Science and Technology Policy
in the United States

Science and Technology Policy in the United States

Open Systems in Action

SYLVIA KRAEMER

RUTGERS UNIVERSITY PRESS

NEW BRUNSWICK, NEW JERSEY, AND LONDON

10095376

Library of Congress Cataloging-in-Publication Data

Kraemer, Sylvia K.
  Science and technology policy in the United States : open systems in action /
Sylvia Katharine Kraemer
      p. cm.
  Includes bbliographical references and index.
  ISBN-13: 978-0-8135-3826-6 (hardcover : alk. paper)
  ISBN-13: 978-0-8135-3827-3 (pbk. : alk. paper)
  1. Science and state–United States–History. 2. Technology and state–United States–History.
I. Title.
  Q127.U6K73   2006
  338.973′06–dc22

                                                                                2005028100

A British Cataloging-in-Publication record for this book is available from the British Library

CONTENTS

LIST OF TABLES

PREFACE

Between 1955 and 1970 federal funding in the United States for scientific research and development (R&D) increased sixfold, from about $6 billion to $36 billion.[1] This relatively rapid infusion of public funds primarily into university research centers, enlarging some and creating others, was accompanied by the emergence of a new academic specialty: science and technology policy studies. Given the importance of government influence on the nature and volume of scientific research being conducted during the first two decades of the Cold War, it is not surprising that the fields of political science and public administration would respond by laying the foundations of our appreciation of the dynamics and consequences of the interactions between public policy and research and development.

If there was a pioneer in the literature generated by the new inquiry into the interplay between science and politics, it was Don K. Price. Price—who would become dean of the Kennedy School of Government at Harvard University—had served as a trusted aide to James Webb, director of the Bureau of the Budget (BOB; to be renamed Office of Management and Budget in 1970) during the creation of the National Science Foundation (NSF). In Price's *Government and Science* (1954) and *The Scientific Estate* (1965) he acquainted a new generation of readers and scholars with the fact that scientists had become one of the four "estates"—scientists, professionals, administrators, and politicians—competing for power and influence in constitutional government in the United States. Scientists and other professionals, no less than administrators and politicians, when acting in political roles are not, nor can they be, guided by what their learning has taught them.

Price's *Government and Science* was followed in 1957 by A. Hunter Dupree's *Science in the Federal Government: A History of Policies and Activities* and, during the 1960s, the publication of what have become classics in the literature of science and technology policy: Derek J. De Solla Price's *Little Science, Big Science* (1963), Daniel Greenberg's *The Politics of Pure Science* (1967), and Harvey Brooks's *The Government of Science* (1968). During the more than three decades that followed, political scientists, economists, historians, public and private research administrators, and scientists themselves have built on the foundations laid in the 1960s. In the process they have focused on special aspects of the science-technology-politics relationship: the roles of, and consequences for universities; the challenges of converting scientific findings to successful technologies; international comparisons of policies to promote research and development; and the economics of innovation, to name only a few. An overview of this rich literature, which provides a scholarly context for the following chapters, is the subject of the essay on sources following chapter 10.

From time to time, as we learn more and more about complex policy issues, it can be useful to step back and increase the angle of our lens, enabling us to appreciate anew the no less complex frames of reference in which those issues occur and will inevitably help to shape their outcome. Those complementary frames of reference will enable us to recognize the ongoing institutional, ideological, and legal traditions that may render policy issues seemingly intractable, and suggest new ways of resolving them or, if resolution seems unlikely, adapting to them. This widening of the angle of our lens, as we focus on the national policy challenges posed by advances in science and technology, has revealed in the background an idea, one that has animated much of modern history, including the history of science and technology: the concept of "open systems."

As the economist Robert B. Reich demonstrated in *The Power of Public Ideas* (1988), the instrumentalism that dominates the prevailing wisdom about politics neglects the comparably important role of ideas in shaping democratic deliberation about "what is good for *society*" and how best to achieve our common aspirations.[2] The purpose of this book is to explore the ways in which a proclivity toward open systems—whether as an idea, an objective, or an approach toward the design of institutions and things—has shaped, and continues to shape, the relationships between science, technology, and policy in the United States.

It is impossible to have toiled with any success in the vineyards of policy development in our nation's capital without learning from others. Colleagues at the National Aeronautics and Space Administration, the National Science Foundation, and the White House Office of Science and Technology Policy especially sharpened my sense of the challenges and possibilities of influencing public policies to advance science and technology, while being sensitive to their social, political, and economic ramifications. This is especially true of John D. Schumacher at the National Aeronautics and Space Administration; Sybil Francis, formerly of the Office of Science and Technology Policy; and Robert Hardy, formerly of the National Science Foundation. The ideas that shape this book were first introduced to, and then developed with the help of, colleagues and students at the School of Public Policy at George Mason University and Colby College in Waterville, Maine; I am especially indebted to Roger Stough, Kingsley E. Haynes, James R. Fleming, and Leonard Reich for making those opportunities possible. Finally—but by no means least in importance—is the skilled and thoughtful editor who serves even the most experienced writer like a pilot bringing a ship into harbor. Audra J. Wolfe at Rutgers University Press has piloted this book around many a shoal unseen by its author, who is immeasurably in her debt.

Science and Technology Policy
in the United States

1

Introduction

Open Systems

Many of us associate "open systems" with computers and industrial products designed for interoperability and standardized interfaces among their components. The architecture of the Internet, for example, is an open system, a characteristic that has enabled its global spread during the last decades of the twentieth century, accompanied by social, economic, and political changes we are only beginning to appreciate. Open systems also play important roles elsewhere in the modern technology-based economy—for example, in the U.S. military's approach to weapons systems procurement, which historically has had a powerful influence on the course of this country's industrial development.[1]

The quality that makes systems "open" is their capacity to absorb novel or improved features with little or no disruption to the original systems themselves. The "plug and play" components now ubiquitous in consumer entertainment electronics and personal computers are examples of open systems at work. Closed systems, in contrast, are everything that open systems are not. They are composed of unique and often proprietary features and connections whose operations require firm control over all their aspects, especially the technical standards used in their design, manufacture, and operation.

In spite of its prevalence in the modern world of electronics and computers, the principle of open systems has been in use in human society for over two millennia. It is a wine of ancient vintage, circulated in periodically newer bottles. Understood as an essential characteristic of a large variety of historical and contemporary ideological and institutional proclivities, the principle of open systems (and the tension between it and its opposite, closed systems) can help to explain a broad array of institutional variations, from the most quotidian to the most public and exalted.

In the realms of ideas and social organization, attitudes and ideologies that are typically said to be liberal or conservative can be understood as manifestations of the natural tension between open and closed systems. Social liberalism as well as economic conservatism embody distrust of government controls over private behavior in the social as well as economic spheres. The degree to which ideologies and institutions display proclivities toward open or closed systems also helps to account for social arrangements that are elitist or exclusive, and those that are pluralistic and inclusive. Oligarchies or class-based political systems are relatively closed, while democratic or representative governments are relatively open. Similarly, selective organizations are closed, while unrestricted membership organizations are open (see Table I.I).

Whether individuals or communities are predisposed toward open or closed systems is likely to be a function of their circumstances and moments in historical time. It is a truism that those who are confident in their capabilities and beliefs will tend to be relatively open to new people, new ideas, and new cultural practices. Authoritarian regimes are more likely to appear in societies that feel threatened (perhaps justifiably) by intrusions of foreign or "outsider" ideas and practices across boundaries of traditional culture and political organization. In the Soviet Union, for example, during the immediate post–World War II period, expressions of intellectual life—scientific as well as literary— were closely policed by the Communist Party's cultural secretary Andrey Aleksandrovich Zhdanov in order to extinguish vestiges of cosmopolitan, or western, thought and writing.[2]

Expanding societies, by comparison, have found openness essential to their capacity for growth. As early as the second century A.D. Rome had begun to offer universal Roman citizenship to colonized areas of the Mediterranean as an instrument of imperial expansion. The history of the United States in the nineteenth century provides an example of geographic expansion and economic growth powered by a population experiencing unprecedented social mobility. Modern science originated and has evolved within this same western social, intellectual, and political universe. In this universe a growing predisposition toward open systems has served as a powerful force for the expansion of both political liberty and standards of truth based on reasoning from individual observations and socially verifiable evidence. Hence the rhetorical distinction often drawn between science and politics must be challenged not only for its mistaken claim of disinterestedness in the human pursuit of scientific investigation, but for its more serious denial of the shared intellectual foundations of progress in scientific knowledge *and* progress toward a wide prevalence of open—transparent, pluralistic, and publicly accountable—political institutions.

A widely publicized instance of this rhetorical contrast between science and politics occurred during public response to the Food and Drug

TABLE 1.1

Open Systems

"Open systems" are any social arrangements, political practices and policies, intellectual premises or outlooks, or technological systems access to which (or participation in) is limited only by individual interest, necessary abilities, and/or commonly available standardized equipment. In the list of illustrations below, the items on the left are examples of "open systems" while the items on the right are corresponding examples of closed systems.

Open System:	*Closed System:*
Popular democracy ..	.. Rule by monarchy, oligarchy
Free market competition ..	.. Market domination by single vendor
Decentralized horizontal management ..	.. Centralized hierarchical management
"One person, one vote" ..	.. Votes weighted by group membership
Public golf course ..	.. Members-only golf course
Barrier-free building access ..	.. Building access by stairs only
Standard gauge railroad tracks ..	.. Variable width railroad tracks
Standardized automobile operator .. interfaces	.. "How do you drive these cars?"
Freedom of the press ..	.. State-censored press
Religious pluralism ..	.. State religion
Ethnic pluralism ..	.. Politically dominant ethnic group
Competitively awarded contract ..	.. "We'll get my brother-in-law's company to do it."
"Anybody's browser will work with my .. operating system."	.. "You use my operating system, you run only my browser."
"My electric razor will plug into any .. outlet in the world."	.. "In England I need this kind of plug, in Italy that kind of plug, in France this other plug."
Competitively awarded research grants ..	.. "Scientists from Super U. are the best, so we'll give them the grant for this research."
Transparency in government .. decisionmaking	.. Government "behind closed doors"
Standardized parts manufacture ..	.. Hand-tooled production

3

Administration's (FDA) September 2000 approval of the use of RU-486, the "abortion pill." "I'm delighted," said Gloria Feldt, president of Planned Parenthood, on an evening network television news broadcast. "I am pleased that the science has finally triumphed over the politics."[3] "Science trumps politics" is a more recent variant of Ms. Feldt's pronouncement, this time occasioned by the FDA's anticipated approval of the "over the counter" sale of the morning-after contraceptive pill.[4]

Such statements promote the view that scientists can and should insulate themselves from the uncertainties and messiness of politics, pursuing aims that are either morally neutral, or will surely improve the human condition through an increased understanding of nature, thus enabling others to apply that understanding to human betterment. But science is messy too, because modern science is a social and institutional activity, propelled forward by intellectual doubts and professional controversies. This very openness to doubt and controversy is more vital to its nourishment than all the world's research grants and international accolades combined, for science is also a creative intellectual pursuit—and such pursuits can survive all manner of deprivation, except the failure to question received wisdom.

THE CONCEPT OF open systems in its cultural and civic expression—or cosmopolitanism—can be found in the philosophy of the Stoics (ca. 300 B.C.–100 B.C.), who opposed the traditional Greek distrust of non-Greeks. Stoics saw themselves as citizens of the world, their membership in the human family being more essential to their civic virtue than their membership in the city-state. The Stoics' belief that the capacity for moral judgment and right conduct was what distinguished human kind, was closely linked to their most enduring contribution to western thought: the idea of natural law, or the belief that there is a law of nature (in contrast to laws established by society, or "positive" law) that can be universally understood through human reasoning. That being so, men and women everywhere could know what was true and just, and thereby question the moral authority of prevailing political institutions. How this ancient Greek idea became a part of early Christian theology is beyond the scope of this book; suffice it to say that the cosmopolitanism of early European Christendom, which enabled it to expand as it did, was one of Stoicism's legacies. Not until the challenge of the Reformation would the Roman Catholic Church begin to close in on itself with the zealous pursuit of doctrinal and ecclesiastical command over all believers.

The belief that natural phenomena are governed by universal principles was a great step forward out of the capricious world of ancient mythology, and a step toward what we now consider modern science. Making good use of the new technology of optical magnifying instruments Nicolas Copernicus (1473–1543), Johannes Kepler (1571–1630), and Galileo Galilei (1564–1642) demonstrated that the heavens and the Earth functioned according to uniform laws that could be articulated through geometry and mathematics. In so doing, they lifted physics and astronomy off of their Ptolemaic and Aristotelean geocentric foundations of different worlds and platonic essences, and set them on the modern ground of a universal material reality, one that could be understood through human reason, observation, and measurement. In the process, they left to metaphysics and religious doctrine such questions as human purpose and destiny, creating instead the first truly cosmopolitan vocation to compete with early Roman Catholicism—that of natural philosophy, or what we today call "science."

Isaac Newton (1642–1747) would unite the quantitative physics of Galileo and the mathematical astronomy of Copernicus and Kepler in a common principle, the law of gravity, which governed matter and motion throughout the universe.[5] Given the concurrent spread of Protestantism and the process of state building in northern Europe, the analogous principle—that natural mechanisms guided human social and political arrangements—offered a compelling alternative to notions of divinely ordained rule.

In recognizing this transformation in western intellectual life it would be difficult to overstate the importance of the role of numbers. Increasingly complex measurements made with telescopes, thermometers, clocks, and barometers required a system of numbers that could be used to calculate quantities in a way that roman numerals could not. The Arabic system of numeric notation based on powers of ten served this purpose, while more advanced mathematics, with its fractions, logarithms, analytical geometry, and calculus would, by the end of the eighteenth century, provide a universal language scientists could use to examine their findings with other scientists.

Just as Latin had served as the language that enabled the Roman Empire and the Roman Catholic Church to expand to its farthest reaches, modern numerical calculation provided the cosmopolitan language by which the scientific revolution could become a universally accessible legacy of the seventeenth century. Two centuries later the digitization of electronic communications would serve an analogous purpose, ensuring that what was observed in New York could be learned and discussed—with near simultaneity—in Prague or Djakarta or anywhere where one had a computer connected to a telecommunications network.

The gradual spread of the belief throughout the North Atlantic region that natural laws—automatic mechanisms that could be described and managed quantitatively—guided human social and political arrangements, brings us to the mechanism of "supply and demand" in Adam Smith's *The Wealth of Nations* (1776). This mechanism, an "invisible hand," would have its parallel in the "balance of powers" in the design of the U.S. government. Tyranny and oppression could be prevented if the different components of political rule were open to public view, disaggregated, and allowed to check the oppressive tendencies in the others.

Opposed to political and religious oppression, both scornful and fearful of dogma, the decidedly secular and inquisitive outlook of the Enlightenment bequeathed to the new republic a political culture that favored open systems in public policy and administration. This propensity has had—and continues to have—consequences wherever the U.S. government engages other forces in society, not the least being science and technology. It is animated by the assumption that human reason is most likely to arrive at the best answers to the important questions of life—be they questions about nature, or the improvement of human social and civic life—if permitted to function and express itself freely. Within the realm of politics open systems prevail when participation is pluralistic, and the exercise of power is both transparent and publicly accountable.

The generation that authored the U.S. Constitution understood freedom of speech and the press to consist of more than the absence of censorship. The Continental Congress spoke of freedom of the press as necessary for "the advancement of truth, science, morality and arts in general, ensuring diffusion of liberal sentiment on the administration of government . . . and its consequential promotion of union among them, whereby oppressive officials are shamed or intimidated into more honorable and just modes of conducting affairs."[6] As the nineteenth-century jurist and constitutional scholar Thomas M. Cooley, commenting on the First Amendment, observed: "The evils to be prevented were not the censorship of the press merely, but any action of the government by means of which it might prevent such free and general discussion of public matters as seems absolutely essential to prepare the people for an intelligent exercise of their rights as citizens."[7]

One of the better known statements of the importance of freedom of speech for the viability of representative government was that made by Justice Oliver Wendell Holmes, Jr. in his 1919 dissent in the case of *Abrams v. United States*. "When men have realized that time has upset many fighting faiths," wrote Holmes, who was joined in his dissent by Justice Louis D. Brandeis, "they may come to believe even more than they believe the very foundations of their own conduct that the ultimate good desired is better reached by *free trade in ideas*—that the best test of truth is the power of the thought to get itself accepted in the competition of the market, and that truth is the only ground upon which their wishes safely can be carried out. That at any rate is the theory of our Constitution. It is an experiment, as all life is an experiment" (italics added).[8] Holmes's allusion to the essential principle of Adam Smith's *Wealth of Nations*—that is, that a freely operating market place is the best arbiter of value—along with the pragmatism of the outlook reflected in this and his other writings, captures so well the extent to which a belief in the necessity of open systems to ensure freedom and progress informs the very core of U.S. public policy.

The late Richard Feynman described the essence of science in this way: "[Science textbooks offer] some kind of distorted distillation and watered-down and mixed-up words of Francis Bacon from some centuries ago. . . . But one of the greatest experimental scientists of the time . . . William Harvey, said that what Bacon said science was, was the science that a lord chancellor would do. He spoke of making observations, but omitted the vital factor of judgment about what to observe and what to pay attention to."[9] The greatest peril facing science in any age, continued Feynman, is the transmission of "mistaken ideas . . . the only remedy" for which is to doubt that what is being passed on from the past is in fact true, and to "try to find out *ab initio,* again, from experience, what the situation is, rather than trusting the experience of the past in the form in which it is passed down. . . . Each generation [must] . . . pass on the accumulated wisdom, *plus the wisdom that it may not be wisdom. . . .* [I]t is of great value to realize that we do not know the answers to different questions. This attitude of mind—this attitude of uncertainty—is vital to the scientist."[10]

What gives open systems in cultural and civic life their vitality is the optimistic uncertainty to which Feynman alludes. Each new possibility—whether it comes in the form of new evidence, a new citizen, a new problem, or a new proposal to resolve an old problem—represents an enhanced prospect for the future, a prospect in which failure is temporary at most. No single scientific discovery has had an impact on modern life greater than the scientific method itself, with its reliance upon public verification of evidence and argument. Similarly, free or open political systems rely on a system of jurisprudence that acquires its authority through the public judgment of evidence.

Thus both science and politics entail to a substantial degree the mediation of experience by judgments informed by tacit beliefs. At the same time, neither is immune from the dynamic in human institutions that ensures that "nothing succeeds like success"—what the sociologist Robert Merton characterized as "the Matthew Effect" in science, or its counterpart in politics, the advantage of incumbency.[11] Of these shared characteristics, however, arguably the most important is the capacity of open systems to adapt creatively to newly discovered phenomena and historical circumstances. This capacity is rooted in the confidence shared by both science and open political systems in the ineluctable singularity of the human spirit and intellect, operating within a framework of consensually accepted laws, as the ultimate arbiter of what is reasonable, true, and just.

All of which begs the question of whether a national proclivity toward open systems has been a uniquely essential ingredient of the U.S. success in the late nineteenth and twentieth centuries in exploiting the potential of science and technology to achieve economic success. Individual readers will have their own convictions—some supportable by scholarly work in cultural anthropology and history—about the *relative* importance of natural resources, political systems, cultural and economic practices, and moment in historical time, in accounting for the success of any given society in a major field of endeavor. This author subscribes to the view that, at the scale of national and international trends, change consists of many moving parts and the effort to single out a single cause is more likely to distort than clarify historical dynamics. More useful—because more likely to approximate the way things actually work—is to identify the more significant forces and how they interact with or reinforce one another. That is the approach taken in this book.

THE FOLLOWING CHAPTERS examine the recent history of science, technology, and politics in the United States as the interplay of institutions, ideas, and issues in the ongoing tension between open and closed systems. The underlying framework is one which regards effective policy making as something that occurs within the dynamic "range of play" where three essential sets of variables intersect: *politics* (the struggle for power among individual and institutional interests, usually status and economic interests); *ideology* (which can invoke religious, intellectual, and cultural precepts in articulating the substance of a policy issue); and *law* (which, in our constitutional system, sets the ever evolving boundaries of policy making). Depending on historical circumstances, one or more of these sets of variables may predominate in shaping the resolution of a policy issue, but all three sets must be engaged in that process for the resolution to prevail.

This examination proceeds with the premise that the creative social enterprise that enfolds modern science, technology, and the American experiment in representative constitutional government is a single national enterprise (albeit with international wellsprings and consequences)—one that thrives best when it proceeds openly, faithful to the principles of transparency, pluralism, public questioning of claims to truth and right, and optimistic skepticism.

The following chapter introduces the conception of the relationship of technology to national wealth contained in Adam Smith's *The Wealth of Nations*, a conception that was introduced into this country by Alexander Hamilton, who, along with Robert Morris, was the architect of the nation's original economic system. The vision of a political economy driven forward by the engine of free-market capitalism exemplifies the concept of open systems. Adapted by Alexander Hamilton to the new nation's first economic policy, Smith's representation of the natural dynamics of a wealth-creating laissez faire economy was among the earliest economic theories to acknowledge the contribution of technology to economic productivity. As for the Hamiltonian policy, it would prove to be the model for U.S. science and technology policy. The natural occurrence of a national and *open* commercial system would be ensured by the elimination of local or regional barriers to commerce; encouragement of domestic manufacturing and technological innovation would provide the endless stuff of commerce; and combinations of policies to stimulate industry and commerce would allow the natural dynamics of the market to generate wealth more certainly than a single, centrally directed economy.

During the nineteenth century, while the actual history of American technological development affirmed Hamilton's design (which received legal support from the U.S. Supreme Court under Chief Justice John Marshall), economic writing shifted to an interest in theories of value and distribution, tending to regard technology as exogenous to economic systems. By the early twentieth century, however, thanks to the work of economists Joseph Schumpeter, Simon Kuznets, and Jacob Schmookler, technology came to be appreciated not only for its critical and endogenous role in economic growth, but as an enterprise that proceeded independently of advances in scientific research.

In chapter 3 I explore the sources of the ideology of science in the United States, initially the platonic idealism transmitted to nineteenth-century America in the works of Plutarch (A.D. 46–120) and his transcendentalist followers, with its veneration of abstract philosophy and disdain for practical or manual work. This legacy from Greco-Roman antiquity not only helped to form the institutions that shape, and have been shaped by, science policy in the United States—most notably the National Academy of Sciences (NAS) and the National Science Foundation (NSF)—but it provided the principal theme for Vannevar Bush's seminal *Science—The Endless Frontier* (1945), as well as for its critics. This theme—which I describe as the Bush paradigm—asserts a necessary and linear relationship between advances in basic science, most typically pursued in universities, technological innovation, and economic growth. The chapter concludes with a discussion of the contemporary status of the paradigm in the study and practice of science and technology policy.

The fourth chapter examines the various ways the U.S. government makes science and technology policy, and the administrative devices it uses to implement the policies that result. Both legislative and executive branch processes for policy making and public administration in the realm of science and technology reflect an open systems approach to policy and public administration. The U.S. science and technology policy toolkit serves as this country's answer to the experiments with centralized industrial policy that have been attempted elsewhere—most notably in Great Britain and Japan. After surveying the entire array of means by which the federal government attempts to implement general policies, the chapter focuses in on federal procurement (how federal dollars are distributed for scientific research and technological development) and intellectual property policy, the two tools that have the most impact on the way federal policy influences this country's research and development enterprise.

The common currency of all effective open systems—whether technological, political, or social—is universal access to information that is complete, verifiable, and appropriately used. The critical political function of reliable information is to ensure that authority is accompanied by public accountability. Chapter 5 explores the uses of scientific and technical authority in the course of policy making, as well as in jurisprudence, which sets the boundaries within which policy can be made and carried out. The relationship of scientific and technical information and authority to the preservation of open government in the United States is not an untroubled one. To the extent it has been a positive relationship, it has most often hinged on the outcome of issues surrounding transparency in the conduct of scientific research, and the reliability of its results. We will consider, as examples, recent issues involving the Freedom of Information Act (FOIA) and the Federal Advisory Committee Act (FACA), the standing of scientific and engineering information in the federal judiciary, and the controversy over the executive branch's implementation of the Information Quality Act of 2000 (also referred to as the Data Quality Act).

Of all the U.S. government's ventures in promoting scientific and technological progress, the creation of the Internet has been among its most successful. Chapter 6 examines how the Internet became the most outstanding example in the United States of the contribution of open systems to the technological enterprise to which that concept was applied. Moreover, the fact that the Internet was developed as an open system enabled it to become a global force for the opening of social, economic, and political organizations to cosmopolitan influences.

The policy issues arising from the expanding use of the Internet are also variations on the classic tension between those things that flourish as a result of open systems—new ideas, public criticism of those in power, and business competition in a free market—and the impulse to control those same things to the continuing advantage of incumbents. These issues cluster around the extension of the U.S. Constitution's First and Fourth Amendment freedoms to cyberspace, the public management of national telecommunication systems, and efforts to limit the accretion of monopoly power in the software industry.

In chapter 7 we observe the U.S. propensity toward open systems influencing not only the shaping of U.S. space policy in the late 1950s, but the administrative strategy the federal government would use in marshalling the country's resources to send satellites, instruments, and humans into space as well as to the moon. At the same time, however, the evolution of the space program illustrates the capacity of publicly funded large-scale technological enterprise to evolve into a countervailing impediment to openness. This occurs because continuing participation in publicly funded science and technology programs confers on private sector contractors and grantees the competitive advantage of accumulated expertise. The advantage of this expertise is not limited to meeting the technical requirements of the sponsored research and development programs; it extends as well to political networks in the Congress and the White House, the arcane workings of federal procurement policy, and the preferences of the officials in the sponsoring agencies.

In chapter 8 we turn to the "health care crisis" in the United States, which can be understood as a struggle between the open systems principle of universal access to an essential public good, and the monopolistic tendencies of scientific and medical authority and their associated institutional interests. These interests—e.g., pharmaceutical firms, health insurance organizations, and for-profit hospitals and medical centers—have a large economic stake in the enormous annual expenditures for health care in the United States. In chapter 8 we consider not only this struggle, and how federal policies have exacerbated the high costs of prescription drugs and the diminishing access of all Americans to adequate medical care, but possibilities for genuine reform.

The open system captured in Adam Smith's vision of the laissez faire economy presupposes the undistorted functioning in a free market of supply and demand, regulated by the costs (to the seller) and prices (to the buyer) of goods and services. Such a system, to operate as expected, requires that costs and profits be accurately allocated among producers, sellers, and buyers. Applied to policy issues that entail interlacing "public goods," which are usually subsidized to some extent by the government, and clearly identifiable private gains, the Smithian model is difficult to achieve.

Nowhere is this more true than in the closely interconnected realms of environmental and energy policy, which are informed not only by the same economic and scientific data used in other areas of scientific and technological policy making, but conflicting concepts of nature and the extent to which nature functions as an open biologic, geophysical, and meteorological system. In the case of fossil fuels (which dominate U.S. and international supplies of fuel and synthetic raw materials), determining how best, and to what extent, to regulate energy production to reduce environmental costs can become legitimately contentious, for regulatory compliance will entail uncertain costs as well. In chapter 9 we examine the extent to which reliance on the open system of a free market has served or hindered efforts to secure adequate near- and long-term energy supplies, while minimizing if not eliminating the cumulative environmental damages from the production and use of hydrocarbon fuels and synthetic materials.

The final chapter gathers up the several narrative threads laid down in the preceding chapters and considers the whole once again. It concludes with reflections on the capacity of open systems—in practice as well as principle—to ensure continued scientific and technological progress *as well as* essential political transparency, pluralism, and accountability. Transparency, pluralism, and accountability are essential because they constitute the critical ingredients of the productive civil society that the United States continues to represent to much of the western world.

2

Technology and the Ideology of Free Markets

It is impossible for the arts and sciences to arise, at first, among any people, unless that people enjoy the blessings of a free government.

–David Hume, "Of the Rise and Progress of the Arts and Sciences"

Neither scientific discovery nor technological innovation could have seized hold of the modern imagination without a profound change in widely held levels of confidence in the power of individuals to comprehend the natural world, and to apply the laws of nature to the shaping of man's social and moral universe. That change consisted less in the sequential replacement of one widely shared perception of the world by another, than in the emergence of persuasive competitors to metaphysical explanations of human destiny. Such explanations saw human souls at the mercy of a single willful divinity, one who ruled a world in which obedience to divine commandment, hierarchical order, and teleological purpose was all that stood between ourselves, utter chaos, and eternal damnation. As the illustrious Congregationalist minister Jonathan Edwards famously admonished unbelievers in 1741, God "holds you over the pit of hell, much as one holds a spider, or some loathsome insect over the fire."[1]

Competition for metaphysical constraints on individual human agency came to the British colonies in North America from emissaries of the European Enlightenment, of whom the Scotsmen Adam Smith (1723–1790), author of *The Theory of Moral Sentiments* and *The Wealth of Nations*, and John Witherspoon (1723–1794), president of the College of New Jersey (called Princeton University after 1896) and tutor to James Madison, were arguably the most influential.[2] They planted in new soil a different set of instructions for achieving a benign, virtuous, and *prosperous* society, one that could function as an open system relying on the dynamic interplay of natural human tendencies to achieve human happiness.

Smith's contribution began with his explication of human psychology as the essential force in shaping human society, which he laid out in *The Theory of*

Moral Sentiments (1759). The easily accessible prose of Smith's work may help to account for its popularity: *Moral Sentiments* has appeared in no less than forty editions, at least eight of which have been American, and was available in French by 1764 and German by 1770.[3] For Smith, man was uniquely equipped by nature with a capacity for moral sentiments (that is, prudence, propriety, benevolence, and justice), a desire to better himself, and a conscience—"the man within the breast, the supposed impartial spectator, the great judge and arbiter of our conduct." Relying on the voice of individual consciences informed by experience, "we shall stand in need of no casuistic rules to direct our conduct." All that is further required for us to lead moral lives and thus promote civil society is "self-command." Just as the escapement of a mechanical clock allows the clock's gears to ratchet forward with each swing of the pendulum, so self-command guided by the "man within the breast" would regulate the interaction of individual personality with the exigencies of family, community, and social experience.

For Smith, the reformer's zeal for superimposing a new political system on society was more likely to do society harm than good. "The man of system," wrote Smith, "seems to imagine that he can arrange the different members of a great society with as much ease as the hand arranges the different pieces upon a chess board." But "in the great chess board of human society, every single piece has a principle of motion of its own." If those who govern would allow the principles that legislatures prefer to "coincide and act in the same direction" as the principles of motion governing society's members, "the game of human society will go on easily and harmoniously, and is very likely to be happy and successful."[4]

What was remarkable about Smith's vision was its naturally occurring dynamism. But this characteristic, which transformed a conception of an open social system from a prescription for chaos to a prescription for orderly growth, was not unique to Smith. We find it as well in James Madison's tenth *Federalist* paper, where he observed that, because of the "diversity in the faculties of men," the forming of factions is "sown into the nature of man." If factions, being natural, are inevitable, how can we prevent the violence to which factional conflict can lead? Democracies, in which every individual is a "judge in his own cause," Madison wrote, cannot control the worst effects of faction. In a republic, however, "the public voice, pronounced by the representatives of the people, will be more consonant to the public good than if pronounced by the people themselves." Further, both "extensive" and "small" republics have their limits: "By enlarging too much the number of electors, you render the representative too little acquainted with all their local circumstances and lesser interests." But if a republic is too small, representatives will be "unduly attached" to their parochial concerns, and "too little fit to comprehend and pursue great and national objects. The federal constitution "forms a happy combination in this respect; the great and aggregate interests being referred to the national, the local and particular to the State legislatures."[5] Harmony and stability could be achieved by applying regulating mechanisms to political life that employ, rather than frustrate, natural human inclinations.

The Wealth of Nations appeared in 1776, the same year as the U.S. declaration of independence. Mention this classic of economic theory to most informed people and the most likely association they will make will be: *laissez faire*. To its detractors laissez faire means an ideology of minimalist government in the service of capitalist greed. But for laissez faire's proponents, minimalist government allows the free market to function as it should, namely to enable prosperity to rise out of the maelstrom economic competition.[6] An appreciation of what Smith actually wrote in *The Wealth of Nations*, and what his followers—most notably Alexander Hamilton—thought they found in its pages, matters if one is to understand the foundations of modern U.S. technology policy.

Let us dispense first with the notion that Smith's prescription for a freely operating market place precluded active government on the public's behalf. The four books of the eminently readable *Wealth of Nations* culminate with this opening to Book IV ("Of Systems of Political Economy"): "Political economy, considered as a branch of the science of a statesman or legislator, proposes two distinct objects: *first, to provide a plentiful revenue or subsistence for the people, or properly to enable them to provide such a revenue or subsistence for themselves; and secondly, to supply the state or commonwealth with a revenue sufficient for the public services*" (italics added).[7]

The final pages of Book IV caution *not* against government action on behalf of the nation's economy, but against government action on behalf of particular industries, what a commentator today might call "picking winners and losers." "Every system which endeavours," wrote Smith,

> either by extraordinary encouragements to draw towards a particular species of industry a greater share of the capital of the society than what would naturally go to it, or, by extraordinary restraints, force from a particular species of industry some share of the capital which would otherwise be employed in it, is in reality subversive of the great purpose which it means to promote. It retards, instead of accelerating, the progress of the society towards real wealth and greatness; and diminishes, instead of increasing, the real value of the annual produce of its land and labour. . . .
>
> All systems either of preference or of restraint, therefore, being thus completely taken away, the obvious and simple system of natural liberty establishes itself of its own accord. . . . The sovereign has only three

duties to attend to; . . . first, the duty of protecting the society from the violence and invasion of other independent societies; secondly, the duty of . . . establishing an exact administration of justice; and, thirdly, the duty of erecting and maintaining certain public works and certain public institutions which it can never be for the interest of any individual, or small number of individuals, to erect and maintain.[8]

What Smith rejected was the centrally directed and administered economy, one in which the government attempts to manage all aspects of manufacture and trade, which in his era was known as *mercantilism*—the economic philosophy that prevailed throughout major European capitals in the sixteenth to eighteenth centuries. Practiced most successfully by France's Jean-Baptiste Colbert, economics minister to Louis the XIV, mercantilism's principal objective was to obtain for the national treasury as much money—or bullion—as possible in order to support the regime and the armies that sustained its power.[9] This required the maximum export of goods manufactured with raw materials obtained as cheaply as possible. Minute aspects of national manufacture and commerce were regulated in the quest for the larger prize, which did not necessarily include the happiness and well-being of the nation's people—though it did include their being actively engaged in commerce, which would increase the amount of money in circulation.[10]

Smith's answer to mercantilism reflected not only the genius of one man, but also the genius of a new age. His response challenged the premises of mercantilism in three fundamental ways: first, the nation's wealth was to be measured not in the wealth of the sovereign, but in "the real quantity of industry, the number of productive hands, and consequently the exchangeable value of the annual produce of the land and labour of the country, the real wealth and revenue of all its inhabitants."[11] Lurking behind this different conception of what constitutes a nation's wealth are the politically revolutionary notions that a society's ultimate lawful sovereign is the citizen, and that its strength lies less in its military or diplomatic power than in the wealth of its people.

Second, while mercantilists had acknowledged the importance of manufactures—most manufactured goods were cheaper to export (by virtue of relative cost per unit volume) than were agricultural goods—Smith was the first eminent political economist to integrate *technology* into his principle that specialization of labor is primarily responsible for increased productivity. Specialization (or "division") of labor, which for Smith was one of the chief prerequisites for "civilized and thriving nations," was due to "three different circumstances; first, to the increase of dexterity in every particular workman; secondly, to the saving of the time which is commonly lost in passing from one species of work to another; and lastly, to the invention of a great number of machines which facilitate and abridge labour, and enable one man to do the work of many."[12]

Third, Smith's assault on mercantilism was possible not only because merchants had grown weary with micro-management by royal functionaries, but also because the age in which Smith lived had conceived of an alternative avenue to civic peace, equity, and prosperity. That avenue began with an accurate understanding of the natural human tendencies of people, and then invited the most desirable of those tendencies to take their course. Needless to say, if one believed that human nature was in essence evil or "sinful," then all aspects of human society required strict rule by religious and secular sovereigns to enable life and society to prevail. But if one believed, as Smith and many of his era believed, that all of nature is comparable to a self-regulating mechanism, then one needed only to ascertain the mechanism to understand and predict the "natural course of events." This view of nature loosened the bonds of physical and natural science to religious doctrine, while its extension to human affairs would release the bonds of social and political thought from the doctrine of the "divine right of kings."

Just what is the natural tendency of economic man, which, if left to itself, would ensure a "general opulence?" It is not any special "human wisdom," but rather "a certain propensity in human nature" which is "the propensity to truck, barter, and exchange one thing for another . . . which originally gives occasion to the division [specialization] of labor."[13] The division of labor, in turn, enables men to produce more than what they need of any particular good, which they can then use as a commodity for exchange in the market place for goods that they desire. Here, very simply, is the natural foundation of industry and trade, without which "general opulence" is impossible. Add to this the "naturally regulated" prices of commodities in the market place—the regulating mechanism being the tendency of demand to diminish as prices rise—and we find ourselves in a world of pure competition in a freely operating market place. Thus we have what Adam Smith called "the system of natural liberty," one in which "every man, as long as he does not violate the laws of justice, is left perfectly free to pursue his own interest his own way, and to bring both his industry and capital into competition with those of any other man, or order of men."[14] Smith's perfect market place is an *open system*, a system much to be preferred over "the policy of Europe" which, "by not leaving things at perfect liberty, occasions other inequalities of much greater importance. It does this chiefly . . . by restraining the competition in some employments to a smaller number than would otherwise be disposed to enter into them; secondly, by increasing it in others beyond what it naturally would be; and thirdly, by *obstructing the free circulation of labour and stock, both from employment to employment and from place to place*" (italics added).[15]

An Open System for the New Republic

The Wealth of Nations was translated into Danish, French, German, Italian, and Spanish and would become one of the mostly widely read books of its time. One of its many readers was the architect of national economic policy for the new American republic, Alexander Hamilton (1755–1804), whose *Report on Manufactures* (1791) contains sections paraphrased from portions of Smith's classic. The creation of the new American republic represented a historic opportunity to experiment de novo with economic policy, and in one of the more fortuitous accidents of history, Hamilton was well prepared and positioned to guide the experiment.

Bright, astute, and ambitious, the autodidact Hamilton escaped his troubled origins on the island of Nevis in the British West Indies by clerking for the international export-import house of Beekman and Kruger (after 1769 Kortright and Kruger) on St. Croix.[16] He learned first-hand the roles played by credit, uniform currencies, national banking institutions, and open markets where great varieties of goods were traded, in securing economic prosperity. He also saw the slave trade at its worst, an experience that would make of him an early abolitionist. An assiduous reader, Hamilton relied on a trio of "texts" to furnish the intellectual structure of his experiences: Malachy Postlethwayt's *Universal Dictionary of Trade and Commerce* (1751), Smith's *The Wealth of Nations,* and the charter of the Bank of England (est. 1694). So impressive was Hamilton's growing mastery of the complexities of finance that Beekman and Kruger sent him to New York, where he began to impress others as well—most notably General George Washington and Robert Morris (1734–1806), superintendent of finance under the Articles of Confederation (1781–1789). Together, Morris and Hamilton, who would become secretary of the treasury (1789–1795), laid the foundations of a national economic policy that would play a fundamental role in the subsequent industrial development of the United States.

That policy would also serve as the paradigm for U.S. science and technology policy, in that it was designed to (a) prevent the creation of local or regional barriers to the spread of domestic industries, thereby creating a national and *open* commercial system; (b) promote the growth of domestic manufacturing and technological innovation; and (c) create the conditions necessary for national economic growth and expansion through an aggregate of policies aimed at the *components* of a modern industrial economy, rather than a single, centrally directed economy that Smith, Morris, and Hamilton had experienced with British mercantilism.[17]

Because the political struggle between Federalists and Democratic Republicans over the nature of the U.S. Constitution hinged to a large extent on the question of the size and power of the federal (or "national") government, it is easy to overlook Hamilton's belief in the economic power of a truly national market place, one in which the "natural liberty" of Adam Smith's *Wealth of Nations* could function. That meant the elimination of local, state, or regional barriers to commerce, which the Federalists achieved with the Constitution's commerce clause reserving to the Congress the power over interstate commerce, a power reaffirmed in Chief Justice John Marshall's decision in *Gibbons v. Ogden* (1824).[18] Hamilton expressed his confidence in the power of a vast national market to ensure the future prosperity of the country in the eleventh Federalist paper, where he wrote: "If we continue united, we may . . . oblige foreign countries to bid against each other, for the privileges of our markets. This assertion will not appear chimerical to those who are able to appreciate the importance of the markets of three millions of people . . . to any manufacturing nation. . . . Let the thirteen States, bound together in a strict and indissoluble Union, concur in erecting one great American system, superior to the control of all transatlantic force or influence, and able to dictate the terms of the connection between the old and the new world!"[19]

Beyond creating a "strict and indissoluble Union," the next most important achievement of the new federal government was to ensure that the "veins of commerce in every part" be invigorated by "free circulation of the commodities of every part." This aim would be served best by a national monetary system backed by the "full faith and credit" of the *United* States.[20] Few could know better than Hamilton, or his predecessor Robert Morris (who financed the Revolutionary War largely with money borrowed from abroad), the importance of the availability of credit to any economic enterprise.[21] Adam Smith had also noted the importance of credit and a national banking system to ensure its stability and flexibility, when he wrote "It is not by augmenting the capital of the country, but by rendering a greater part of that capital active and productive than would otherwise be so, that the most judicious operations of banking can increase the industry of the country."[22]

Hence the critical decision, advocated in Morris's and Hamilton's successive reports on the public credit (1782, 1790) that the federal government would fund the nation's *as well as the states'* Revolutionary War debts, charter a national bank to serve as a depository for government funds and manage the government's revenues, and use the nation's public debt—that is, bonds—to serve as a basis for a uniform and elastic currency.[23] The new government would also establish a system of revenues (such as import tariffs, land taxes, poll taxes, and excise taxes on liquor) to meet its regular public expenses. Had the United States government failed to establish its credit, the country's independence would have been in jeopardy, and its economic future in doubt.

Technological Innovation in "One Great American System"

The Wealth of Nations had extolled the value of technological innovation and its author, characteristically, sought to explain what brings about technological innovation. "The invention of all those machines by which labour is so much facilitated and abridged," wrote Adam Smith, "seems to have been originally owing to the division [specialization] of labour. Men are much more likely to discover easier and readier methods of attaining any object when the whole attention of their minds is directed towards that single object than when it is dissipated among a great variety of things."[24]

For Smith technological innovation occurred as the incidental by-product of a skilled workman's ingenuity in trying to make his task easier, experiments of machine makers, or a natural philosopher's (progenitor of today's scientist) observations about previously unnoticed and unexploited relationships between physical objects and their motions. None of the three explanations Smith offered relied on government action to encourage them to happen. Nor was Smith, given his low expectations of governments "attempting to perform" functions in which they "must always be exposed to innumerable delusions," likely to have recommended such a policy.[25]

Hamilton, however, was willing to venture beyond Smith by recommending that either import duties or "bounties" be instituted to "encourage" various industries still in their "infancy." Industries that should be so favored included goods made from iron and steel, copper and lead; coal mining; wood products from cabinets to ship timbers; leather and fur goods; grain spirits and liquors; cloth made from flax, hemp, cotton, and wool; paper and printed books; and refined sugars and chocolate. Hamilton envisioned that duties, or protective tariffs, would serve to raise national revenue. Hence when such duties were lifted—and Hamilton can be faulted for underestimating the political difficulty of lifting duties—a substitute source of comparable revenue had to be established. That alternative source was a special fund financed by the surplus resulting from duties already collected.

This fund was to be used for paying the bounties (that is, subsidies) to the industries Hamilton singled out, and to support "the operations of a board to be established, for promoting arts, agriculture, manufactures, and commerce," the members of which would be government officials serving as board commissioners. The board would offer financial inducements to immigrants having certain critical skills as well as "manufacturers in particular branches of extraordinary importance." It would also fund what today would be called research and development: "Let these commissioners be empowered to apply the fund . . . to induce the prosecution and introduction of useful discoveries, inventions, and improvements, by proportionate rewards, judiciously held out and applied." Hamilton hastened to add, however, that the fund was not to be used to support speculative research, but "invention and . . . useful improvements" in very specific instances of need.[26]

While Hamilton's program reached beyond the view of the government's limited role in the nation's economy found in *The Wealth of Nations*, it did not extend to the comprehensive and centrally conceived and administered economic program for which mercantilism was known. Moreover, as he cautioned his readers, the "public encouragements" he recommended were but temporary expedients, tolerable in a young country just then setting out on its own. As for promoting technological innovations, "private societies" such as the Pennsylvania Society for the Promotion of Manufactures and Useful Arts, "are truly invaluable."[27] Hamilton's approach is embodied in the patent clause of the Constitution, which authorizes the Congress to "Promote the Progress of Science and useful Arts, by securing for limited Times to Authors and Inventors the exclusive Right to their respective Writings and Discoveries." The government can encourage economically important invention and "discoveries," but it cannot, and ought not, attempt to bring them into being itself. Hamilton also appreciated that government, be it federal or state, would have to promote the building of a national infrastructure to support commerce. By chartering quasi-public corporations and issuing bonds to domestic and overseas investors, individual states enabled the construction of roads, canals, bridges, harbors, and, in time, railroads. This, too, as we have seen, was well within the vision of *The Wealth of Nations*. On the other hand, the protective tariffs that Hamilton advocated to foster the new republic's nascent industries were mercantilist measures through and through.

Thus was born what came to be known as "the American system:" the pairing of internal improvements with protective tariffs.[28] Combined with the vast continent's treasure of natural resources and a population schooled for enterprise by its inherited culture, the American system made possible the phenomenal economic growth of the United States in the nineteenth century. As sometimes happens, however, the success of the American system generated its own enemies. Banking and businesses began to flourish while farmers on the expanding frontier struggled with erratic prices for their crops and what they believed were artificially high prices for manufactured goods. As a result, the Democratic Republican party of Jefferson mounted an effective opposition to the Hamiltonian program culminating in the election of Andrew Jackson in 1832 and Jackson's veto of legislation to renew the charter of the Bank of the United States.

THE ENLIGHTENMENT legacy brought to the former British colonies by its founding generation included not only the first expression of an economic philosophy that treated technology as an endogenous factor in the generation of national wealth, but the first expression of an open systems approach toward technology itself. In his classic *From the American System to Mass Production, 1800–1932*, historian of technology David A. Hounshell describes the transmission to the new United States of the concept of interchangeable parts manufacture from France, where the Enlightenment rationalism of military planners bore fruit in the recognition that firearms made of standardized parts could be more readily repaired and interchanged among soldiers in the field. Standardization—for example, the "opening" of the system of the firearm's components—did not necessarily entail mechanization. It did entail the specialization of labor, with the attendant possibility that specialized workers would introduce improvements into their particular tasks. This was a phenomenon that Smith recognized in his account of pin making in the opening pages of *The Wealth of Nations*. The principle was undoubtedly recognized as well by the French arms maker Honore Blanc, contemporary of both Smith and Thomas Jefferson, who recommended Blanc's ideal of interchangeable parts to the U.S. secretary of war Henry Knox.

As Hounshell takes care to point out, in the early nineteenth century interchangeable parts manufacture was not necessarily cheaper; nor did it necessarily lend itself to mechanization. "It is important to keep in mind," writes Hounshell, "that throughout much of the nineteenth century the ideal of interchangeability, despite its powerful appeal and seeming rationality, was considered a somewhat irrational pursuit. . . . Yet in the United States, mechanics [e.g., firearms contractors to the War Department] continued to pursue the dream, and the government allowed them generous financial support for their efforts."[29] The innovation that enabled the principle of interchangeable parts to become the foundation of modern mass production in the United States was *intellectual*: the realization that attempting to achieve a uniformity of parts by replicating a series of identical components individually, successor by successor, invites the accumulation of errors into the series. Instead, the principle behind true interchangeability was manufacture of many components to a *single standard*, which could be confirmed by the use of standard gauges for measuring the accuracy of various components' dimensions and setting up machine tools. Second, precision throughout a series of uniform parts required that those parts, during the process of manufacture, be held in the fixtures that secured them to machining equipment in exactly the same way each time. This could be achieved through the use of a consistent "bearing point" for all fixtures used in working on a single part. Both the use of standard gauges and bearing points were achieved during the 1820s in (John) Hall's Rifle Works at Harpers Ferry arsenal, whence these practices spread throughout the nineteenth-century enterprises in the United States that produced clocks, sewing machines, and eventually all mechanized production.

Thus it was that the foundations of American economic policy *and* what became known as the "American system" of mass production shared a common intellectual ancestor: the Enlightenment belief that the principles that govern the natural (that is, material) world are universally applicable; that once identified such principles must be allowed to function "naturally" or without interference—be they a clock escapement, the laissez faire market place, a standard manufacturing gauge, or a system of checks and balances—in order to achieve "the best of all possible worlds."[30]

WITH THE MATERIAL extravagances that came with the rise of big business and the appearance of robber barons in the second half of the expansionary nineteenth century, laissez faire acquired a new meaning. From the mouths and pens of apologists for unfettered business enterprise, laissez faire came to signify Social Darwinism, or the belief that only by the survival of the fittest in pure economic competition (a condition that never existed, given that the public provided most of the social capital on which business relied) could the country continue to thrive—or even prevail over what were then considered inferior peoples around the globe.

But Adam Smith's work in the United States had already been done. It had provided those responsible for the financial future of the fragile and bankrupt republic a model for national economic policy that would endure to this day. That model rejected the comprehensive and centrally conceived and administered economic policies that had evolved with the new nation-states of seventeenth-century Europe. In place of the manipulations of government ministers and functionaries, *The Wealth of Nations* urged confidence in the natural course of the market place, which, thanks to human nature's inclination to "truck, barter, and exchange," would ultimately produce a more secure prosperity for the nation's citizens.

Unlike Smith, however, Robert Morris and Alexander Hamilton were faced with at best the rudiments of a national economy. The system they proposed to deal with it was decentralized, pluralistic, and pragmatic. Internal improvements and a variety of what Hamilton envisioned as temporary measures to promote or protect new industries was as far as the American faith in cumulative individual actions would go. Their prescription prevailed as a model for public policy in the United States well into the twentieth century. Just as federal technology policy would be embedded in an open economic system, so also would federal policy to promote scientific discovery be embedded in an open system of public policy and administration, as we shall see in chapter 4.

Hamilton's contribution to the development of the U.S. economy was not limited to treatises—the Jeffersonians ensured that the appearance of *The Report on Manufacturers* would fall far short of an enthusiastic congressional reception—but extended as well to the creation of a laboratory in which his economic policy might be tested. The lessons of Britain's economic advantage, thanks to her industrial prowess, were not lost on Hamilton, who engineered in 1789 the formation of the New York Manufacturing Society. He then joined his assistant treasury secretary Tench Coxe, another strong advocate of manufacturing, in creating the Society for Useful Manufacturers (SUM). The society settled on the Great Falls of the Passaic River in northern New Jersey as the site for its new industrial city, one that would exploit the abundant waterpower at the site to operate mills and factories producing all manner of goods. New Jersey governor William Paterson repaid the compliment of having the new town named after him with a monopoly charter and ten-year tax exemption. Unfortunately the venture suffered an early setback when several of the society's directors managed to lose most of the society's funding by speculating with it, and then losing it in the financial panic of 1791. Paterson itself would recover in the nineteenth century, however, as a center of textile, locomotive, and firearms manufacture.[31]

Some unfinished business remained for the twentieth century, however: to fully integrate technological innovation into the general understanding of the *dynamics* of economic growth. Throughout much of the nineteenth century economists remained preoccupied with the categories of labor, land, capital, profits, rents, and—for policy purposes, tariffs and monetary and fiscal policy. Technology was something exogenous, occurring outside of normal economic considerations, though it might indeed influence those considerations. Adam Smith and his acolytes in the new American republic recognized that technological innovation would play an integral role in the process of producing wealth, and acknowledged as much in their forays into political economics. But a full empirical understanding of the ways that technology might play that role was necessary before it could become a well-integrated component of public policy. Building that understanding was the achievement of Joseph A. Schumpeter, Simon Kuznets, and Jacob Schmookler.

Technology: A Critical Role in Economic Productivity

Joseph A. Schumpeter (1883–1950) was not the first economist to recognize technological innovation as an essential ingredient in economic change, but in the United States he was by far the most influential. Born and educated in Austria, Schumpeter served briefly as minister of finance for the Austrian government immediately after World War I, and lost his own fortune when the Biedermarkbank (of which he was president) failed in 1924. These experiences no doubt contributed to the sociological realism with which he pursued the study of economics and finance, which he taught at the universities of Graz and Bonn before settling down to teach and influence a new generation of economists at Harvard University, from 1933 until his death in 1950.

Schumpeter dismissed a linear conception of economic progress, propelled forward by a freely operating market place, as existing only in an imaginary world. In its place he offered the concept of economic life as a state of imperfect equilibrium as producers and consumers continuously determine what shall be produced and how much of it will be sold, at what price—all under whatever circumstances may exist at any point in time. This state of imperfect equilibrium, characterized by constant adaptations to changing economic data, was likened by Schumpeter to a "circular flow." Economic development occurs when there is a "spontaneous and discontinuous change in the channels of the flow, disturbance of the equilibrium, which forever alters and displaces the equilibrium state previously existing."[32] The primacy of technological innovation then follows from Schumpeter's insight that "innovations in the economic system do not as a rule take place in such a way that first new wants arise spontaneously in consumers and then the productive apparatus swings round through their pressure. . . . It is . . . the producer who as a rule initiates economic change, and consumers are educated by him if necessary; they are, as it were, taught to want new things, or things which differ in some respect or other from those which they have been in the habit of using."[33] To early twenty-first-century readers only too familiar with planned product obsolescence and commercial media advertising, this is a noteworthy observation for a work first appearing in 1912.

Once we accept Schumpeter's axiom that production involves the combination of materials and forces, then it is only a short step toward technological innovation as the principal mover in economic development. True change in the channel of "circular flow," he argued, results not from incremental adjustments in the combinations used to produce things, but when genuinely novel combinations of productive means are carried out. "This concept," he continues, "covers the following five cases: . . . the introduction of a new good—that is one with which consumers are not yet familiar—or of a new quality of a good; the introduction of a new method of production . . . which need by no means be founded upon a discovery scientifically new, and can also exist in a new way of handling a commodity commercially; the opening of a new market . . . ; the conquest of a new source of supply of raw materials or half-manufactured goods . . . [and] the carrying out of the new organisation of any industry, like the creation of a monopoly position (for example through trustification) or the breaking up of a monopoly position."[34] With the exception of Schumpeter's third case, and the arguable exception of his fifth case, all of his new combinations are varieties of technological innovation.

Schumpeter elaborated on this concept (as well as other themes) in probably his best-known work, *Capitalism, Socialism and Democracy*, first published in 1942. Since the late 1800s the industrialized economies had experienced not only the industrial revolution "dear to the heart of textbook writers," but three similar "long waves of economic activity," of which the most recent, when Schumpeter wrote, had peaked in 1911. Not only do these revolutions constitute the key forces behind economic development, they are also responsible for cycles of economic prosperity and depression. Industrial revolutions "periodically reshape the existing structure of industry by introducing new methods of production—the mechanized factory, the electrified factory, chemical synthesis and the like; new commodities, such as railroad service, motorcars, electrical appliances; new forms of organization—the merger movement; new sources of supply—La Plata wool, American cotton, Katanga copper; new trade routes and markets to sell in and so on." This process of industrial renewal did not happen without cost, for it left in its wake, each time it occurred, "antiquated elements of the industrial structure," resulting in "prolonged periods of rising and falling prices, interest rates, employment and so on, which phenomena constitute parts of the mechanism of this process of recurrent rejuvenation of the productive apparatus."[35] This process Schumpeter would characterize as "creative destruction," and it would be a process that not even capitalism itself could escape.

While it was not difficult to accept Schumpeter's view of the importance of technological and business innovation to economic change (especially considering his broad-ranging experience, erudition, and persuasiveness as a writer) its importance was confirmed by the later work of two economists whose paths would also cross at Harvard: Simon Kuznets and Jacob Schmookler. Kuznets was among the first economists to explore the question of the relationship of technology and economic change using a combination of quantitative historical data and patent statistics. He shared Schumpeter's belief in the importance of innovation, which he referred to as "technique," applicable to improvements in business organization as well as engineering. Technique was, for Kuznets, one of the three most dynamic forces in economic history, the other two being population and demand. Examining the history of U.S. economic development from the 1830s through the 1920s, he found that periods of increasing economic productivity did indeed ebb and flow. Why was this so?

New technologies, Kuznets concluded from his research, do not revolutionize an industry indefinitely. "As time goes on . . . only a very small percentage of the patents issued represents changes that have been or will be actually introduced. What the number of patents really indicates is the amount of attention that the inventive capacities of the nation pay to problems connected with a specific mechanism or process."[36] From this and other data Kuznets concluded that no industry benefits from technological innovation for long on its own; it relies on subsidiary industries to supply necessary raw materials for the products being manufactured. Unless the rate at which these other industries produce keeps pace with the innovating industry, they will "act as a check on the growth of the industries favored by the advantages of technical development."[37]

The supply of capital available for investment in innovation is an important variable as well. "The rise of a new industry or the revolutionary expansion of an old one implies a considerable new investment. Capital must be provided either from the returns of the industry concerned, or from the returns of the other industries." For example, early steam engines had been financed with money from the iron industry and hardware manufacture, as well as British and Dutch banks, while the early automobile industry was financed by capital from entrepreneurs in other industries, credit extended by parts makers, and reinvested profits. Capital markets did not assist the industry until it was established and "had achieved comparative technical stability."[38]

Building on the work of his predecessors, Jacob Schmookler stimulated more than a decade of work at Harvard in quantitative studies of the sources and economic consequences of technological innovation. Like Schumpeter and Kuznets, Schmookler questioned the preoccupation of "classical and neoclassical" economists with tariffs, monetary and fiscal policy, wages, and prices. His own work taught him that *intellectual capital* or technological capacity in advanced countries, "reflected in the production of better products and the use of better methods . . . has been much more important than the accumulation of physical capital in explaining the rise of output per worker . . . when the period studied covers several decades."[39]

Schmookler examined both U.S. patents issued since 1874 and grouped by their Standard Industrial Code (SIC) classification, and 934 inventions made during the period 1800–1957 that he judged important on the basis of their descriptions in technical and trade journals and technological and economic histories.[40] Having done so, he found that in the majority of cases, no stimulus to the invention was identified.

> [But in] a significant minority of cases, the stimulus is identified, and *for almost all of these* that stimulus *is a technical problem or opportunity conceived by the inventor largely in economic terms.* . . . When the inventions themselves are examined in their historical context, in most instances either the inventions contain no identifiable scientific component, or the science that they embody is at least twenty years old . . . scientific discovery is seldom a sufficient condition for invention, either in the short run or the long. . . . Particular scientific discoveries are seldom even necessary conditions for later inventions if we think of the latter in terms of rough substitutes.[41]

Instead, the documents Schmookler examined indicated that the most prevalent "initiating stimulus" for invention was the inventor's "economic valuation of a technical problem or opportunity."[42]

FROM THE FIRST HOURS of the new republic, the reach as well as the boundaries of government action to promote and regulate technology had begun to take their place in the configuration of American politics. Centrally conceived and administered direction of the nation's economy and institutions had been clearly rejected by Adam Smith and the Federalists—most notably Alexander Hamilton—as well as their Jeffersonian opponents, who feared central government even more. Experience among the colonists and new states, whose geographic distances could be breached initially only by horse and stream, conspired with a diffuse belief in the beneficence of human nature, freely pursuing its rightful ends, to create a national virtue out of necessity.

Meanwhile, the integration of technology into the political economy of the new republic did not extend to a corresponding integration of research in the natural sciences. This was due partly to the fact that scientific inquiry in the eighteenth century was undertaken as a branch of philosophy—natural philosophy—and thus largely (though not exclusively) the preoccupation of academics, as indeed science would remain for much of the following two centuries. But nothing served to isolate the natural sciences from the changing winds of political practice and ideals more than the exceptionalism of their shared ideology, an ideology that was inherited from the Greek and Roman authors whose works were staples of a formal education in America during the eighteenth and nineteenth centuries. That ideology, and its consequences for the political arrangements for science in modern America, are the subject of the next chapter.

3

The Ideologies of Science

In the last year of the William Clinton administration (1992–2000) the President's Committee of Advisors on Science and Technology (PCAST) issued an attractively printed and illustrated report entitled "Wellspring of Prosperity: Science and Technology in the U.S. Economy."[1] Like all such reports it was a political document, designed to secure support for the administration in exchange for continuing support for the policies advocated in the report. The report appeared in the eighth and last year of a Democratic administration whose vice president (Al Gore) was campaigning to become Clinton's successor. An ardent advocate for the protection of the natural environment, Gore had shown a greater interest in science and technology as a national resource than many of his predecessors. "Wellspring of Prosperity," as its extended subtitle promised, would tell us how investments in scientific research are improving lives.

For the Clinton PCAST investing in discovery meant "sustaining and nurturing U.S. scientific leadership across the *frontiers* of scientific knowledge," which is "not merely a cultural tradition of our nation, [but] . . . an economic and security imperative," and ensuring that "America remains at the forefront of scientific capability by sustaining our *investments in basic research*, thereby enhancing our ability to shape a more prosperous future for ourselves, our children, and future generations while building a better America for the twenty-first century."[2] Investments in basic research, however, are not like other forms of public sector investment (e.g., education, highways and airports, defense facilities), which can be readily appreciated as tangible things or specific public services. As the report noted, it is "*impossible to accurately predict which areas of science and engineering will yield ground-breaking discoveries, what those inventions will be, how they will impact other scientific disciplines and, eventually, benefit our daily lives.*"[3]

To the uninitiated, this might seem an extraordinary claim: We should invest in an undertaking that will have unpredictable consequences and whose only certainty is novelty—presumably a universally desired attribute. However, if we adjust the focal length of this image of "new frontiers" and "new discoveries," the rhetoric resolves itself into the reality of federal funding for basic research and scientific disciplines. Fundamental science, basic research, scientific disciplines—these things are not only important, reported the PCAST; they are critical to "America's record-breaking performance in the current world economy [of the 1990s]." "As experts seek explanations for America's record-breaking performance in the current world economy, it is tempting to credit our bountiful natural resources and our diverse, hardworking population. But . . . the credit for our recent success really goes to the powerful system we have generated to create new knowledge and develop it into technologies that drive our economy, guarantee our national security, and improve our health and quality of life."[4]

When the Clinton administration's PCAST turned to elaborating what it meant by *support*, it relied on investment principles it had adopted earlier in the administration:

- Science and technology are major determinants of the American economy and quality of life.
- *Public* support of science and technology is *an investment in the future.*
- Education and training are crucial to America's future.
- The federal *government should continue to support strong research institutions and infrastructure.*
- The federal *investment in science and technology* must support a diverse portfolio of research, *including both basic and applied science.*
- Stability of funding is essential.[5]

The extent to which an ideology of science had secured its claim on U.S. policy making is reflected in the principal argument the Clinton administration's panel of eminent figures in the world of science used to justify indefinite amounts of federal funding for scientific research conducted primarily in the nation's universities. Though first introduced into the canon of bipartisan domestic policy in the United States at the end of World War II, a war in which technological innovation played a critical role, the litany acquired its presumptive truth as a result of its deep ideological roots in American history.

Since congressional authorization is necessary for all federal spending, statutory language and legislative histories are our most reliable guide to what the Congress understood to be the purpose of its successive appropriations of federal dollars for scientific research. The closest we can come to a statutory definition of basic research can be found in the Code of Federal Regulations, where it is described as "that research directed toward increasing knowledge in science. The primary aim of basic research is a fuller knowledge or understanding of the subject under study, rather than any practical application of that knowledge."[6]

Meanwhile, the National Science Foundation publishes annually a statistical compilation and analysis of quantitative data—e.g., dollars spent, degrees awarded, salaries earned, patents awarded—by which policy makers and observers can gauge comparative trends in national and international science and technology. The taxonomy the NSF uses in collecting and interpreting its data is critical to any inferences policy makers might draw from it. The NSF parses the federal research and development (R&D) budget into five categories (including R&D itself):[7]

> R&D. According to international guidelines for conducting R&D surveys, research and development, also called research and experimental development, comprises *creative work that is undertaken on a systematic basis.* R&D is performed *for the purpose of "increasing the stock of knowledge, including knowledge about humanity, culture, and society," and using "this stock of knowledge to devise new applications."* (Organisation for Economic Co-operation and Development, 1994).
>
> Basic research. The *objective* of basic research is to gain more comprehensive knowledge or understanding of the subject under study *without specific applications in mind.* In industry, basic research is defined as research that advances scientific knowledge but *does not have specific immediate commercial objectives*, although it may be in fields of present or potential commercial interest.
>
> Applied research. Applied research is *aimed at* gaining the knowledge or understanding to meet a specific, recognized need. In industry, applied research includes *investigations oriented to discovering new scientific knowledge that has specific commercial objectives* with respect to products, processes or services.
>
> Development. Development is the systematic use of the knowledge or understanding gained from research directed toward the production of useful materials, devices, systems, or methods, including the design and development of prototypes and processes.
>
> R&D Plant. R&D plant includes the acquisition of, construction of, major repairs to, or alterations in structures, works, equipment, facilities, or land for use in R&D activities [italics added].[8]

The NSF's *Science and Engineering Indicators 2004* acknowledges that these definitions are imperfect and "subject to reporting complexities."[9] But complexity is not their problem; their problem lies in the fact that they rely for their distinguishing characteristic on the subjective *intent* or motive with which research is pursued, rather than its results.

Attempting to bring some clarity to this question, *Science and Engineering Indicators 2004* devotes a separate chapter to "academic R&D." Here, at least, we have a category that relies on objective characteristics for its meaning: Thus academic R&D is conducted in "an institution that has a doctoral program in science or engineering . . . [and] that expends any amount of separately budgeted R&D in S&E [science and engineering]."[10] Academic R&D is what some people employed in certain educational institutions *do.* But this is as clear as *Science and Engineering Indicators* can be, for "academic R&D activities are concentrated at the research (basic and applied) end of the R&D spectrum and do not include much development activity. . . . Despite this delineation, the term 'R&D' (rather than just 'research') is primarily used throughout this discussion because data collected on academic R&D often do not differentiate between research and development. Moreover, it is often difficult to make clear distinctions among basic research, applied research, and development."[11] In other words, we cannot really know whether we are funding basic or applied research or any other kind of investigation. The only thing we can be sure of is that when the Congress funds academic research, we are spending public funds on activities that colleges and universities define as research.

Since the White House Office of Management and Budget (OMB) grapples every working day with accounting categories for federal dollars, one can expect to find in OMB's own directives a more specific discussion of what the federal government considers to be scientific research. Whereas NSF relies on the subjective intentions with which research is pursued to distinguish basic from applied and other forms of research, the OMB looks, in addition, to the general methodology used in the research being done, as well as where it is being done.[12] Research, to the OMB, is "*a systematic study* directed toward fuller scientific knowledge" while development "is the *systematic use* of knowledge and understanding gained from research."[13] Since the expectation of systematic inquiry is implicit in the notion of a scientific discipline, the OMB's addition of the descriptor "systematic" contributes little to the precision of the official definition of scientific research.

While the PCAST's "Wellspring" report calls on the federal government to fund fundamental science, basic research, and scientific disciplines because they are responsible for "America's record-breaking performance in the current world economy," the government has been unable to agree on a concrete definition of what these things are, other than that they refer to scientific research conducted in universities, research hospitals, and non-profit laboratories. The conclusion we must draw is that what the government is actually funding is a class of institutions, whose claim to special treatment is that they conduct research in a variety of academic disciplines that specialize in knowledge of the natural world. Moreover, the special treatment claimed by these non-profit research-performing institutions is typically characterized as arising from a metaphorical social compact or partnership between government and universities. Until the Congress passes a law authorizing such agreements, they must remain metaphors—the central metaphors of the rhetoric that has evolved since World War II to explain the exceptionalism of this country's research universities and their claims to federal funding. Which brings us to Vannevar Bush and his *Science—The Endless Frontier* (1945).

The Bush Report

It is difficult to imagine a discussion of public funding for scientific research without coming upon a reference to *Science—The Endless Frontier*. One commentator has described it as "a kind of creation myth, a founding story about the new world conceived by the union of science and government during the war [World War II]."[14] Its author was Vannevar Bush, director of the federal Office of Scientific Research and Development (OSRD), created in June of 1941 at Bush's behest (Bush was then head of the National Defense Research Council, or NDRC) to coordinate the World War II research efforts of the military services and university scientists already at work on improved detection, targeting, communications, and weapons systems.

Bush's prominence as a standard bearer of federally supported scientific research and godfather of the post–World War II scientific elite is rife with irony. Grandson of a fishing captain and son of a Universalist minister who was happiest mixing with "all kinds," Bush grew up in "the middle-class Irish and Yankee part of Chelsea [Massachusetts] . . . across the Boston & Albany and Boston & Maine railroad tracks from Jewish immigrants and impoverished newcomers."[15] He graduated from Chelsea's public high school and worked his way through Tufts University, eventually joining the faculty of the Massachusetts Institute of Technology (MIT).

As a sickly child and social outsider, Bush entertained himself inventing and tinkering with gadgets, good preparation for a life as the successful inventor and engineer that he would become. He was most proud of being an engineer, whom he saw as one who "needs to pay attention to men as well as to things . . . [who] utilizes science to produce results that the public needs or wants. He stands between science and industry in many cases. He has to deal with men in companies, on boards of directors, in government, and the ability to do so is just as essential to his success as it is that he should be able to apply his technical skills." Bush might have thought he was describing himself.[16]

Until the professionalization of science toward the end of the nineteenth century, exploring the natural world—or "natural philosophy"—was a leisure activity accessible only to those whose means allowed them to spend their days in intellectual pursuits, or to those who would become teachers. Teaching for gentlemen scholars served, like the ministry, as a respectable occupation for the excess sons of the gentry and urban merchant class, much preferable to slipping into a class of tradesmen and shopkeepers. The serious pursuit of science tended to center in the northeast, where trade and manufacturing had for decades enabled the accumulation of sufficient capital in families to support learning for its own sake. Meanwhile, the transportation needs of the expanding country in the nineteenth century—whether for land surveys, laying roads and railroad beds, or coastal navigation—opened up new occupational avenues. With the growth of industrialization and large-scale agriculture in the U.S., scientific occupations could be found not only in colleges and the new universities, but also in industrial and agricultural laboratories. Throughout it all, the government of the expanding nation had numerous needs that could best be met with individuals familiar with geology, astronomy, scientific agriculture, land surveying, and the like.[17] In funding the work of scientists and engineers familiar with these disciplines, it helped to create a class of individuals who came to depend upon the federal government for funding which, if well used, could become the basis of successful careers.

As the growing numbers of scientists in government, industry, and university laboratories sought to distinguish themselves as a way of ensuring greater public influence and support, they began to parse the meaning of science into moral gradations to justify their evolving expectations for U.S. science policy. *Pure* science, *basic* science, *fundamental* science, *abstract* science, and *curiosity-driven* science—these qualifiers are exogenous to both the philosophical realism and the positivism of modern laboratory science. Rather, they come from the world of human sentiment and ideals that transcend the material concerns of ordinary people, representing instead the qualitative aspirations of exceptional individuals. That is why, try as they might, the authors of definitions at the NSF and OMB could not arrive at concrete distinctions between basic and applied research, and OMB abandoned the attempt at definition to the universities seeking federal funds for research. That is also why *Science—The Endless Frontier* has proven such a politically emblematic document.

The Bush report has been memorialized in countless science policy discussions for its linear argument for the public support of scientific research. This is the view that science, uniquely and necessarily, generates the knowledge which engineers, in turn, draw upon to invent new technological devices and processes. Upon these new devices and processes depend economic growth and human betterment. This is the "Bush paradigm." It was neither original with Bush—Francis Bacon laid out the same argument for the pursuit of science over three centuries earlier—nor is it likely that Bush himself held such a simplistic view.[18]

Given the long history of federally funded research in the United States, it is highly unlikely that the federal government would have abandoned support for science after World War II. Indeed, while technology has played a role in nearly every major war, few would question the extent to which World War II was as much an engine of technological advance as of social change. Radar, sonar, improved aircraft performance and range, missile guidance, the atomic bomb, and advances in antibiotics such as penicillin are only few of the strategically significant technologies that resulted from the mobilization of scientists and engineers for World War II. What was historically significant about *Science—The Endless Frontier* was Bush's unrelenting insistence that the government fund not merely research, but basic research in U.S. universities and endowed (non-profit) research institutes. (In his veneration of basic research Bush was not original either. Joseph Henry [1797–1878]—discoverer of electromagnetic induction, first secretary of the Smithsonian Institution, and first president of the American Association for the Advancement of Science—insisted on distinguishing between abstract and practical science as well.)[19]

As we saw with the Clinton administration's PCAST report, reports designed to shape policy are typically the work of committees convened to issue a consensus statement on the need for, and virtues of, the recommended policies. The committee assembled in 1944 to produce what became the report requested by President Roosevelt (in a letter undoubtedly written by Bush) consisted of fifty-six individuals formed into four subcommittees corresponding to the four issues upon which Roosevelt requested recommendations: the release of scientific information developed for the military during the war, continuing research in medical and related sciences, the proper roles of public and private research organizations, and "discovering and developing scientific talent in American youth." The composition of these committees is not surprising, considering the era. High-ranking executive officers (that is, presidents, deans) and faculty of well-established and mostly private research universities predominated among the committee's members. Among those universities, Columbia, Harvard, Massachusetts Institute of Technology (MIT), Johns Hopkins, and the University of Minnesota had more than one member; Harvard had four. A handful of private foundations and industrial research laboratories was represented.[20]

The political heft of these institutions' members was augmented by a senior staff member each from the National Research Council (NRC) and the American Institute of Physics (AIP), organizations that were strong advocates for scientific research. Responsibility for representing the world beyond academia and the laboratory fell to the two members from the Brookings Institution and the Institute for Current World Affairs. The federal government had two representatives—the deputy administrator of the Foreign Economic Administration and the director of the U.S. Geological Survey. But the country's precollegiate schools, whose efforts and goodwill would be critical to "discovering and developing scientific talent in American youth," were represented only by the executive secretary of the National Association of Secondary School Principals. The military services (which in two years would be wrestled into the Department of Defense [DOD] that we know today) were not represented at all. Thus, whatever else one can say about the significance of *Science—The Endless Frontier*, we cannot say that it documented a social compact agreed to by the scientific community and the American people or their government.

What the report *did* document was an appeal that the federal government grant the nation's universities the funding and authority necessary for them to meet their responsibility to (a) train young scientists, and (b) conduct the basic research necessary to ensure the "advances in science [which] when put to practical use mean more jobs, higher wages, shorter hours, more abundant crops, more leisure for recreation, for study, for learning how to live without the deadening drudgery which has been the burden of the common man for ages past . . . [for] higher standards of living . . . the prevention or cure of diseases . . . conservation of our limited national resources, and . . . [to] assure means of defense against aggression."[21]

World War II funding for scientific research already provided a precedent, so Bush needed only to specify a particular level of post-war funding, which he ventured to be $10 million at the outset, "and may rise to about 50 million dollars annually when fully underway at the end of perhaps 5 years." How these figures were arrived at, he did not say. Moreover, since by definition "basic research . . . ceases to be basic if immediate results are expected on short-term support," unlike any other form of government spending, programs to support university researchers should not have to account for themselves in the annual congressional budget process. The continuity in income necessary to sustain their efforts required, instead, that they should be able to enjoy "commitments of funds from current appropriations for programs of five years duration or longer."[22] This begged the question of whether most other government programs would not also benefit from continuity and stability in funding.

Years of experience in Washington had undoubtedly taught Bush that there would need to be some kind of organization to orchestrate the government's support for scientific research. None existing would serve the purpose. A new organization should be established, Bush urged, a National Research Foundation, the purpose of which would be to "develop and promote a national policy for scientific research and education . . . support basic research in nonprofit organizations, . . . develop scientific talent in American youth by means of scholarships and fellowships, and . . . contract and otherwise support long-range research on military matters."[23]

Undoubtedly as a result of his own wartime experience, Bush envisioned an important role for the new foundation in shaping military R&D. While it is true that the military in the U.S. is ultimately subordinate to civilian branches of government, Bush's proposal would transfer to scientists this constitutional function. He recommended that "as a permanent measure, it would be appropriate to add to the agency . . . the responsibilities for civilian-initiated and civilian-controlled military research. The function of such a civilian group would be primarily to conduct long-range scientific research on military problems—leaving to the Services research on the improvement of existing weapons."

This notion is especially interesting for two reasons. First, it overlooks the close interconnection between the development of military weapons systems planning and operational experience. Second, it assumes that "long-range scientific research on military problems" can be segregated from lessons learned during incremental military systems development. But Bush had little admiration for the military's capacity for worthwhile long-range research:

> It is the primary responsibility of the Army and Navy to train the men, make available the weapons, and employ the strategy that will bring victory in combat. The armed services cannot be expected to be experts in all of the complicated fields which make it possible for a great nation to fight successfully in total war. There are certain kinds of research—such as research on the improvement of existing weapons—which can best be done within the military establishment. However, the job of long-range research involving application of the newest scientific discoveries to military needs should be the responsibility of those civilian scientists in the universities and in industry who are best trained to discharge it thoroughly and successfully.[24]

Furthermore, the government should not only defer to scientists in its ordinary oversight responsibility (inherent in the budget authorization and appropriations process); it should also defer to them in its responsibility for allocating funds among various scientific projects. This was because the ordinary principle of federal administrative procedure—citizens should be treated equally under the law—was not suitable for science. Federal procurement laws and regulations, which governed the ways public funds could be spent, were generally guided by the principle of "full and open" competition for awards given on the basis of explicit and objective measures. The same principle resulted in elaborate federal accounting requirements which, though despised by most recipients of grants and contracts, provided an objective means of ensuring that public funds were spent as intended. Science was different, according to Bush's committee. It should be exempt from such onerous requirements:

> Since research does not fall within the category of normal commercial or procurement operations which are easily covered by the usual contractual relations, it is essential that certain statutory and regulatory fiscal

requirements be waived in the case of research contractors. . . . Similarly, advance payments should be allowed in the discretion of the Director of the Foundation. . . . Adherence to the usual procedures in the case of research contracts will impair the efficiency of research operations and will needlessly increase the cost of the work to the Government.[25]

The politically ambitious agenda laid out for Bush's National Research Foundation would have been less striking were it not for the proposed composition and nature of the foundation's leadership. Two centuries of U.S. constitutional experience and practice had reinforced the principle that reliable accountability to "the people" requires that the highest ranking officials of the executive branch be held answerable to the Congress through the Senate nomination and approval process. Similarly, the best assurance of fidelity to the law in the workings of government can be had through a civil service answerable to the president and the Congress. But Bush and his committee preferred an alternative principle: "Responsibility to the people, through the President and the Congress, should be placed in the hands of say [sic] nine Members, who shall be persons not otherwise connected with the Government and not representative of any special interest, who should be known as National Research Foundation Members, selected by the President on the basis of their interest in and capacity to promote the purposes of the Foundation."[26] Bush's definition of "special interest" seemed not to include university faculty and presidents.

Once appointed by the president on the basis of their scientific eminence, the "Members" would be free to run things as they wished, including appointing themselves, setting policy and regulations, administering the foundation, conducting its procurements, and reallocating funds appropriated by the Congress. For his pains, and in order "to attract an outstanding man to the post," the director could command a salary that was not subject to current civil service salary scales.[27]

The Bush committee report insisted repeatedly throughout *Science—The Endless Frontier* that the benefits of science were best achieved by enabling scientists at federally subsidized educational and private research institutions "to explore any natural phenomena without regard to possible economic applications." However, at the same time, Bush and his committee argued that these same researchers and their institutions might require special incentives, above and beyond the grants and contracts they received, to pursue their work. The most important incentive would be "patent rights resulting from work financed by the foundation." Though any financing the foundation could offer scientists would actually come from congressionally appropriated funds, the government, in Bush's view, should defer to the foundation in the disposition of intellectual property stemming from foundation-supported research. Bush proposed that "legislation . . . leave to the Members of the Foundation discretion as to its patent policy in order that patent arrangements may be adjusted as circumstances and the public interest require." Just what the circumstances and public interest might be would be left for the foundation to decide.[28]

The Ideology of Science

The importance the Bush report gives to "basic research," and the exceptional political powers it envisions for its most eminent practitioners, compel us to look closely at the layers of meaning that underlie the concept of "basic research." The report's language contains ample evidence that "basic research" signified something more than the disinterested inquiry Bush insisted it was. Passages referring to basic research in *Science—The Endless Frontier* describe it this way: Basic research is (italics added):

—[the] *creation* of new scientific knowledge, *least under pressure for immediate, tangible* results
—expanding the *frontiers* of knowledge
—*free to explore any natural* phenomena without regard to possible economic applications
—free to pursue the *truth wherever it may lead*
—the free play of *free intellects*, working on subjects of their own choice, in the manner dictated by their *curiosity for exploration of the unknown*
—[the] free, *untrammeled study of nature*, in the directions and by the methods suggested by [the researcher's] . . . interests, curiosity, and imagination
—[that which has produced] many of the most important discoveries [that] have come as a result of *experiments undertaken with very different purposes in mind* . . . the results of any one particular investigation cannot be predicted with accuracy
—research in the *purest realms of science*
—[contrasted to] Yankee mechanical *ingenuity* [which] *building largely upon the basic discoveries of European scientists*, could greatly advance the *technical arts*
—free from the *adverse pressure of convention, prejudice, or commercial necessity*
—*fundamental* knowledge
—the *creation* of new scientific knowledge [which] rests on that *small body of men and women who understand the fundamental laws of nature* and are skilled in the techniques of scientific research [29]

As the phrases emphasized above reveal, the Bush report's definition of true science echoes the notion of the intellectual life praised by the New England Transcendentalists, a life that disdains material, immediate, and commercial concerns. It also regards nature—the object of scientific inquiry—in platonic fashion as best understood on the ideal plane of a unifying, rational theory (for example, "fundamental laws").[30] So, too, do these passages describe the highest form of knowledge as something true and infinite which is *created* by our intellects, and not simply a mirror of the material world to be known primarily through the evidence of our senses. Thus, for Bush, the misapplied metaphor of the geographic frontier of American history served as a ready substitute for a concept that had been captured for American letters by another prominent son of New England, the essayist, Transcendentalist leader, and Unitarian minister, Ralph Waldo Emerson (1803–1882). The Transcendentalists had drunk deeply from the ancient vintage of platonic idealism, and especially from the writings of one of Emerson's favorite authors, Plutarch (A.D. 46–120), whose platonic idealism suffused his histories and moral essays.

According to platonic tradition, in a democracy power is achieved by quantitative means, while in an oligarchy of the wise, it is achieved by qualitative means—an assessment of the merit of an individual person's knowledge. And not simply knowledge, but theoretical knowledge, which ranked highest in the classical hierarchy of three kinds of knowledge. Ranking beneath theoretical knowledge was the *techne* of mechanics and hand-workers, while beneath *techne* functioned ordinary sense perception. This ranking undergirds modern scientists' argument for peer review as the appropriate means of identifying science deserving of public support. It also undergirds the argument that scientists working with public funds should retain the independence to select, direct, and evaluate their work. Thus the resistance articulated in *Science—The Endless Frontier* to requiring scientists to defer to the public ("the crowd" in Plutarch's essays) in allocating funding for activities identified as pure science draws from a deep taproot in western thought.

When composing the argument they would use to attempt to persuade the White House to champion university-based science, Bush and his colleagues needed a vocabulary to distinguish their vision of science as pursued in the elite universities from science as pursued elsewhere. The intellectual wellsprings of Cambridge, Massachusetts, provided them one. But why was it so important to support science *in universities*? Because universities were where fundamental (or academic) knowledge was preserved, created, and transmitted to subsequent generations. In the platonic scheme of things, academic knowledge is antecedent and morally superior to practical or technical knowledge. It is also where gentlemen aspiring to positions in life that would involve them neither in husbandry or trade could be educated among those sharing similar aspirations and destinies. While industrial or government research laboratories certainly offered employment to the scientifically inclined, they could not substitute for the universities in the Boston-Washington axis where, not by accident, the World War II science establishment felt most at home.[31]

Universities receiving government research and development contracts during World War II had undergone considerable expansion in facilities and faculty; these needed to be maintained. Quite aside from the fact that Bush himself had been on the MIT engineering faculty from 1919 to 1932, he served as dean of engineering under physicist and president of MIT, Karl T. Compton. Compton had joined the Washington science establishment as chair of the Science Advisory Board created by President Roosevelt in 1933, which functioned under the aegis of the National Academy of Science. In 1931 Compton had not only been named president of MIT (he then named Bush to join him as MIT's vice president), but chairman of the governing board of the American Association for the Advancement of Science (AAAS). Not surprisingly, Compton became a tireless advocate for the public support of scientific research in the best universities. Compton was a close colleague and friend of Bush, who named him to the committee he assembled to prepare what became *Science—The Endless Frontier*. So also was James B. Conant—who received his doctorate in chemistry from Harvard in 1916, the same year that Bush received his doctorate in electrical engineering, jointly granted by Harvard and MIT. Conant was elected president of Harvard in 1933, and served as Bush's closest and most trusted colleague during their Washington years. When Bush persuaded the White House to create the OSRD in 1941, which he would head as director, Bush saw to it that Conant was named to his former post of chair of the NDRC (est. 1940). Later, when Bush grew uncomfortable carrying the entire weight of OSRD on his own shoulders, he asked Roosevelt to approve a "Top Policy Committee" for the organization. Roosevelt did; Bush and Conant were the only non-government members.[32]

It is no wonder that the three men—Compton, Conant, and Bush—built close working relationships rooted in the academic life of two of New England's leading (and neighboring) universities, later to become powerful advocates for a strong post-war commitment for federal government funding for research universities. They were joined by Isaiah Bowman, J. Hugh O'Donnell, and Walter C. Coffey, presidents respectively of The Johns Hopkins University, University of Notre Dame, and the University of Minnesota. During the debates over the bills to create a National Science Foundation, Bowman and O'Donnell would be among the staunchest advocates for giving the board of the proposed foundation full authority over its governance and policy.

Once the public accepted the idea of basic science as a professional pursuit, one which by its very nature required almost total public or private subvention, there was still the matter of public accountability in a constitutional republic. The exercise of authority in such societies requires at the very least the passive consent of the majority of citizens. However, advocates of government support for university research wanted to secure not only funding, but authority over what was done with the funding—that is, to be accountable not to the Congress which, however imperfectly, responds to the electorate in dispensing public funds, but only to themselves. How could the public be persuaded to allow scientists to decide for themselves how to spend the people's money? For this, too, New England offered a deep ideological well from which to draw.

Creating the National Science Foundation: A Closed System at Bay

As for *Science—The Endless Frontier*, President Harry S Truman, formerly of Missouri was having none of it. It was not that Truman was indifferent to modern science. But Roosevelt, his health failing through much of his second term, died on April 12, 1945, leaving a vice president with whom he had shared virtually nothing of the deliberations and decisions of his presidency during World War II. Truman was not part of any high-level consensus about the merits of a post-war research program, and so was free to form his own opinion on the future relations between the federal government and the research institutions it had enlisted to help win World War II. In this he was helped by Senator Harley M. Kilgore, Democrat from West Virginia, lawyer and son of an oil prospector. Missouri and West Virginia were far away from Cambridge, Mass. in more ways than one—culturally, economically, and ideologically. This distance was reflected in the debates over bills to establish a national science foundation led by Senator Kilgore, who introduced the Truman administration's bill, and Senator Warren G. Magnuson (D-Washington), who was working closely with Bush. Before Truman could be presented a bill that he could sign, these issues had to be resolved:

- Should funds be limited to basic science, that is, projects defined by the researchers themselves, or should they also be extended to applied science and technologies, that is, funding for projects defined by the government to address politically identified need?
- Should funds be distributed through some geographic formula that would ensure that state universities and land-grant colleges received "their share" of funding, or should funds be distributed solely on the basis of scientific merit?
- Who should control the foundation—scientists, or officials of the federal government?
- Should the social sciences receive federal support as well, and within the framework of a national science foundation?
- Who should retain the intellectual property rights to inventions resulting from federally funded research?

These issues were a natural outgrowth of the conflict between the ideology of basic science and the American ideology of popular government—as in the question of whether scientists or the government should control policy for the foundation, and whether funds should be distributed by geographic formula or on the basis of scientific "excellence." The issue of control was fiercely argued. Johns Hopkins University president Isaiah Bowman, who had served on Bush's committee, was blunt in voicing his and Bush's opposition to entrusting U.S. scientific research to a government organization rather than an autonomous (but publicly funded) foundation governed by scientists. Demonstrating what good bedfellows ignorance and arrogance can be, Bowman warned a Senate committee: "Do not open doors for untrained and worse than worthless [government] employees who may creep into positions of control and attempt to pass themselves off as wise administrators who understand better than so-called fuzzy scientists how the job should be done."[33] Bowman's and Bush's view that the direction of scientific research must be left entirely to academic scientists in order to retain its purity and promise was echoed throughout the debates over national science foundation bills introduced between 1945 and 1950.

Vannevar Bush's opposition to including the social sciences (which Kilgore and some other congressional Democrats favored) was predictable, given his conservatism. What many physicists and engineers may not have appreciated was that sociology, economics, history, and political science also had distinguished intellectual pedigrees, traceable back to the mid-eighteenth century, if not earlier. But intellectual pedigree was not the real issue. The trouble with the social sciences was that they attempted to marry the scientific method, with its emphasis on systematic data collection and abstract reasoning, to the study of human society. That is, the social sciences engaged with the world of human experience rather than nature, and thus risked being compromised in truth-seeking by the "nail of pain and pleasure which fastens the body to the mind."[34] They were liable to such un-republican uses as liberal social reform; and, of course, they would be competitors for a research budget that might never be large enough for everyone.

The debate also questioned whether contracts or grants were the better legal instruments for funding scientific research. This was no minor technical matter. A standard government procurement contract requires a well-defined product or service in return for specified payments. But basic research is not supposed to have a practical purpose or outcome for which researchers must answer to someone else, much less the government. Hence research done under a federal *contract* is assumed to be applied research, in which science is applied to the development of a commodity or process for the government. On the other hand, basic research must be funded on terms that allow scientists to go about their work more or less as they please—or to conduct "curiosity-driven" research. Hence the appeal of grants—funds awarded outright to certain categories of individuals or organizations statutorily eligible to receive them. That being so, why not simply grant to all universities awarding science degrees a certain amount of funding based on such factors as the number of students enrolled in the natural sciences, or the amount of funding raised by state and private sources for scientific research? This, in fact, is what Senator Kilgore and other advocates of distributing research funds on a geographic basis were proposing.

However, receiving public funding because one's institution belonged to a particular category would be inconsistent with the ideological principle that public subvention should follow scientific merit. Moreover, the leading research universities, having acquired much of their preeminence from government work during World War II, favored funding instruments that would help to preserve that preeminence. Advocates for the distribution of funding on the basis of excellence, they argued for the use of peer review to determine relative excellence. The peers doing the reviewing, of course, would by definition be other scientists working in similar fields. The awarding of basic research grants would remain under the control of scientists in the disciplines being funded. Thus the question of how the new foundation's funds were to be distributed was entangled in such issues as the procedures by which funds were allocated, whether the recipients were obligated to the government to account for the results of the funded research, and whether funding would continue to flow to those who had become a part of a familiar group of established peers.[35]

After five years of legislative wrangling over the twenty-one different NSF bills introduced into Congress during 1945–1950, the Congress was able to pass a bill that President Truman could sign on May 10, 1950. The final bill created a National Science Foundation that was to provide support for basic research in the natural sciences, mathematics, and engineering. It could, but was not directed to, support the social sciences.[36] And (in a defeat for the Bush committee's proposal) the foundation would be governed by a policy-making board and full-time director, all appointed by the president and confirmed by the Senate. It would dispense its funds through project grants (rather than to individuals) awarded largely on the basis of peer review—in this respect following a path set by the Office of Naval Research (ONR), established in 1946. The foundation was authorized to award scholarships and fellowships, as well as disseminate scientific information. Intellectual property (patent) issues were deferred to the NSF's governing board.

While the foundation was not required to distribute funds by a geographic formula, over time it did begin to award program grants to promote science education and science and mathematics in minority institutions. During 1971–1978 the NSF attempted to fulfill a 1968 change to its charter mandating it to conduct applied research into various socioeconomic problems. Its effort was the Research Applied to National Needs (RANN) program addressing such problems as pollution, energy sufficiency, transportation, and urban decay. Complaints from the natural science disciplines, fearing competition for their budgets, and from other agencies, fearing encroachments into their own missions, led to the gradual phasing out of the program. The NSF was also charged to evaluate and coordinate the intramural research performed by federal agencies. This would prove an exceptionally challenging task, given that research done elsewhere throughout the government is carried out primarily to serve the sponsoring agencies' missions (e.g., NASA, Department of Defense). Those agencies and their constituencies and congressional supporters were unlikely to welcome "second guessing" by the federal government's basic research agency.

Some of the alignments surrounding the creation of the NSF might have been partisan, but it would be a mistake to characterize the five-year struggle as the result solely of a natural fault line between American populism and the elitism of a small group of eminent scientists. The most serious divide was between the scientist-administrators from the elite private universities and the most senior civil servants in the Truman administration. This latter group included budget directors Harold D. Smith and his successor James E. Webb (who served as budget director from 1946 to 1949 and would become NASA administrator in 1961); Don K. Price, political scientist and staff to the budget director; sociologist, economist, and presidential assistant John R. Steelman; and Henry A. Wallace, an agronomist, former secretary of agriculture and vice president under Roosevelt and now secretary of commerce, whose department included the U.S. Patent and Trademark Office.[37]

Without exception this group opposed the Bush committee's insistence that the new foundation be governed by a board which, while appointed by the president, would not be associated with the government and would select a foundation head answerable only to themselves. Budget director Webb spoke for the Bureau of the Budget and the White House staff when he argued in a letter to Bush: "Persons with responsibility for the disbursement of public funds should not be actively associated with the beneficiaries of those funds. . . . Only an official with undivided allegiance and fully supported by a responsible relation to the President can safely make such determinations."[38] Nothing could have cast a brighter glare on the divide that separated the Bush committee from the senior people in the Truman White House than Truman's reply, three years later, to the news that the newly named board of the NSF was at work framing criteria for the selection of the foundation's director: "Oh, don't bother with that. There's only one criterion. He must get along with me, and I already have my man picked." (The man Truman had picked was Alan T. Waterman, a Yale physicist and, since 1946, deputy director and chief scientist at the ONR.)[39]

THE CONCEPT OF BASIC or pure science remains an idea with a long history of political significance. Like notions of patriotism and piety, when used to legitimate a claim to moral and political superiority the idea is not, because of its subjective quality, easily challenged by outsiders. Meanwhile, setting intellectual and institutional boundaries and devising secular rationales for influence and funding was a natural component of the process of professionalization which all of the intellectual disciplines underwent during the nineteenth century.[40]

While *Science—The Endless Frontier* has become an icon of academic science, the Bush paradigm had only one modest success in the creation of the National Science Foundation: the dedication of a single federal agency to the support of "basic" research. Even this small victory was short-lived, however. In 1968, during the administration of another Democratic president, Lyndon B. Johnson, and under the guidance of a Democratic congressman from Connecticut, Emilio Q. Daddario of the House Committee on Science and Technology chaired by George R. Brown of California, the NSF's charter was substantially broadened, putting the narrow focus on academic basic research in jeopardy.

Hearings on a new NSF bill surfaced four issues: the foundation's failure to support the social sciences, as it had been authorized to do in its enabling legislation; the need for applied research support for industries trying to cope with the growing political momentum favoring regulation to improve environmental protection and the sustainable use of natural resources; its failure to play any significant role in the development of comprehensive national science policy; and its passivity in allowing others to decide for it what it should fund.

The legislative proposal to revise NSF's charter authorized funding for applied research, and thus provoked renewed calls for the protection of basic research lest the nation's long-term future be jeopardized. The National Academy of Sciences weighed in with its warning that "the support of applied research at academic institutions may be the thin edge of the wedge which could ultimately result in pressures to support mission-oriented work at the expense of basic research." This would have "a damaging effect upon science, upon our academic institutions and, in the long run, upon our economic well being."

The foundation was also to establish a division for the social sciences, and, in addition to collecting data on scientific research around the country, it would have to interpret the data and report annually to the Congress. Also disconcerting to the foundation's academic constituency was the creation of an NSF deputy director and four new directorships, all to be appointed by the president and confirmed by the Senate. As a consequence, the foundation would have an unusually large (for its size) number of political appointees in its top management. New programs would require authorization from NSF's congressional oversight committees as well. These changes to NSF's charter brought the foundation about as far away as one could get from Bush's vision of a publicly funded but scientist-controlled organization to support basic science.[41]

What evolved instead was a policy best characterized as pluralism in both the conduct and direction of scientific and engineering research in the United States. The Atomic Energy Commission (AEC) and the ONR, both established in 1946, provided new sources of research funding for physicists, geologists, chemists, and engineers (and numerous sub-disciplines that matured during mid-century), as did the National Aeronautics and Space Administration (NASA, est. 1958) and the National Oceanic and Atmospheric Administration (NOAA, est. 1970).[42] Biologists and chemists of all varieties might find research support at the Department of Agriculture, the Food and Drug Administration, and the National Institutes of Health (NIH, created in 1930 out of several older federal medical services, and authorized to fund research in basic biological and medical problems). While the predecessor organizations of these federal research agencies performed most of their research with in-house scientists, contracts and grants to external research institutions prevailed after World War II. The military services, organized into a single Department of Defense in 1947, continued to be important sources of funding for research in engineering-related disciplines.

Pluralism rather than centralization in federally sponsored research has won out as the most favored approach to the public support of science by researchers and R&D agencies alike, with the OMB and occasional experts in public administration remaining the only significant advocates for the efficiency that is supposed to result from centralization. Among university researchers, having a variety of federal organizations with differing missions from which to seek support has a singular advantage. Programs dispersed throughout many federal agencies fall under the protective patronage of numerous committees in the Congress and their constituencies.

Meanwhile, relatively small research programs arrayed throughout a number of agencies are a more difficult target for budget cutters than a single more visible research budget. One can also argue that two scientific investigations of similar research questions are not necessarily duplicative. Different investigators and different research designs tackling similar problems may produce different, but no less worthwhile, results. From a broader perspective, research institutions benefit as well from the dispersion of funding sources between the private and public sectors, with the government's share of the national total of $265.5 billion for research and development declining to slightly over 25 percent in 2000. Periodically, as during the Ford and Clinton administrations, official Washington continues to be stirred by proposals to create a single Department of Science to impose order and efficiency on the federal science enterprise. Just as predictably, such proposals fail.

A Partial Victory for the Bush Paradigm: The National Academy of Sciences

A small group of scientists and engineers largely from New England's leading academic institutions had dominated the management of wartime research for the Roosevelt administration, just as it had dominated federally sponsored research during the second half of the nineteenth century. The vision reflected in Bush's 1945 report was of the institutionalization of that elite. In modern parlance, the distance between Vannevar Bush and Harley Kilgore was the distance between *ex*clusionary (oligarchical) and *in*clusionary (democratic) government. Granting the critical contribution of Bush's elite to the allied victory over the Axis powers, we have no way of knowing whether a science policy after the war based faithfully on Bush's program would have produced more or better research—even if we could agree on how we would measure "more" or "better" research.

The closed system that lay at the heart of *Science—The Endless Frontier* was also not original to Bush. Creating such an establishment, one dominated by a Boston-Washington axis of gentlemanly influence—had been proposed during the Civil War when the southern states, having seceded, were no longer represented in the Congress. The proposal for what became in 1863 the National Academy of Sciences originated with a group of about two dozen men employed at one time or another by the federal government in scientific mapping of the continental United States.[43]

Sharing ties as well to Harvard University and the Smithsonian Institution, this self-consciously elite group identified fifty men like themselves and persuaded Senator Henry Wilson of Massachusetts—himself not known for scientific interests—to propose legislation in the waning hours of the lame-duck Congress of 1863 to charter their own elite society, like the Royal Society of London for improving Natural Knowledge (est. 1660). The proposed National Academy of Sciences would be chartered by the state (as the Royal Society had been chartered by King Charles II). Also like the Royal Society, it would be independent and self-governing—except that it would look to the Congress to support it by requesting studies of scientific matters. While Senator Wilson may have managed to hustle the legislation creating the National Academy of Sciences through the Congress, and the legislation may have been signed by an otherwise preoccupied President Lincoln, the elitist nature of the academy soon attracted enough critics that the academy felt obliged to increase its membership to seventy-five.

Thanks to a bequest from one of its founding members, the academy managed to survive, its members regularly commending their researches to one another. Such scientific or technical work as the government needed during the next war (World War I) was financed through the professionally administered research of the National Research Council, organized for that purpose in 1916. Neither President Roosevelt nor President Truman would approve appropriations for it. And, as we have seen, Truman was not enamored of a similarly exclusive arrangement in Vannevar Bush's proposal for a national science foundation. Today the National Academies of Science and Engineering have fared somewhat better. While executive branch agencies occasionally look to the academies to produce studies of complex or controversial issues besetting their programs, the Congress will often ask the academies to evaluate the same programs, or render expert opinions about other matters before them. Studies, and periodic gatherings of various boards representing the interests and views of elite research organizations, form the academies' principal work, which is paid for by earmarked funds in department and agency budgets.

THE SEEDS OF policy that would nurture the phenomenal growth of a U.S. technology-driven economy in the nineteenth and twentieth centuries were not planted primarily by the closed circle of scientists traveling back and forth between New England's elite campuses and Washington between 1850 and 1950. While their understanding of scientific excellence and its importance to the American prospect was well-founded, the propensity toward open systems in American culture prevailed in the pluralism that characterized federal support for science and technology in the second half of the twentieth century. The conventional pairing of policies to promote science and technology as "science and technology policy" formalizes a fiction, namely, that technological and economic progress *necessarily* depend on scientific progress.

Ideology has combined with historical experience in the United States to produce a policy that has shaped the contours of this country's political relationship with science and technology: the policy of using government resources to build into the nation's industry and academic institutions a capacity to produce, on call, world-class scientific research and a technologically advanced industry second to none. "Preeminence" is the often cited objective of this policy, an objective which critics might rather describe as one designed for global military and economic hegemony. Such a policy, operating in mutual dependency with communities of expertise, could and did exacerbate the critique during the 1960s of modern technology among those who (with some historical irony) could be described as later day Transcendentalists. Certainly for many of those critics, Henry David Thoreau's *Walden; Or Life in the Woods* (1854) was a virtual Bible.

Two of the most astute and literate of these critics have been Lewis Mumford and Theodore Roszak, whose writings reached far beyond the "counter culture" of the 1960s and continue to have a wide readership, judging by the numerous editions readily available from Internet book sellers.[44] "The roots of technocracy," wrote Roszak,

> reach deep into our cultural past and are ultimately entangled in the scientific world-view of the Western tradition. . . . [I] define the technocracy as that society in which those who govern justify themselves by appeal to

technical experts who, in turn, justify themselves by appeal to scientific forms of knowledge. And beyond the authority of science, there is no appeal. . . . Within such a society, the citizen, confronted by bewildering bigness and complexity, finds it necessary to defer on all matters to those who know better . . . since it is universally agreed that the prime goal of the society is to keep the productive apparatus turning over efficiently.[45]

Broad and sometimes exaggerated generalizations are staples of cultural criticism, whether from the political left or right. Roszak, however, was giving voice to a fear that had been expressed, albeit in more measured terms, by the outgoing president Dwight D. Eisenhower in January of 1961. This was a new fear arising from the growing influence in Washington of the nation's leading scientists and research institutions as a result of their substantial contributions to winning World War II and waging the Cold War. "In holding scientific research and discovery in respect," warned Eisenhower, "we must also be alert to the equal and opposite danger that public policy could itself become the captive of a scientific-technological elite."[46]

During the same decade, while Jacob Schmookler demonstrated the frequent independence of technological innovation from prior scientific discoveries, college and university campuses began to subject the intellectual and institutional production of scientific knowledge (as well as the origins and varieties of technological practice) to an unprecedented degree of sociological, philosophical, and historical examination. The most striking result of this examination was the challenge to the ideology that had propelled an unprecedented period of federal funding for, and deference to, academic science. The notion that science was somehow immune from the human foibles and organizational dynamics that shape other categories of social, political, and intellectual activity—or the belief in the intellectual and moral exceptionalism of science—lay in tatters, resurrected only on celebratory occasions calling for its liturgical affirmation.

Some of this critical analytical interest was undoubtedly attributable to the anti-establishment temper of a generation that came to maturity wrestling with a perplexing and costly war in Vietnam, the apparent captivity of post–World War II research to military and industrial interests, and the discovery that civil rights for all Americans remained more promise than fact. A discussion of the extensive and varied literature produced by scholars committed to the sociology, philosophy, and history of science is beyond the scope of this book. However, three figures stand out among those who laid the intellectual foundations of the social studies of science (and technology), and also managed to reach audiences beyond specialist readers in the new academic discipline: Robert K. Merton (1910–2003), Karl Popper (1902–1994), and Thomas S.

Kuhn (1922–1996). Each in his own way challenged—albeit somewhat indirectly—the notion that science and politics can be understood, much less pursued, as completely separate vocations.

Whether Vannevar Bush was aware of Merton's *Science, Technology and Society in Seventeenth Century England*, first published in 1938, we cannot say.[47] What we can say is that Merton, who led contemporary sociology from the varieties of determinism that characterized its origins in the United States toward a functional interpretation of social organization and behavior, also isolated the ideology of science from the platonic exceptionalism that animated the political vision in *Science—The Endless Frontier*. Merton did not deny that scientific knowledge itself was exceptional; but under his scrutiny it ceased to be the product of exceptional individuals who inhabited the highest plane of intellectual and moral virtue, entitling them to special authority in political society. What sets science apart, for Merton, is its institutional norms, norms which enable its practitioners to confirm empirically and logically observations about regularities in nature.[48] These norms are communalism, disinterestedness, skepticism—or the suspension of judgment pending persuasive evidence and argument—and universalism. The last of these norms, which arguably are observed only imperfectly, affirms the importance of the principle of openness toward new and initially foreign ideas as essential to the accumulation of scientific knowledge. While Merton's acceptance of scientists' claims to valid knowledge did not go unchallenged, the principle of openness as essential to institutional creativity has achieved the status of popular wisdom.[49]

Proceeding from a discipline and intellectual milieu far different from Merton's, the Austrian born philosopher Karl Popper voiced a comparable skepticism about the notion that scientists' statements about the natural world could be proven to be indisputably valid or true.[50] While not rejecting scientific investigation as the most probable means of accumulating an increased understanding of the material world outside ourselves, Popper nonetheless cautioned that our understanding of the material world would always be imperfect, and all the experimentation and evidence in the world would not make it perfect. This is so because there will forever remain the possibility that tomorrow will bring with it contrary evidence to undermine what was today thought to be a certain scientific fact or fully proven theory. Accordingly, the work of science moves forward not by cumulatively verifying empirical observations of natural phenomena and thus confirming scientific theories to explain them. Instead, scientific knowledge progresses when evidence is found erroneous or misleading at some future time, and scientific explanations disproven or falsified. To the extent that science is a social or institutional process, if the community or institutions of science do not positively invite skepticism and disproof, they will fail.

While Popper's philosophical examination of science was published in Vienna within a few years of Merton's path-breaking *Science, Technology and Society in Seventeenth Century England* (1938), it was not translated into English until 1959 and thus became, for American readers, an important critical work of the 1960s. Popper's inquiry was soon joined in this role by Thomas S. Kuhn's *The Structure of Scientific Revolutions* (1962). A physicist at Princeton University, Kuhn may have been especially well prepared to notice that science did not seem to be the progressive linear accumulation of discovery of truths about nature that it was said to be in celebratory observations. Contrary to what we may have been taught about the fruits of scientists' faithful application of the scientific method, argued Kuhn, "there is no scientifically or empirically neutral system of language or concepts" with which scientists can represent what is "out there" in the material universe.[51] What we like to think of as scientific progress cannot be accounted for by observation and experiment alone. Personal and historic accidents come into play because they force individuals to challenge the accepted theories and practices of "normal" science.

And what is normal science? Normal science is a shared tradition of research practices, observations, and experiment that have arisen from the acceptance of an explanatory theory or "paradigm." A scientific revolution occurs when an observational or experimental problem becomes intractable, and individuals, often younger scientists, must, as a result, question the previously accepted paradigm. Proofs cannot resolve competing paradigms, wrote Kuhn, because paradigms are "incommensurable; . . . the transition between competing paradigms cannot be made a step at a time, forced by logic and neutral experience." "Copernicanism made few converts for almost a century after Copernicus' death," remarked Kuhn; "Newton's work was not generally accepted, particularly on the Continent, for more than half a century after the *Principia* appeared. . . . Max Planck, surveying his own career in his *Scientific Autobiography*, sadly observed that 'a new scientific truth does not triumph by convincing its opponents and making them see the light, but rather because its opponents eventually die, and a new generation grows up that is familiar with it.'"[52]

Kuhn's notion of a paradigm became something of a paradigm itself, and persons in many different walks of life could be heard using the phrase "paradigm shift" to characterize changes in artistic taste, management philosophies, or other ways of looking at the world that predominated in various institutions or society. Meanwhile, the scientist-philosopher Michael Polyani (1891–1976) at the University of Chicago and physicist Gerald Holton (Ph.D. 1948) at Harvard University wrote, respectively, of the "tacit knowledge" and "themata" that tend to provide common navigational instructions for communities of researchers. Each of these scholar-observers of modern science—Merton, Popper, Kuhn, Polyani, and Holton—in his own way, dislodged the vocation of natural scientist from the moral and intellectual Olympus it had occupied among platonists and their modern acolytes.

DURING THE BLEAK DAYS of World War II, Karl Popper turned from his meditations on the nature of science and the scientific method to the philosophical foundations of totalitarianism.[53] Those foundations lay, he concluded, in the attraction among those fearful of change to prophetic views of history requiring a collectivist allegiance, and their attempts to limit the uncertainties of political life with rigid tribal social structures rationalized from metaphysical ideals of perfection. The notion of justice in Plato's *Republic*, the historical idealism of Georg W. F. Hegel (1770–1831), and the historicism of Karl Marx (1818–1883) exemplified for Popper reactionary ideologies that emerge when the possible opening of society to individual expression, criticism, reason, and moral judgment threatens the fixed social rankings of closed societies and, as an inevitable consequence, their political stability.

The fulcrum of the open society for Popper was the free individual, for whom the capacity for intellectual and moral agency entails personal responsibility for his own moral judgments, and for reasoned, intelligent, and critical discussion of policy. As were all educated Europeans of his generation, Popper was well read in ancient history. In *The Open Society and Its Enemies* (1945) he wrote of the principal figures of Athenian history and philosophy with an immediacy suggesting a conversation with one's contemporary. There, in ancient Athens, Popper found in Socrates the champion of the free individual—a philosopher who redefined the human soul in intellectual and moral, rather than metaphysical terms, so that when Socrates enjoined his hearers to "care for your souls," he was urging them to preserve their intellectual and moral integrity, which is the essence of what separates us from beasts. Such individuals are not mere egoists; indeed, they are capable of great loyalties—but the institutions and undertakings to which they are loyal are those that enable other free individuals and open society to flourish. One such undertaking is scientific inquiry, which is both cause and consequence of the intellectual activity of free individuals.[54] Thus did Popper recapture for his readers the common heritage and shared interest of science and politics in open societies, democratic politics, and free individuals. These are essential to the continuation of a civilization based on reason and a shared humanity, a civilization of which modern science is an inseparable part.

4

The Science and Technology
Policy Toolkit

However widespread the appeal of principles of openness and transparency in government, ideologies and principles in themselves are only abstractions. They must acquire traction in daily experience to breathe and endure. In the arena of policy it is the often maligned business of public administration (that is, what government bureaucracies do) that will prove—or disprove—the viability of political ideals. Among the industrialized democracies, variations are due largely to the extent to which each country's political system supports centralized or decentralized approaches to policy making, as well as a larger or lesser role of government in its economy.

In the United States, where the Congress (with or without the active support of the executive branch) has periodically created agencies or programs aimed at promoting generic or specific technologies (for example, synthetic fuels), such efforts may not have nearly the same effect as the combined impact of a host of other federal regulatory or programmatic functions the ostensible purpose of which is only indirectly related to science or technology. The U.S. science and technology policy "toolkit" thus consists of a decentralized and pluralistic array—some might call it a hodge-podge—of funding and regulatory policy devices, not all of which could be ascribed to science and technology policy objectives per se. Both federal and state governments have created programs and agencies with the specific purpose of fostering scientific research, technological innovation, and its commercialization.[1]

Federal programs are necessarily national in scope and benefits, though the facilities and employment they generate may benefit individual states, thanks to the intervention of members of Congress representing constituencies in those states (otherwise known as "pork barrel" politics). State programs tend to take two forms—incentives to high-technology firms to locate facilities within the state, or the creation of research and special trade areas which may be subsidized through such means as special state tax treatment or low-cost long-term lease of state-owned property, for example, state-owned institutions of higher education.

In the case of federal programs and agencies instituted specifically to increase scientific research and technological innovation, some were created to accelerate the development of technologies necessitated by national emergency (such as nuclear weapons during World War II). Others were created in the belief that the technologies they spawned would prove economically important, but the private sector on its own would not invest in their development because any returns on such investments would be too far downstream to be appropriated by current investors.

Any reasonable categorization one might use to differentiate among the various ways public policy influences the nature and rate of scientific research and technological innovation is helpful for analytic purposes. But categorizations are likely to obscure the way such policies have their effect, which is most typically through the interactions among them. This is so because science, engineering, and innovation occur in institutions, which are subject to more than one government policy at any given time. For example, the ability of the National Science Foundation to shape the nature of scientific research on university campuses results not only from budget decisions made in the Congress, but from the way the White House Office of Management administers federal information and procurement policies. Or, as another example, the research NASA supports in space science disciplines can be influenced not only by what the Congress and NASA science administrators consider essential for a coherent program of planetary astronomy, but by the way members of NASA's advisory committees and committees of the National Academy of Sciences influence the agency's long-range planning.

Moreover, when institutions that conduct research and technological innovation rely on federal dollars, they become susceptible to political pressures that come with government oversight. Such pressures can make themselves felt through any number of the variety of policy tools the U.S. government has at its disposal. Some of those pressures may be benign—such as efforts to increase accountability in all federal activities through reform measures like the Government Performance and Results Act (1993). But political pressures can also be exerted through measures that may seem largely ideological or economic in inspiration, such as restrictions on federally funded genetic research, or congressional appropriations "earmarked" for facilities located in congressional districts.

The relative importance of the various policy tools the United States uses to promote and shape the course of scientific research and technological innovation will vary, then, according to the nature of their interactions, the amount of resources consumed, the importance of their results, and the degree of political controversy that has surrounded them. Of all these scientific and technology policy tools U.S. patent policy and federal procurement policy and administration are the most important, because they singularly determine the means of distribution and recovery of public capital investment in scientific research and technology.

General Observations

Some general observations about the making and execution of science and technology policy in the U.S. will help us to appreciate the composition of its component parts. First, the process is highly pluralistic. Numerous institutions—especially universities, private laboratories, high technology industries, and government R&D organizations—may have an interest in a particular policy or program, and how such a program is implemented (whether by a federal mandate or tax incentives, for example). Moreover, for every member of Congress there is a potential economic interest in that member's district—typically that interest is employment or increased revenues (such as lease payments for drilling in publicly owned lands). For every president there may be a host of generous campaign donors whose businesses or organizations favor or oppose research in particular areas, such as birth control or global warming.

If we liken the process to a vast machine, there are numerous levers that can be pulled and dials that can be adjusted to accommodate changing circumstances and interests. All of these interests can reduce the possibility that a few organizations or individuals seize enough influence to institute a radical change in policy that does not have at least the tacit acceptance of the others. "Pluralism," in short, ensures that the process remains open to the interacting influences of many players, while the more open the process, the more likely it is that a creative or at least broadly acceptable solution to a policy issue is found. Thus pluralism, by its nature, is an open systems approach to policy making. And if we accept Joseph Schumpeter's view that modern capitalist economies proceed through stages of "creative destruction" as new firms based on new technologies supplant once-new industries, then the open system of U.S. science and technology policy can assist, through its adaptations to changing circumstances, Schumpeter's "creative destruction."[2]

To its beneficiaries, the pluralism of the system is one of its most attractive characteristics. Federal funding programs scattered here and there throughout the government—under the watchful eyes of their various constituencies—are more difficult for their opponents to target for reduction or termination than are funding programs aggregated into a single agency. Nor are such programs in direct competition with one another in the same organization or its budget. For this reason initiatives to aggregate most federal science and technology programs into a government-wide department have thus far been unsuccessful, and are likely to remain so.[3]

Critics may argue that the pluralistic or open system of science and technology policy in the United States is *too* adaptable. Responding as it must to the proclivities of members of Congress whose priorities change with the concerns of their constituents, federal science and technology programs fail to provide the multi-year funding stability necessary to sustain long-term research and development undertakings. What's more, the very exposure of such programs to political influence—e.g., earmarking federal scientific research funds, or the pork barrel allocation by congressional districts of large research and development projects—may frustrate the system's adaptation to Schumpeter's creative destruction and, in that sense, prove the deficit of their virtue.

Ever since the New Deal, when comprehensive planning to promote the nation's industrial and economic well-being seemed a defensible expedient, the notion of industrial policy has attracted some enthusiasm in national politics.[4] Measures such as the creation of the Federal Communications Commission (FCC, 1933), Securities and Exchange Commission (1934), Federal Deposit Insurance Corporation (1934), and Social Security Act (1935) served to ameliorate some of the worst excesses of the laissez faire economy. Since the end of World War II, which transformed the federal government into the largest purchaser in the United States (and the world) of scientific and engineering research and their products, there have been three periods of increased political tolerance for government intervention in the process of technological innovation in U.S. industry. Each was a consequence of what appeared at the time to be an international challenge to the global industrial hegemony of the United States: the Cold War that followed World War II, the energy crisis of the 1970s brought upon by extraordinary high oil prices, and the global economic competitiveness crisis of the 1980s.

The response of the Congress and the two-term Reagan administration of the 1980s to what was widely perceived as a national economic emergency bore fruit in a number of initiatives designed to jump-restart a technology-driven U.S. economy. Nonetheless, since that decade the most notable survivor of half-a-century of science and technology policy in the United States has been a bipartisan policy consensus surrounding its principal strategies and limits. These are:

(1) Federal government support for scientific research and technological innovation and development should occur as part of the funding for federal agencies whose missions entail extensive use of targeted research and

development—principally the National Aeronautics and Space Administration, the National Institutes of Health, the Department of Energy (DOE), and the Department of Defense.

(2) Technology policy will continue to be a component of economic policy.

(3) The federal government will continue to support basic research, principally but not exclusively through the National Science Foundation, but it will not rely on funding for basic research alone to generate technological innovation and new national industries.[5]

(4) The conviction that the federal government should not compete in the market place beyond its necessary role as buyer of the goods and services required to conduct its congressionally directed activities.

Public encouragement of scientific research and technological innovation has not been a serious political issue in the United States since the Federalist era, when Benjamin Franklin and Thomas Jefferson questioned the wisdom of the patent clause of the Constitution, arguing that all invention relies to some extent on the discoveries of others. They were outnumbered, and the Federalist vision of a government-fostered national economy strengthened by domestic manufactures prevailed. To the extent that differences have survived, they have been at the margins, with latter day Democrats more willing to countenance federal programs to push certain technological innovations into the market place, and Republicans favoring allowing the market place to decide for itself what innovations it wants—a market "pull" strategy. Within these broad outlines individual programs to commercialize particular products of federally funded research and development projects rise and fall, their fate in the Congress frequently having less to do with a project's predicted economic benefit or the members' ideology and party affiliations than with the presence in their electoral districts of contractors supported by the project.[6]

The Toolkit

Mission Agencies

The lion's share by far of federal support for scientific research and research and development is administered through federal mission agencies, or agencies whose missions depend upon their effective use of science and technology.[7] As we can see from Table 4.1 the research and development responsibilities lodged in these agencies have evolved over time in response to a discrete set of national needs considered politically urgent at the time: the Department of Defense, with significant increases in the early period of the Cold War, and then another surge during the strongly pro-defense Reagan administration, which also championed a ballistic missile defense system; the initiation of nuclear energy R&D during World War II with the Manhattan Project, followed by a surge in energy research during the energy crisis of the mid-1970s; aeronautics and space R&D, which grew tenfold during the 1960s to fund the Apollo moon landing program, and saw lesser growth in the 1980s and early 1990s to support the R&D for the International Space Station program; and the steady growth in funding for biomedical research through the National Institutes of Health since the 1960s, a decade that saw a resurgence of social welfare legislation stimulated by the Johnson administration's "Great Society" programs.[8]

Since 1950 funding for the National Science Foundation—sole inheritor of Vannevar Bush's mantle as the agency responsible for federal support of basic scientific research in all research disciplines—has grown from less than 1 percent of total federal obligations for research and development when the NSF was created, to 3.8 percent of the same total. However, DOD, DOE, NASA, and NIH also award basic research grants and contracts in disciplines associated with their own programs, with the result that only about 20 percent of federal support to academic institutions for basic research comes through NSF, which ranks fifth among federal agencies receiving appropriations to support science and technology. (The NIH, by contrast, receives over three times as much funding for research awards as does NSF.)

Nonetheless, the NSF stands apart as the single federal institution whose mission is support primarily for academic research. For this reason college and university scientists, especially in the natural sciences, look to the foundation for the grants that enable them to conduct the research that they must do to advance in their fields. Moreover, the NSF is the only federal agency today that attempts to support scientific research across a broad array of disciplines.[9]

Intramural Laboratories

The oldest established research and development institutions in the United States are the federal armories created in the early eighteenth century to supply armaments for the fledgling nation. Today their legacy is being maintained in the dozens of research and development facilities operated by, or under private sector contract to, the departments of Energy, Defense, the National Aeronautics and Space Administration, the National Oceanographic and Atmospheric Administration (a part of the Department of Commerce), and the Federal Aviation Administration (a part of the Department of Transportation). The departments of Energy and Defense and NASA, which normally receive more than 20 percent of the federal R&D budget, account for over 80 percent of the federal government's funds allocated to its intramural laboratories.

While federal intramural research laboratories may appear to compete with university laboratories for federal support, a handful of large universities actually operate federally owned laboratories under contracts from the agencies to which those laboratories belong. These include NASA's Jet Propulsion Laboratory, operated by the California Institute of Technology; the Department

TABLE 4.1

Federal Obligations for Research and Development, 1940–2000*
(in $ millions, real-year dollars; n/a = data not available)

Agency	1940	1945	1950	1955	1960	1965	1970	1975	1980	1985	1990	1995	2000
Department of Agriculture	29.1	33.7	53.0	72.0	131.4	224.5	281.2	420.8	688	943	1108	1380	1747
Atomic Energy Commission (after 1947)	...	...	221.4	289.8	989.5	1241.0	1345.9	...	...	...	...	...	...
Manhattan Engineer District (1943–1947)	...	859.0	...	...	...	...	...	...	...	...	...	...	...
Energy R&D Adm. (1974–1977)	...	...	...	...	...	...	...	2047.3	...	...	...	...	...
Nuclear Regulatory Commission (est. 1974)	...	...	...	...	...	...	...	...	183	150	218	88	53
Department of Energy (after 1977)	...	...	...	...	...	...	...	...	4754	4966	5631	6145	6063
Department of Commerce [4]	3.3	5.0	12.0	9.9	33.1	61.3	121.6	215.3	343	399	438	1136	1037
Department of Defense [1] [2]	26.4	513.0	652.3	1555.9	5653.8	6796.5	7360.3	9012.4	13981	29792	37268	34346	33167
Department of the Interior	7.9	18.0	32.1	31.9	65.3	113.2	156.8	303.3	411	392	509	562	580
Environmental Protection Agency (est. 1970)	...	...	...	...	...	...	89.0	258.0	345	320	420	558	502
National Advisory Committee for Aeronautics	2.2	24.1	54.5	73.8	...	...	...	...	...	...	...	...	...
National Aeronautics and Space Administration	...	...	...	...	401.1	4951.5	3799.9	3064.4	3234	3327	6533	9015	6882
National Science Foundation (est. 1950)	...	...	...	**8.5**	**58.0**	**187.2**	**288.9**	**585.0**	**882**	**1346**	**1690**	**2149**	**2726**
Office of Scientific Research and Development	114.5	...	...	...	...	...	...	...	...	...	...	...	...

Public Health Service	2.8	3.4	39.6	67.9	...	...	...	...	...	...	...	...	...
Department of Health, Education and Welfare/HHS [3]	...	...	...	...	324.2	346.4	347.9	551.4	598	623	1269	1158	1508
National Institutes of Health	...	...	...	...	...	523.0	873.0	1846.0	3182	4828	7137	10299	16918
Department of Education (est. 1980)	...	...	...	...	...	...	...	...	139	125	170	176	247
Department of Transportation	...	...	...	...	...	64.4	327.8	311.5	361	429	366	604	467
Veterans Administration/Department of Veterans Affairs	...	...	...	...	...	37.4	58.6	94.8	133	227	238	238	342
US Agency for International Development (est. 1979)	...	...	...	...	...	...	...	...	149	220	335	303	218
Total for Agencies Listed [5]:	186	1456	1065	2110	7656	14546	15051	18710	29383	48087	63330	68156	72457
Difference over preceding 5 years:		1270	(391)	1045	5546	6890	505	3659	10673	18704	15243	4826	4301

* Sources: U.S. Bureau of the Census, *The Statistical History of the United States from Colonial Times to the Present* (Stamford, Conn.:1960); U.S. Bureau of the Census, *Statistical Abstract of the United States, 1979* (Washington, D.C.: Hoover's Business Press, 1999); National Science Foundation, *Federal Funds for Research and Development, Historical Tables: Fiscal Years 1951–2002* and *Federal Funds for Total Research and Development by Agency: Fiscal Years 1995, 1996, and 1997*, http://www.nsf.gov/sbe/srs.

[1] Before 1947 combines the departments of the Army, Navy, and Air Force; excludes military pay.

[2] 1955 and later, includes military pay, which adds approximately 10 percent to the annual R&D total for the Department of Defense.

[3] In 1980 the Department of Health, Education, and Welfare was split into the Department of Education and the Department of Health and Human Services.

[4] Includes National Oceanic and Atmospheric Administration (est. 1970) and the National Institute of Standards and Technology (NIST, est. 1901 as the National Bureau of Standards, renamed NIST in 1988).

[5] Actual federal government-wide totals would include a number of smaller agencies and commissions with relatively negligible R&D budgets and which are not specifically identified in all National Science Foundation or Office of Management and Budget data. Added to the totals in this table, these entities would increase the annual totals by about 10 percent.

of Energy's Los Alamos National Laboratory and Lawrence Livermore National Laboratory, both operated under contract to the department by the University of California; and the Department of Energy's Argonne National Laboratory, operated by the University of Chicago.

As a type, these research and development institutions have a long and distinguished history beginning with Springfield Armory in Massachusetts (est. 1794) and Harpers Ferry Armory (est. 1799) at the confluence of the Potomac and Shenandoah Rivers. There, under the direction of the U.S. Ordinance Department, federally directed work on firearms joined that of private makers in establishing the American system of manufacturing (e.g., mechanization and interchangeable parts manufacture), a key element in the United States' successful industrialization during the nineteenth century.[10]

In the twentieth century a few of these institutions became so prominent that they acquired the identities of their most notable programs, such as Los Alamos National Laboratory and the building of the first U.S. atomic bomb, and Johnson Space Center in Houston, Texas, institutional home of NASA's human space flight program. Less widely recognized, but no less important for their contributions to the nation's ability to produce the technologies necessary to carry out World War II and prevail during the Cold War, are such installations as Wright-Patterson Air Force Base in Ohio and the Army Ballistic Missile Agency in Huntsville Alabama (transferred to NASA on its creation in 1958, it was renamed the George C. Marshall Space Flight Center).

Periodically advocates of more efficient government—which usually entails reducing duplication in functions or programs—call for the consolidation or elimination of some federal laboratories, much as the Department of Defense is periodically summoned to consolidate or close some of its bases. Such initiatives tend to be marginally successful, if only because each federal laboratory represents jobs and the economic well-being of someone's congressional district. During the Clinton administration, Vice President Gore adopted for his own agenda government "reinvention," an initiative influenced by notions of personal and organizational transformation then much in vogue among self-improvement and management consultants. In an effort to bring about government "that works better and costs less," Gore launched a National Performance Review of the federal departments and agencies which included, in 1995, a Federal Laboratory Review orchestrated by the White House National Science and Technology Council and Office of Science and Technology Policy. A few months later the review concluded that the Department of Defense, the Department of Energy, and the National Aeronautics and Space Administration laboratories might benefit from clearer goal setting, reduction in "onerous" self-generated regulations and paperwork, reduced staffing levels, improved inter-laboratory coordination, and reduced audits of contractor compliance. Nonetheless, and not surprisingly, the review concluded that these agencies' "laboratory systems . . . provide essential service to the Nation in fundamental science, national security, environmental protection, energy, aerospace, and technologies that contribute to industrial competitiveness."[11]

Federally Funded Research and Development Centers

A much smaller group of research organizations, representing about 3 percent of the nation's research and development expenditures, are federally funded research and development corporations, or FFRDCs. These organizations, which can be administered by universities and industries as well as stand alone as non-profit corporations, receive 70 percent or more of their funding from federal agencies, and are permitted to do some research for other clients provided that research is not otherwise available from the private sector.

The FFRDCs originated as "think tanks" established during the Cold War by the Department of Defense, which wanted to be able to obtain on short notice—that is, without the months' delay that ordinary procurement procedures could involve—and without repeated and detailed congressional scrutiny, defense systems analysis and evaluation and occasional technical assessments and management support. Among the better known think tanks is the Rand Corporation, but they also include such familiar names in federal R&D circles as the Mitre Corporation, the Software Engineering Institute, the Institute for Defense Analysis, ANSWER, the Center for Naval Analysis, and the Aerospace Corporation. Today more than half of the primary funding for all FFRDCs comes from the Department of Energy. The Department of Defense continues to make significant use of these organizations, as does the National Aeronautics and Space Administration.

Because FFRDCs are able to offer substantial (non-civil service) salaries to senior-level staff, and are not bound by federal procurement regulations which normally require competitive bidding, they have acquired for themselves the unenviable nickname, "beltway bandits." In 1978 the Congress responded with the Ethics in Government Act to mounting complaints of an apparent "revolving door" through which senior-level civil servants passed (as employees) back and forth among think tanks and other large federal contractors. In addition to defining unethical conduct for government officials acting in the public trust, the act created the Office of Government Ethics, which oversees enforcement of federal ethics regulations designed to minimize conflicts of interest among federal officials often working in tandem with industrial firms.

National Infrastructure and Regulatory Agencies

The federal science and technology policy tools we have considered thus far were created to enable the federal government to design and carry out the research and development programs necessary to meet specific policy objectives more than to promote the advance of science and technology itself. While it would be difficult to trace the influence and measure the specific effects of these tools, the value of their combined effect is surely enormous. But any account of the U.S. science and technology policy toolkit would be incomplete without acknowledging federal strategies used to foster the conditions that enable scientific and engineering progress in the national economy. As we have seen, the belief that innovation is critical to an expanding economy dates back to the very origins of the United States, has been demonstrated by economists, and seems warranted in retrospect by two centuries of experience.

The most basic need for a thriving modern economy is an infrastructure, not only of roads, navigable waterways, and airways, but of the financial institutions that enable the expansion of credit and investment. The Hamiltonian program to ensure that the new nation's emerging currency could be backed by "the full faith and credit of the [government of] the United States" was as critical to the economic development of the country as any other single measure the federal government could have taken. The Supreme Court under Chief Justice John Marshall affirmed in *McCulloch v. Maryland* (1819) the constitutionality of a congressionally created national bank, as well as congressional regulation of interstate banking. Likewise, a legal system that treated lawful contracts as inviolable, and allowed for the chartering of limited liability corporations, provided the institutional vehicles without which large-scale private enterprise would have been difficult. Technological innovation was an inseparable part of this process, for it could not proceed without new methods of manufacturing, industrial engineering, business management, finance, and the construction of technological systems such as the railways and rural as well as urban electrification.[12]

From the earliest days of the republic U.S. public policy (at both the state and federal level) has provided some form of subsidy to promote the building of a nationwide infrastructure, whether as help in financing canals and roadways, dredging harbors and internal waterways, partial funding for telegraph lines, surveying railroad routes, tax exemptions, issuing guaranteed railroad mortgage bonds, supplying timber and stone from government lands, or granting lands for railroad rights of way. The first grant of federal land to promote railroad construction was made in 1850 to the Illinois Central, and the last was in 1871, when public criticism of the "give away" program forced the Congress to terminate the railroad land grants. The criticism proved to be somewhat unjust, for the railroads, in exchange for their grants of federal land, had to transport at reduced rates federal property, government mail, and troops—a not negligible requirement during the Civil War. By one calculation the value of the reduced rates to the government in 1945, when the federal requirement ended, was $500 million, or about what lands granted in the nineteenth century were then worth.[13]

Once automobiles and trucks replaced horses and wagons as the principal means of transporting people and light cargo, both state and federal governments built new roads, paved old ones, and connected major arteries—the best known examples of which are U.S. routes 1, 40, and 66. The road-building triumph of the post–World War II period has been the Interstate Highway System, launched with the 1956 Interstate and National Defense Highway Act. The importance of a good transportation network for transporting military material was underscored by U.S. involvement in two world wars.

As commercial aviation became a serious possibility, the Congress passed the Kelley Act (1925), which authorized the U.S. postal service to contract airmail service with commercial airlines, while local and state governments began to build and maintain airports. The creation in 1915 of the National Advisory Committee for Aeronautics (NACA), designed to function as an aeronautical research adjunct to the nascent aircraft industry and Army Air and Signal Corps, completed the structure of government support for an aircraft industry that grew to be one of this country's leading export sectors.

The broad interpretation of Congress's jurisdiction over the nation's interstate commerce given by Supreme Court Justice John Marshall in *Gibbons v. Ogden* (1824) ensured not only that individual states could not erect barriers to national commerce, but that the federal government could exercise regulatory authority over interstate commerce in the interests of public safety. This it began to do as early as 1838 when horrific steam boat accidents in interstate waterways led the Congress to institute a system of boiler inspections financed by steamboat operators and overseen by federal judges. After the defeat of the south in the Civil War and subsequent passage of the Fourteenth Amendment, with its guarantee that no person could be deprived of protection of "life, liberty, or property, without due process of law," corporations argued that federal regulation of their business was tantamount to the seizure of private property and thus unconstitutional. This claim was firmly refuted, at least in the case of businesses which, like the railroads and public utilities, were "clothed with a public interest," in the Supreme Court's 1877 ruling in *Munn v. Illinois,* a case that arose from discriminatory railroad freight rates. The Interstate Commerce Commission, created the same year, was authorized to regulate railroad shipping rates to prevent such abuses as rates that favored long-haul shippers over short-haul shippers, and pooled routes to limit rate competition.

The twentieth century brought with it a flowering (or a plague, depending on one's perspective) of federal regulatory agencies and commissions, beginning in 1906 with the addition to the Department of Agriculture's functions administration of the Pure Food and Drug Act to prevent adulterated food from reaching markets; the creation in 1914 of the Federal Trade Commission to oversee compliance with the Clayton Antitrust Act, passed in the same year to clarify the practices meant by the Sherman Antitrust Act's (1890) prohibition of "combinations in restraint of trade"; the Federal Communications Commission

(1933) to ensure orderly and publicly responsible use of the airwaves and the nation's communications networks; the National Air Pollution Control Act of 1955 to set emissions standards; the National Highway Traffic Safety Administration (1966), the Environmental Protection Agency (1970), and the Consumer Product Safety Commission (1972).

Federal regulation of technology-intensive industries may not appear to be part of an overall strategy to promote technological innovation. In fact, however, it has played an important role in three ways, all of which have favored industries for which investment in managerial as well as technological innovation is essential to survival. First, especially in the case of the railroads, federal regulation reduced the uncertainties of the market place and leant some predictability to the competitive environment. Second, federal regulation, by limiting the worst abuses and imposing a reasonable degree of public accountability in industries sustaining essential public services, probably preempted more radical efforts to control major industries during the Great Depression and World War II. Nationalization, such as occurred in post–World War II Great Britain when the Labour government nationalized the railroads and utilities, as well as the coal, iron, steel, and telecommunications industries, is not easily undone.[14]

However, an occupational hazard of government regulation is excessive influence by the regulated industries on the regulators. Regulatory agencies normally depend on the companies they oversee for the information they need to do their work. The more helpful industry representatives can be to typically understaffed federal officials, the greater their likely influence over those officials' decisions. Meanwhile, industry lobbyists continuously monitor congressional activity that might increase the regulators' authority to investigate, enjoin, or penalize them as they attempt to maintain or increase market share. Such, for example, is the case with the Food and Drug Administration's relationship with the pharmaceutical industry.

Third, and perhaps most importantly when considering policies conducive to technological innovation, federal regulation of industry has been designed to preserve the open system of a genuinely competitive market place. Among the best known of these efforts are the breakup of the Standard Oil trust in 1911 and the breakup in 1982 of AT&T's (American Telephone and Telegraph) conglomerate which, at the time it settled its litigation with the U.S. justice department's antitrust division, owned twenty-two regional and local telephone companies, long-distance lines, Western Electric, and Bell Laboratories research facilities. In times of national emergency, however, the federal government may overlook the onus of monopoly power, if doing so is necessary to ensure that the nation's military has as much of the best equipment possible. During both world wars, for example, the federal government, under threat of patent seizure, directed firms in the aircraft, automobile, and electronics component industries to suspend patent litigation and form patent pools, so that all could benefit from the innovations produced by some. If federal efforts to minimize monopoly power within large American industries have been imperfect, that does not negate the fact that the policy principle is there, and today is honored to a greater or lesser degree by both major parties.

Targeted Programs

Of all the devices in the science and technology policy toolkit, federal programs targeted at stimulating industrial research and innovation, while highly touted during periods when economy watchers proclaim that the United States has lost its competitive edge, are probably the least significant in their effects. This is so because there is a strong ideological resistance evident in both major parties to creating any program (or significantly supporting one, once created) that has the appearance of an "industrial policy."[15] The Kennedy administration, for example, in 1962 proposed the creation of a Civilian Industrial Technology Program on the recommendation of the President's Science Advisory Council, Office of Science and Technology, and Council of Economic Advisors. The initiative relied heavily on a report prepared by its proponents entitled *Technology and Economic Prosperity*, which argued that "spillover" effects from heavily funded defense and aerospace research and development could not be relied upon to promote civilian technology. The proposal was defeated in the Congress. During the Johnson and Nixon administrations as well as the Kennedy administration, the Bureau of the Budget successfully resisted additional efforts to institute federal programs to promote technology on the grounds that not only was there no evidence that the private sector underinvests in technological innovation, but such programs risk evolving into government bureaus that distort the natural functioning of the market place.[16]

Two economic crises of the last quarter of the twentieth century re-ignited White House interest in actively stimulating the technology sector of the economy: the energy crisis of the 1970s, brought on by a sustained world rise in the cost of oil, and the global economic competitiveness crisis of the 1980s. Neither President Jimmy Carter nor the Congress could remain politically unscathed without responding in some way to the oil crisis which produced television images of endless lines of automobiles waiting at gas pumps, or elderly and low-income citizens shivering at home through winters in their coats and blankets. They responded with the Federal Non-nuclear Energy Research and Development Act of 1974, which abolished the Atomic Energy Commission, created the Nuclear Regulatory Commission, and launched the Energy Research and Development Administration (re-created as the Department of Energy in 1977). The Ocean Thermal Energy Conversion R&D and Demonstration Act and Magnetic Fusion Engineering Act followed in 1980, as did the Stevenson-Wydler Technology Innovation Act. What became known simply as "Stevenson-

Wydler" directed that federal R&D agencies take explicit measures (such as creating special programs) to promote the adoption by private companies of technological innovations occurring in the agencies' own research facilities, or under one of their grants or contracts. (The term "technology transfer" was coined to characterize any systematic effort to promote the movement of new technologies among public and private sectors.) Soon thereafter federal R&D mission agencies created organizations to promote the adoption by industry of their agencies' technologies.

Also passed during the last year of the Carter administration, Bayh-Dole (1980 Amendments to the Patent and Trademark Laws named for their sponsors, Senators Birch Bayh and Robert Dole) authorized federal agencies to grant to small businesses and non-profit organizations (which would include virtually all colleges and universities) exclusive or partially exclusive licenses to any patented federal inventions—provided such licenses would not confer a monopoly advantage on the recipient business. Moreover, small businesses and non-profits could elect to patent any innovations they made while working under a federal grant or contract.

Though a Republican president, Carter's successor Ronald Reagan could not reject the findings of his own Commission on Economic Competitiveness, which confirmed what the business press had been reporting, namely that the United States was losing ground on nearly every international indicator of economic strength, including per capita income, the merchandize trade balance, and the country's share of high-technology world trade. A flurry of legislation ensued, all designed to "push" the results of federal R&D programs into the arms of waiting American industries, not the least of which were the small businesses reputed to be more innovative than the lumbering giants that dominated Wall Street. Reagan's 1983 Memorandum on Government Patent Policy extended to *any* federal contractor the same right to elect to patent inventions made under federal grants or contracts that Bayh-Dole had extended to educational institutions, non-profits, and small businesses.

Two further examples suffice to illustrate the approach taken by federal policy makers to subsidize technological innovation in the private sector. The first, initiated in 1986, was the authorization given to federal mission agencies to enter into Cooperative Research and Development Agreements (CRADAs), which allowed government-owned laboratories to provide personnel, facilities, and equipment—but *not* funding—for joint projects with industry. Within a few years around 3,000 CRADAs were in effect, primarily with laboratories owned by the Departments of Defense (47 percent) and Energy (23 percent).[17] Early in the administration of President George H. W. Bush, in 1990, the Congress authorized the Department of Commerce to establish within its National Institute of Science and Technology an Advanced Technology Program to fund projects to develop generic "high-risk enabling technologies." Projects were to be chosen on the basis of technical merit as well as economic criteria, and were to entail some cost-sharing. During the following decade around 1,100 companies and universities received a total of over $3 billion in awards that funded projects in biotechnology, electronics, information technology, advanced materials and chemistry, and manufacturing.[18]

How effective are such programs? One can only guess, because any measurement would require factoring in proprietary business information which would not be consistently maintained across firms—even if such firms were willing to release it. Some economists question whether such public programs do not compete with lines of research that businesses might otherwise pursue independently, or whether the economic gains to any industry are greater than the loss of revenue represented by the programs' funding. In any event, the amounts involved constitute such a small share of an industry's or the government's R&D budgets that a significant impact is hard to imagine.[19]

What these illustrations do tell us, however, is that there is a middle ground into which both major parties are willing to venture in the name of advancing technological innovation for economic growth. This is a middle ground in which the government can do what is administratively feasible to ensure that the private sector benefits from the elaborate institutional structure it has built to perform its own research and development, *short of participating in the commercialization of those innovations itself.* Commercial activity must remain the exclusive province of privately owned and operated firms. The sanctity of the market place remains protected not only by ideology, but by the belief that the government has constitutional responsibilities that do not necessarily promote business efficiency, for it serves citizens, not customers.

Finally, most industrialized economies, including the United States, use some form of reduced tax or low-interest credit schemes as incentives to industry to invest in research and development. Allowing industries to deduct R&D expenses from taxable profits, a tax incentive available in the United States for decades, has a certain appeal to policy makers. They are relieved of the burden of selecting technologies or markets for special treatment to promote research and development, and because the corresponding revenue losses do not appear in the federal budget as an outlay, the tax incentives are relatively invisible. In 1981 the Congress added an additional research and engineering (R&E) tax credit of 20 percent of qualified research above a base amount based on the ratio of research expenses to gross receipts.

The extent to which such credits make a significant difference in the amount and quality of research and development conducted by industry is debatable. A 1987 study of the costs and benefits of R&D tax incentives used in other industrialized economies done by the United Kingdom's office of Inland Revenue and HM Treasury concluded that added investment in R&D does not compensate for the revenue losses. Other studies question how much increased R&D is actually new, or a reclassification of activities that would be carried out anyway. Meanwhile, we know that in the United States the credit is used largely by a group of around 10,000 companies with assets worth over a quarter of a billion dollars and concentrated in the pharmaceutical, aircraft, computer, automotive, and electronics industries. As for smaller companies, it is uncertain whether those claiming the R&D and R&E tax credits would conduct any R&D beyond what they can claim under the credit.[20]

Federal Procurement Policy

In the use of its various policy tools, the U.S. government spends money in four ways: (1) as the employer of civil servants and military personnel, and purchaser of equipment and facilities it needs for governmental functions such as military bases; (2) through "assistance" programs—such as school lunches for children from low-income families—by which it gives congressionally stipulated amounts of monetary assistance to statutorily specified categories of recipients; (3) through grants for activities like research and cultural activities—such as funding for the National Science Foundation or the National Endowment for the Humanities—the Congress has determined to be of public benefit; and (4) by awarding grants and contracts to private sector concerns to perform various functions deemed not to be "inherently governmental." Federal support for scientific research and technological innovation occurs principally through two of these four means—either as the employer of scientists and engineers, or as the source of R&D grants or contracts. Some of these grants and contracts are awarded to private as well as state-owned academic institutions to operate government-owned research facilities. Two examples of the latter are NASA's $1.5 billion contract (fiscal year 2002) with the California Institute of Technology to operate the Jet Propulsion Laboratory in Pasadena, CA, which manages NASA's planetary research missions, and the Department of Energy's $3.8 billion contract (fiscal year 2002) with the University of California System to operate the Los Alamos, Lawrence Livermore, and Lawrence Berkeley national laboratories.

Federal mission agencies spend at least 75 percent of their R&D budgets externally on grants and contracts to industrial firms, academic institutions, and non-profit organizations (see Table 4.2). In a representative year, 2001, NSF turned 99 percent of its budget into research grants and contracts primarily with academic institutions. The NSF's share of total federal R&D procurements that year was only slightly more than 5 percent, while the National Institutes of Health's share was over 25 percent. By far the biggest single source of federal R&D dollars transferred to the private sector through procurements was the Department of Defense, which obligated 47 percent, or $27.8 billion. These figures are significant because they indicate the extent to which federal pro-

TABLE 4.2

Mission Agency R&D Funding and Procurements, Fiscal Year 2001
(in $ millions)*

Agency	R&D Obligations	R&D Procurements (%)	TOP 15 R&D Contractors/Grantees
Department of Defense	$39,396	$27,818 (76%)	Lockheed Martin Corp.
			Boeing Co.
			United Technologies Corp.
			Northrup Grumman Corp.
			Raytheon Co.
			TRW Inc.
			Aerospace Corp.
			MITRE Corp.
			General Dynamics Corp.
			Science Applications International
			Massachusetts Institute of Technology
			Carlyle Group
			Computer Science Corp.
			ITT Industries
			Washington Group International
NIH	$19,235	$15,556 (81%)	Johns Hopkins University
			University of Pennsylvania
			University of Washington
			University of California, San Francisco
			Washington University
			University of Michigan
			University of California, Los Angeles
			University of Pittsburgh
			Science Applications International Group
			Yale University
			Duke University
			Harvard University

(continued)

TABLE 4.2

Mission Agency R&D Funding and Procurements, Fiscal Year 2001 (in $ millions)* *(continued)*

Agency	R&D Obligations	R&D Procurements (%)	TOP 15 R&D Contractors/Grantees
NASA	$9,602	$7,105 (74%)	Columbia University
			UNC, Chapel Hill
			Baylor College of Medicine
			Boeing Co.
			Lockheed Martin Corp.
			Thiokol Corp.
			Space Gateway Support
			McDonnell Douglas Corp.
			Science Applications International
			Raytheon Co.
			TRW Inc.
			United Technologies
			The Johns Hopkins University
			Assn. University of Research & Astronomy
			California Inst. of Tech.
			Universities Space Research
			Smithsonian Inst.
			Stanford University
			University of Colorado, Boulder
Department of Energy	$6,794	$5,923 (87%)	Bechtel National
			Lockheed Martin
			Fluor Corporation
			Battelle Memorial Inst.
			Brookhaven Science Assoc. LLC
			BNFL Inc.
			Duke Cogema Stone & Webster LLC

(continued)

TABLE 4.2

**Mission Agency R&D Funding and Procurements, Fiscal Year 2001
(in $ millions)* *(continued)***

Agency	R&D Obligations	R&D Procurements (%)	TOP 15 R&D Contractors/Grantees
			Wackenhut Services
			Oak Ridge Associated Universities
			University of California System
			University of Chicago
			University of Tennessee System
			University Research Assn.
			Midwest Research Inst.
			Stanford University
NSF	$3,180	$3,153 (99%)	Raytheon Polar Service Co.
			UCAR
			University of Illinois, Urbana-Champaign
			University of Wisconsin, Madison
			Cornell University
			University of Washington
			University of California, San Diego
			California Inst. of Technology
			University of California, Berkeley
			Columbia University
			MIT
			University of Michigan
			AuI/National Radio Astr. Observ.
			Woods Hole Oceanogr. Inst.
			AURA/National Optic. Astron. Observ.
	$59,555		

(Table notes on following page)

curement policy and practices can impact not only what federally sponsored research and development gets done, but who does it.

The federal government's policy of procuring military systems and other material from private suppliers has two centuries of precedent, beginning with the first federal R&D procurement contract negotiated in 1798 between Eli Whitney and the new U.S. war department. As military requirements for equipment and weapons became more sophisticated and technology-intensive during two world wars, the federal government increasingly relied on industrial suppliers for weapons research and development. The Manhattan Project to build the first U.S. atomic bomb, which resulted from a number of large and complex procurements awarded by the U.S. Army, is the most historic—but by no means the only—example. The same pattern was followed by NASA during the early period of the Cold War, which thereby helped to build the defense/aerospace industry. The Department of Energy, which grew out of the Manhattan Project's successor, the Atomic Energy Commission, followed suit.

Since the 1960s federal procurement policy and regulations have placed increasing emphasis on contracting out any and all activities that did not entail "inherently governmental functions," which the OMB defined as "activities [that] require the exercise of substantial discretion in applying government authority and/or making decisions for the government."[21] Inherently governmental activities normally fall into two categories: the exercise of sovereign government authority or the "establishment of procedures and processes related to the oversight of monetary transactions or entitlements." Both major political parties have found the promise of "reducing government" to have effective public appeal and, as a consequence, every White House since the Nixon administration has sought to achieve the *appearance* of reducing government by reducing the civil service workforce, transferring the eliminated personnel functions to private sector contractors.

The second fundamental element in the U.S. government's procurement policy is its emphasis on awarding contracts and grants fairly and competitively, that is, through an open system by which all those who might be reasonable competitors have an even chance in a transparent process of competing for awards on the basis of standards of selection that correspond to the work to

---

(Table notes from previous page)

* Sources: National Science Board, *Science and Engineering Indicators, 2002* (Washington, D.C.: National Science Foundation, 2002), Vol. 1, 4–14; National Aeronautics and Space Administration, *Annual Procurement Report, 2002* (Washington, D.C.: 2003); *Government Executive: Procurement Review* (August 2002); Department of Energy, *Annual Procurement and Financial Assistance Report, FY 2002* (www.doe.gov; downloaded October 18, 2003); National Science Foundation, *Overview of Grants and Awards* (www.nsf.gov/home/grants.htm; downloaded October 19, 2003).

be done or the product being sought. Much of the system's notorious paperwork, administrative detail, and "red tape" arise from the elaborate data collection and documentation required to record that, in every detail, government procurement officers have made awards on the basis of a "full and open competition." Exceptions exist (and are occasionally abused), primarily for procurements related to national security or some form of national emergency. While federal procurement policy and practices have, on occasion, been a source of contention between recipients of federal R&D funding and the agencies dispensing it, generally the notion of open and fair competition for government dollars has corresponded to academic researchers' notions of recognition and awards based on merit. When disputes arise, they are usually over how merit is to be decided—whether through the process of peer review, or some other measure of a research proposal's worthiness.

Federal procurement procedures share with the Internal Revenue Code the distinction of being one of the two principal means by which the federal government can attempt to achieve particular social or economic policy objectives through the way it collects and distributes federal dollars. Both sets of rules have become voluminous, subjects which few can master in their entirety—those few being tax attorneys, accountants, and procurement specialists. Whether the policy objective is to encourage charitable giving or long-term capital investment or to promote small minority and women-owned businesses through procurement "set-asides," aims arguably extraneous to the business at hand (collecting taxes or buying goods and services for the government) have rendered both systems for getting and spending public dollars highly cumbersome—and some might argue, highly inefficient.

In the area of aerospace and weapons systems procurement, former Lockheed Martin Co. president and chief executive officer Norman Augustine has, with great good humor, dissected the federal acquisitions process (that is, procurement) and identified the systemic shortcomings *in spite of which* working weapons systems get built, and the United States manages not only to land a man on the moon, but to bring him back—not once, but several times. Most individuals who have had to scramble through the briar patch that is the federal procurement process are likely to agree with Augustine's first law of federal systems procurement: "The thickness of the proposal required to win a multimillion dollar contract is about one millimeter per million. If all the proposals conforming to this standard were piled one on top of the other at the bottom of the Grand Canyon, it would probably be a good idea."[22]

Large defense and aerospace corporations, for which federal contracts provide a significant portion of annual revenue, have become expert in managing procurement procedures. They are in business to do business in a way that cannot be said of colleges and universities, which have come to rely on federal dollars to fund the research that helps to attract and retain productive scientists on their faculties. This dependency has taken on the aura of an entitlement, though universities prefer to distinguish themselves from beneficiaries of federal assistance programs. Some observers prefer the characterization of the funding dependency between universities and the federal government as a "social contract" or a "partnership," though one would be at a loss to find such an agreement documented and enshrined in the National Archives. Rather, it is a contract informally assented to by the parties much in the same way that the English common law refers to accumulated precedent and practice. But this genteel characterization cannot always obscure the fact that the dispersal of public funds in the United States is governed by principles that must apply to all equally. Thus a general discussion of the relationship of the federal government to the nation's universities must, inevitably, devolve into discussions of the administration of the same procurement processes that afflict any organization seeking to do business with the federal government.

About the time that President Bill Clinton was re-elected to a second term, "national, political, corporate, and education leaders," along with the President's Committee of Advisors on Science and Technology, appealed to the president to take note that universities were suffering from "a period of stress," and appealed for a White House review of the federal government's policies and administrative practices in its relationship to the nation's universities, "where training of young scientists and engineers is advanced synergistically with the creation of new knowledge." The White House responded by creating an interagency task force under the aegis of the National Science and Technology Council. The task force surveyed the country's research universities and non-profit research institutions to ascertain how the federal government might be more accommodating to their needs.

The resulting report (*Renewing the Federal Government-University Research Partnership for the Twenty-first Century*, April 1999) and recommendations were replete with high-minded sentiments, as most such reports are, which formed the backdrop for ten specific recommendations. Three themes stand out in the report's prologue and permeate the text: (a) scientific research performed in universities continues to be essential to the nation's current and future well-being; (b) the federal government has a presumptive obligation to finance research in universities; and (c) the university's institutional obligations to teach and to conduct research are inseparable.[23] Of the ten recommended actions in the report, five had to do with administrative aspects of the way in which universities receive federal dollars for research, all of which are specified in various OMB circulars.[24] Two were fairly straightforward and to be expected from any such report. University research administrators and research project scientists called for the adoption throughout the federal government of uniform procedures both for making research grant proposals and for administering those grants once awarded. They also called for the increased use of merit review in selecting awardees from competing proposals while reducing the number of awards allocated on the basis of political or agency priorities. This second recommendation incorporated a change in terminology—merit review now replacing the more traditional term, peer review, that deserves a special note.

Since the immediate post–World War II years, when the allocation of extra-mural federal research dollars was dominated by a small elite from a handful of universities clustered in the northeastern United States and California, both peer review and the presumed independence of scientific truth from social, political, or economic influences came under increasing challenge. What was significant about this chipping away at science's ideological ramparts was that it was done not primarily by the populists who had supported Senator Kilgore's vision for the National Science Foundation, but by respected scientists and scholars in the new field of the social studies of science (see chapter 3).

In 1967, for example, the widely respected sociologist Robert K. Merton read a paper at that year's meeting of the American Sociological Association that was published in *Science* a few months later, "The Matthew Effect in Science." Drawing from interviews with Nobel laureates and personal writings of other scientists, Merton argued that scientists distribute professional recognition among their peers on the basis of psychosocial processes, with the result that (as the Bible attributes to Saint Matthew), "For unto every one that hath shall be given, and he shall have abundance; but from him that hath not shall be taken away even that which he hath."[25] Still, four years later, out of the venerable Massachusetts Institute of Technology came a sharp critique by a Cambridge University trained nuclear physicist, Charles W. McCutchen, published in 1991 as "Peer Review: Treacherous Servant, Disastrous Master," which appeared in MIT's journal, *Technology Review*.[26]

Thus it was that, when the Congress passed in 1993 the Government Performance and Results Act (see chapter 5) which imposed unprecedented standards of accountability on all federal agencies (including R&D mission agencies), few could make a convincing argument for deferring to the judgment of external peer-review panels when allocating federal dollars for scientific research. Federal research managers could see the "Matthew Effect" at work in their own programs, as fewer and fewer institutions received more and larger awards. Even if these institutions were in fact the "best" qualified according to published criteria and the judgments of peers, this narrowing tendency suggested—though it would be hard to prove—that a set of elite institutions with a handful of influential scientists was overtaking publicly supported research. The transition from peer review to merit review as the scientific community's response to critics of peer review was reflected in the

Clinton administration's *Partnership* report, where the term "merit" review replaced "peer" review, and might "be used in conjunction with other selection criteria to support agency or program goals."[27]

Priority and Profits: U.S. Patent Policy

Every scientific advance and every new technology begins as an idea. For those who work with their minds rather than their hands, the ability to benefit materially from new scientific insights or new ways of doing things is as important an incentive to continue their work as profits are to industry and commerce. The notion of ownership, and profiting from one's ownership—for good or ill— is a bedrock principle of capitalism. For scientists, thought and experiment are the fuel, and priority in discovery the engine, of advancing careers. Publication in peer-reviewed journals is the normal way scientists establish the priority of their discoveries. For inventors, being able to turn their workable ideas into commercially valuable property—or "intellectual" property—has played a comparable role in advancing their careers. Government protected patents to the fruits of invention—like publication for the scientist—establish personal ownership of the intellectual content of inventions. For both scientists and inventors, copying the expressions of their ideas and publicizing or marketing them as one's own is a form of theft.

As soon as governments recognized the value of inventions and technological innovations to their own economies—a recognition that was a part of the rise of capitalism in the late middle ages—they began to award patents, or legally protected monopolies, to inventors to encourage them to stay and practice within their jurisdictions. The initial purpose of patents in English law (the basis of colonial American law) was to confer monopolies for the production and trade of various items, including land. Such patents ("letters patent" or, literally, "open letters"), were granted by Elizabeth I (1533–1603) and James I (1566–1625) to their favorites until Parliament put an end to this royal prerogative with its 1623 Statute of Monopolies.[28] The earliest British technology patents awarded to colonists were granted in 1715 and 1716, respectively, by Pennsylvania (for a corn cleaning mill and technique for making straw hats) and South Carolina (for an oil and tar composition to prevent river-worm rot on the bottoms of boats).[29]

By 1787 the need for a national policy promoting manufacturing had become clear enough to at least the Federalists, and the Constitutional Convention made sure to include a provision in the Constitution empowering the Congress "to promote the Progress of Science and useful Arts, by securing for limited Times to Authors and Inventors the exclusive Right to their respective Writings and Discoveries."[30] In 1790 the Congress passed the nation's first patent law, which assigned to any two of the secretary of state, attorney general, or secretary of war the authority to award patents for inventions or discoveries that were "sufficiently useful and important." Patentees would have to provide not only detailed descriptions, but working models of their inventions.[31] (When the patent office ran out of storage space, the model requirement was dropped.)

Nearly 10,000 patents had been awarded under the 1790 Act by the time the Congress turned to the subject of patents again in 1836, when it established the U.S. Patent Office and the position of commissioner of patents within the Department of State. Institutionalization of the patenting process was accompanied by a refinement in the reasons for which patents could be given. While importance and utility remained as requirements, the invention in question now had to be *demonstrably* new, never before invented or discovered. If the matter of novelty or importance was contested, a panel of three appointed by the secretary of state—one of which had to be an expert in "the particular art, manufacture, or branch of science to which the alleged invention appertains"—would resolve the dispute.[32]

As one of the principal tools in the country's science and technology policy toolkit, U.S. patents' nationalistic ends have justified their cosmopolitan means: encouraging foreign inventors to develop and market new technologies in the U.S. could only enrich the quantity and mix of commodities flowing through the domestic market place, thus further stimulating economic growth. In the year of its founding (1836), the U.S. Patent Office issued 103 patents for inventions, eight of which were awarded to residents of foreign countries.[33] By 2003 the rate of patenting in the United States had grown so prodigiously that the U.S. Patent and Trademark Office (USPTO) issued over 257,000 patents. And as an indication of the extent to which the high-technology economy is also a global economy, over 50 percent of those patents were issued to foreign owners. (The top three non-U.S. recipients of U.S. patents were Japan, Germany, and the United Kingdom.)[34]

Do numbers of patents tell us about the degree of inventiveness of a society, an industry, or a firm? For reasons we shall note later in this chapter, the temptation to count patents as indicators of inventiveness when making or evaluating technology policy is a temptation to be resisted. But the *content* of the patents themselves can yield critical information about salient areas of innovation in particular industries or regions. Fortunately for not only patent attorneys, but historians of technology, industry analysts, and others concerned with technology and economic policy, the patents themselves by law are open for public inspection, while the U.S. Patent Office has maintained an online patent count and technology classification search capability since the late 1980s.

Issues in Contemporary Patent Policy

While patents have become a common fixture of modern industrial life, they are in fact one of the most problematic tools of national technology policy. Quite aside from the long-standing question raised by Franklin and Jefferson—whether any invention can occur so independently that it justifies granting the inventor a monopoly on the product—the award, enforcement, and commercial trade in patent rights today raise a number of fairly critical policy issues. Their importance arises from the vast sums of money that are consumed by patent litigation, and the fact that in 2001 the value of U.S. exports in royalties and license fees alone was $38.8 billion, twice what it had been ten years earlier.[35] Contemporary patent policy issues are of varying degrees of complexity and some are closely intertwined:[36]

(1) How well do patents granted today conform to the statutory criteria for patentable inventions, namely that they must be demonstrably novel, non-obvious in light of "prior art" and have utility?

(2) In twenty years the number of law suits over patent infringement (often involving the legal validity of the patent in question) has increased by a factor of ten. Patent litigation is extremely costly—which is discouraging to small inventors. It also takes time, which can impact when or whether a new product appears in the market place.

(3) Patents originated in a technological world are dominated by machines and their components. Rapidly unfolding developments in computer software and biotechnologies pose new challenges to our application of the legal definition of a patentable invention.

(4) Since the early 1980s the right to patent inventions made by companies or universities during research funded by the federal government normally belongs to the inventors, rather than to the federal government. Has this policy helped or hindered the commercialization (bringing to market) of products resulting from federally funded research and development?

(5) In the modern global market place, in which patent ownership (and any resulting royalties or licenses) must be valid across international boundaries to be meaningful, how critical is it for the United States and its trading partners to achieve international harmonization (or consistency) of patent laws?

The first three of these issues are closely related, because they are shaped not only by the daily decisions of patent examiners in the USPTO, but by the propensity in any particular industry to litigate questionable patents or possible infringements, as well as trends in the way federal courts have adjudicated issues of patent validity and infringement. Theoretically, the extent to which federal courts hold patents invalid should correspond to some extent to the care patent examiners take to ensure that an invention clearly meets statutory requirements for patentability. But such a relationship is difficult to demonstrate empirically, and few studies have confirmed, even in a preliminary way, a connection between patent examinations and court rulings.[37] What is more certain is the impact of the courts' broad interpretations of the statutory criteria for patentability and the decisions of the USPTO.

The Broadened Scope of Patentability

Until 1982 patent cases worked their way through the federal court system like any other litigation involving federal law. The process took time and money—lots of it—and the outcome of individual cases could vary, depending upon the extent to which individual courts approved of patents in principle. To both expedite the process and standardize patent law across the country, the Congress in 1982 passed, and the pro-business President Reagan signed, the Federal Courts Improvements Act which, among other things, created the Court of Appeals for the Federal Circuit (CAFC) to hear appeals from federal district court patent cases.

To the extent that the 1982 act was a "pro-patent" measure, it was successful. While federal appeals courts before 1982 overturned only 12 percent of district court findings of patent invalidity or non-infringement, since the creation of the CAFC, the overall probability that a litigated patent will be held to be valid has risen to 54 percent.[38] This translates into increased patent litigation, increased court injunctions delaying product commercialization until an infringement case is resolved, and increased monetary awards against infringers.

Initially legal differences over the validity of individual patents involved primarily the traditional issues of novelty, non-obviousness, and utility—criteria which can easily be debated in any given instance. More recently, the patentability of classes of technologies has been the focus of legal and political controversy. For example, patentable items by law can be "any new and useful process, machine, manufacture, or composition of matter, or any new and useful improvement thereof." *Not* patentable are laws of nature, physical phenomena, abstract ideas, and algorithms or mathematical formulas (considered comparable to laws of nature). Is computer software patentable? Many might argue that it is a "useful process" for directing computers to perform certain tasks, while others might insist that software is nothing more than a set of algorithms. However, given the open system character of computer programming—by which computer users actively participated in the early development of software programs, and today many users make individual improvements to such open source code programs as Linux—how does one make a firm proprietary claim to a section of software code?

One does make such claims, of course, when the programmers have been employees of such companies such as Apple Computer or Microsoft. But organizations like the League for Programming Freedom have argued that awarding copyrights or patents for software code will impede, rather than promote, software development. First, writing software code is as much an art as a mechanical process—consider, for example, the importance of look and feel to the usefulness of a software application. Second, important innovative work is often done by individuals or small groups of individuals, who cannot afford the legal services necessary to avoid infringing the patents of others, much less defend against patent infringements of their own programs.

Prior to the 1980s, the USPTO treated software inventions as unpatentable on the grounds that they were merely algorithms which, like other principles of nature, could not be patented. Software producers could treat their codes as trade secrets or copyright them. Then the Supreme Court turned an about face and ruled computer software itself patentable in *Diamond v. Diehr* (1981).[39] The case arose out of a patent application for an industrial rubber curing process that involved the use of a digital computer to register precise temperatures of the curing rubber and release it from its mold at the exact time at which the rubber was perfectly cured. Justice William H. Rehnquist, writing for the majority, argued that a process for curing rubber was patentable subject matter, and while a computer algorithm itself was not patentable, its use in the patent claimant's curing process did indeed result in an innovative and patentable improvement to that process.

This clear broadening of the scope of patentable *enhancements* to otherwise ordinary processes is reflected as well in the Supreme Court's rejection, in *State Street Bank v. Signature Financial Group* (1998), of the doctrine prevalent up to that time that business methods (comparable to other unpatentable thought processes) were not patentable.[40] In this instance the court ruled that the use of *computerized automation* in a procedure for transferring assets among mutual funds is indeed patentable. The patentability of software remains a murky area, however, with the federal courts now more inclined to uphold than to reject the validity of a patent awarded for software code.[41]

The ramifications of national policy extend far beyond the interests of inventors, industry, and patent attorneys. Intellectual property is a form of *personal* property, and when inventions are patented they become private property, protected by the same Fourth Amendment as one's land, home, or capital investments.[42] Thus the seemingly endless expansion by the federal courts of what can be patented to "anything under the sun that is made by man"[43] has become deeply controversial, impinging as it now does on differing cultural views of the sanctity of life.

The landmark Supreme Court case addressing the conversion to private property, and thus into marketable commodity, of a biological life form is *Diamond v. Chakrabarty* (1980), which arose out of the USPTO's denial of a patent for a new microbe created through genetic engineering by General Electric microbiologist Ananda M. Chakrabarty. The microbe had a property existing in no other naturally occurring microbe of being able to break down the components of crude oil—thus making it a potentially important means of cleaning up crude oil spills. The USPTO argued, however, that bacteria, being living organisms, are life forms, and thus not patentable. Attorneys for Chakrabarty argued that, to the contrary, under the terms of the law whether or not microbes were life forms was immaterial; the fact that they had been "made by man" rendered them patentable. The court in a narrow (five-to-four) majority decision agreed.[44]

Because of the depth of the controversy generated by *Chakrabarty*, both the court's majority and minority opinions merit a closer look. References to a pronouncement by the court that patentability extends to "anything under the sun that is made by man" invites the assumption that the Supreme Court was reaching boldly into new ground in *Chakrabarty*. In fact, the phrase comes from the congressional committee reports accompanying 1952 legislation to recodify the patent statutes.[45] Both Chief Justice Warren E. Burger and Justice William J. Brennan, Jr., authors of the majority and minority opinions respectively, agreed that the potential dangers to humanity, animal diversity, and the environment of transforming biological life forms into private and commercial property, were issues to be resolved by the Congress through the legislative process: "That process involves the balancing of competing values and interests, which in our democratic system is the business of elected representatives [and not the courts]."[46] However, it *is* the business of the court to interpret statutes already made, and when the intent of a statute was not unarguably clear, to refer to the legislative history of the statute.

This was the point at which the majority and minority differed. Both cited the 1930 Plant Patent Act and the 1970 Plant Variety Protection Act; but they interpreted the record of Congress's intent with these two statutes differently. When the Congress passed the 1930 Plant Patent Act which allowed the patenting of plants reproduced asexually "in aid of nature," it had an opportunity to distinguish between animate and inanimate things, and to exclude living things from the statute's provisions. The Congress declined that opportunity, recognizing (in the Supreme Court majority's opinion) that "the relevant distinction was not between living and inanimate things, but between products of nature, whether living or not, and human made inventions . . . [such as Chakrabarty's] microorganism [which] is the result of human ingenuity and research."

Justice Brennan's opinion for the minority agreed that awarding Chakrabarty a patent for the *process* he had used to produce the new microbe was not at issue. The minority disagreed, however, that Congress ever "intended that he be able to secure a monopoly on the living organism itself, no matter how produced or how used." Citing "this Nation's deep seated antipathy to monopolies," the minority maintained that the court, in its reading of the pertinent statutes, "extend[s] patent protection . . . further than Congress has provided," especially in the "absence of legislative direction." Instead, "the courts should leave to Congress the decisions whether and how far to extend the patent privilege into areas where the common understanding has been that patents are not available." The majority interpreted the Congress's silence on patentability of living things in the Acts of 1930 and 1970 as indicating approval; the court's minority took the opposite view: "Because Congress thought it had to legislate in order to make agricultural 'human-made inventions' patentable and because the legislation Congress enacted is limited, it follows that Congress never meant to make patentable items outside the scope of the legislation."[47]

The patentability of life forms has not only aroused deep-seated feelings about the sanctity of life, it has also divided the biomedical research community. The worst of these fears—that one day someone will receive a patent for genetically engineering a human being—is unfounded, thanks to the Thirteenth and Fourteenth Amendments to the Constitution, which prohibit the ownership of persons. What researchers in university biology departments fear is the loss of access to biological materials and procedures necessary for their work. At the same time, universities themselves are patenting biological materials either directly, or as partners with firms in the pharmaceutical industry.[48] Monsanto Chemical Co., for example, in 1974 funded a $25 million research project at the Harvard Medical School to find a tumor-blocking agent. Monsanto also agreed to provide cell cultures as well as funds to endow a professorship. Both Harvard and Monsanto would patent their own inventions arising from the project, though only Monsanto would have lucrative licensing rights—if the project succeeded.

Questions surrounding the patentability of software, computer-assisted manufacturing and commerce, business methods, and biotechnology processes all highlight the expansion of notions of legitimate intellectual property ownership from novel substances and devices, to novel methods in "soft" as well as hard technologies. This expansion is perhaps best illustrated by the awarding of patents not only for novel pharmaceuticals and medical devices, but for surgical procedures. The United States is virtually alone in awarding "pure" medical patents—that is, patents for medical procedures involving a novel technique rather than a new drug or device—a practice that is consistent with the broadened notion of patentability embedded in the *Chakrabarty* (1980), *Diehr* (1980), and *State Street Bank* (1998) decisions.

Awarding patents for surgical procedures has engendered considerable controversy, especially among ophthalmic surgeons, who saw laser surgery for improved eyesight transformed into a shopping mall commodity during the late 1990s. Although the USPTO has been awarding medical method patents since the late 1950s, its willingness to do so went relatively unnoticed until 1990, when an ophthalmic surgeon (Dr. Samuel Pallin) filed suit in federal court in Vermont against Hitchcock Associates for infringement of Pallin's patent for a surgical method for removing cataracts. The court ruled in 1993 that parts of Pallin's patent were invalid and prohibited him from enforcing any part of it. Allowing the issue to be resolved one by one in a series of patent infringement cases would have been incalculably costly in time, dollars, and compromised patient care. Thus the American Medical Association, American Academy of Opthalmology, and American Society of Cataract and Refractive Surgery urged the Congress to ban medical and surgical procedure patents. A plastic surgeon as well as member of Congress, Rep. Gary Ganske, M.D. (R-Iowa) introduced a bill with 137 co-sponsors in 1995 to prohibit the issuance of patents "for any invention or discovery of a technique, method, or process for performing a surgical or medical procedure, administering a surgical or medical therapy, or making a medical diagnosis," unless that technique or procedure is a necessary accompaniment to the use of a device or substance that is itself patentable. The bill was referred to the House Committee on the Judiciary, Subcommittee on Courts and Intellectual Property, where it is likely to remain, so long as medical services remain commodities—like pharmaceuticals—marketed by the commercial health services industry.[49]

Federally Funded Inventions

Patentability issues associated with new areas of technology bear on the comparably important national policy issue of the increasing privatization of the public domain. Historically public domain in the United States has referred to the government's principal physical asset—land. With the expansion of government-funded infrastructure of science and technology since World War II, however, federally funded research—whether conducted intramurally or by contractors and grantees—has produced *publicly* created intellectual property of immeasurable value.

Prior to World War II technological innovations embodied in munitions sold to the federal government were developed either in private factories with advance funds from government contracts, or in the government's shipyards and its armories at Springfield, Mass. and Harpers Ferry, Va.[50] The disposal of government intellectual property rights was not a significant issue and, when disputes arose, the courts normally resolved them. Patents issued to the U.S. government were negligible up to 1930, and averaged only forty-two per year between 1931 and 1943—roughly 1 percent of all patents issued during the same period.[51]

At the height of the U.S. involvement in World War II, weapons-related research and development funded by the U.S. military services contributed to a tenfold increase in U.S. government intellectual property during 1944–1955. By 1950 it had become clear that a federal policy governing patentable inventions made with public funds was in order, and President Harry S Truman issued

Executive Order 10096, the first uniform, federal government-wide patent policy. The policy provided that the federal government would be the presumptive intellectual property (IP) owner of any invention made by a federal employee or "on government time." (Contractors for the Department of Defense and the National Advisory Committee for Aeronautics would be the exceptions, usually retaining title to inventions made under their government contracts.) In the case of the Atomic Energy Commission (AEC),[52] the national security ramifications of its work with nuclear materials virtually dictated that the federal government would retain title to all intellectual property created by the AEC—hence the designation of the AEC as a "title" agency.

Eight years later, when the Eisenhower White House drafted legislation to create the National Aeronautics and Space Administration, the administration retained the same principle used for the Atomic Energy Commission, which is, NASA would be a title agency presumptively retaining patent rights to any inventions made by its personnel or with its funds. This time, however, many contractors to the former National Advisory Committee for Aeronautics (NASA's predecessor agency), some of whom were also contractors for the Department of Defense, protested when they learned of the patent provisions in the Eisenhower administration's bill creating a new federal civil space program.[53]

The question of whether NASA would remain, like the AEC, a surrogate owner, on behalf of the American people, of federally "made" inventions, would be determined by how attached members of Congress and their constituents were to a conception of the public good that relied on an unwritten free market canon. An essential principle of that canon was that initiative, industry, and creativity—and by extrapolation, the economic progress of the United States—have all been driven by unfettered competition. The anticommunist fulminations of the 1950s reverberated in the Congress, as did the well-established American conviction that private enterprise was the engine that "made America great." For the government to retain patent rights to inventions made by its R&D contractors (or grantees) would, in this view, not only ensure that those inventions would lie fallow; it was also tantamount to the seizure of the creative fruits of private enterprise.

The political division over this issue occurred fairly consistently along Republican, pro-business and Democratic, rural-populist lines. Pro-business patent policy advocates like the Electronics and Aerospace Industries associations and the National Association of Manufacturers, argued that government retention of patent rights to inventions made by federal contractors was tantamount to theft, and would discourage the best companies from contracting with the government.

The Democrats' position was argued forcefully by Senator Russell B. Long (D-La.), chairman of the Senate Small Business Committee's Subcommittee on

Monopoly, who countered that the overriding public policy objective should be economic progress; economic progress required the rapid dissemination of the results of research—rather than monopoly patent ownership—especially when the U.S. government has paid for that research. By this standard, NASA and AEC patent policies were the right ones, and should be extended throughout the government.[54] Long also questioned his opponents' assertion that the government, if it retained patent rights, would be unable to get the best contractors, or would be charged premium prices by them. Recalling public statements by senior executives from the Douglas Aircraft Company and Raytheon Corporation, he insisted that industrial contractors reaped real benefits from their government contracts—most notably experience and know-how—that could be used on other contracts.[55]

The outcome—the Kennedy administration's patent policy issued October 12, 1963—steered cautiously between a rock and a hard place. The administration's policy directed that the federal government would normally retain patent rights to its contractor's inventions *if* they involved products or methods likely to give the contractor a commercial advantage over its market competition. Patent rights should remain with the government if the contract had been awarded for research in fields "which directly concern the public health or public welfare" or in which the government had been "the principal developer." On the other hand, government contractors would "normally" retain patent rights to inventions in technological fields in which they had previously established technical competence or a market position.[56] Since few federal agencies were likely to award a contract to firms that were *not* technically competent to carry out the work required, the new guidelines seemed to ensure that most contractors would be able to claim rights to any invention made under a government contract.

A succession of Republican presidents during the 1970s and 1980s (excepting only the administration of Jimmy E. Carter, 1977–1981), helped ensure the continuation of a policy that transferred to the private sector, through exclusive licenses and waivers of the government's patent rights, monopoly ownership of intellectual property arising from federally made or funded inventions. Announcing without foundation that "studies and experience over the past seven years" justified such policies, President Richard M. Nixon released in August 1971 his own administration's statement of government patent policy. Much of Nixon's policy simply repeated language from the Kennedy patent policy of 1963. However, Nixon's policy also authorized the heads of government agencies to grant *exclusive* licenses to commercial developers of inventions made under government contracts, subject to a loose array of criteria likely to be met by anyone seeking a license.[57]

The 1980s witnessed the full flowering of the trend toward privatization, or the transfer to the private sector of numerous functions previously conducted by the government—though still at government expense, and often using publicly developed equipment, facilities, and public financing guaranties. The pendulum had begun to swing back from decades of government activism in social and economic policy toward a period of government downsizing and anti-government rhetoric, complaining of endemic bureaucratic incompetence and the perpetuation of government programs for no reason other than to preserve the bureaucracy administering them. The sentiment spread widely on both sides of the Atlantic, resulting in the two-term Republican presidency of Ronald Reagan (1981–1989) and, in Great Britain, the eleven-year term of Conservative prime minister Margaret Thatcher (1979–1990), the only British prime minister in the twentieth century to win three consecutive general elections.[58] This was the policy milieu in which federal patent policy in the 1980s took a sharp turn toward nearly total privatization of publicly funded technological innovation. First in 1980 came Bayh-Dole, and, three years later, President Reagan's Memorandum on Government Patent Policy (see the earlier section on targeted programs).[59]

Since 1987 the Department of Commerce's Office of Technology Policy (OTP) has collected and reported federal agency data on the number of inventions disclosed, patent applications filed, licenses granted, royalty income from licenses, and the number of active federal agency-private sector CRADAs.[60] The department has also asked federal agencies to provide data reflecting the number of patents issued and new CRADAs initiated. The OTP readily acknowledges that such data, however, measures only activity and tells us nothing about "outputs or outcomes." In its most recent report, the department recommends that the federal government "develop the measures needed to monitor the technology transfer process" and institute "a system . . . to collect and disseminate such information."[61]

This last recommendation begs the question of why such a system was not required by the various statutes that comprise federal technology transfer legislation in the first place. Perhaps the conviction of the policies' rightness was so strongly and widely held that no one thought any verification was required. Nonetheless, until the federal government systematically collects and reports comprehensive information about the technological content and industry (by firm) allocations of waived patent rights as well as technology transfer activities, we will be left pretty much where we were in 1965, when President Lyndon Johnson's attorney general Nicholas Katzenbach advised the Senate Small Business Committee that he knew of no "data, studies, or facts of any kind at all which could possibly support" the notion that giving patent rights to contractors would "foster the prompt working of inventions."[62]

One of the difficulties of ascertaining the impact of federal patent policy on any national measure of inventiveness (assuming we could agree on such a thing) is that once federal intellectual property rights are issued to a contractor or grantee, that property ceases to belong to the federal government which, then, can no longer command information on how new private intellectual property owners dispose of their patenting rights. For example, in recent years patents for "drug, bio-affecting, and body treating compositions" as well as "molecular biology and microbiology" have been the largest number of patents issued of any patent class.[63] Some of these patents have been issued to the National Institutes of Health, which applied for patents on 2,500 gene fragments to ensure their availability for researchers.[64] However, the citizen watchdog group Public Citizen lost its 2000 suit (in the Federal District Court for the District of Columbia) for the public release of information on the royalties paid to the NIH by pharmaceutical companies producing and marketing drugs under licenses to patents held by the NIH. The court held that the royalty information was proprietary commercial information legally protected from public disclosure.[65]

Secondly, contractors can opt *not* to apply for patents on certain inventions *at all*, deciding for reasons of business or litigation strategy to protect their monopoly use of that invention as a trade secret. Contractors also can and do treat federal government–originated patent rights as corporate assets, part of the capital value of their company during the course of an industry consolidation, or other form of inter-firm exchange. For example, out of twenty-four known patents resulting from 1993 NASA waivers to contractors of patening rights, twenty-one of the patents themselves were turned into assets transferred to other aerospace or allied industry firms.[66]

Since patent rights or patents themselves have no intrinsic currency value, one can only estimate the value of intellectual property transferred from the federal government to private hands. Perhaps the most useful monetary measure is an indirect one, that is, the potential cumulative dollar value of licenses and royalties associated with particular patents. As with so many other public assets, the greatest value of government held patents is that they are *public,* that is, their existence serves a collective national purpose, which in this case is to increase the store of knowledge from which citizens can draw in fulfilling their own aspirations. Limited, exclusive, and royalty-free licenses to federal patents, which can be cancelled if the license holder fails to develop and market the patented product or process—a policy introduced during the Nixon administration—would preserve the best of both worlds. Firms could enjoy a monopoly right to manufacture patented items for a specific period of time. If they failed to do so, other firms would have an opportunity to step in, thus ensuring genuine competition in the market for new technologies.

International Harmonization of Patent Laws

As a tool of technology policy, the granting of patents by the United States can be characterized as pro-patent, pro-business, and pro–free market. A pro–free market patent policy in a global economy requires at least reciprocity in the treatment of intellectual property. The United States goes one step further by awarding patents only on the basis of patentability, usually without regard to the nationality of the patent applicant. Thus foreign firms or subsidiaries can manufacture in the United States with the same patent protection as their U.S. counterparts.

Not all industrialized countries use their governments' authority to grant patents the same way as the United States. For example, Japanese patent policy favors Japanese industry over patentees, especially foreign patentees. The government assists Japanese companies in obtaining licenses to foreign-owned patents, and requires foreign companies to grant licenses to Japanese firms as a condition of operating in Japan. Before any patent can be valid in Japan, it must survive challenges to its validity, while Japanese courts rarely give injunctive relief against alleged infringers. Critics charge that the Japanese "flood" the USPTO with applications in order to force U.S. companies wanting to do business in Japan to enter into cross-licensing agreements with Japanese partners as a condition of operating in that country. And indeed, in 2001 the USPTO issued 34,891 patents to Japanese applicants, or three times the amount issued to the next largest foreign country receiving U.S. patents that year, Germany (which received 11,894).

As a different example, Taiwan received the third largest number of patents (6,544) issued by the USPTO in 2001. This Pacific rim country lacks the strong tradition of technological innovation necessary to sustain a patent system, but has grown in its capacity for research and development as well as its marketing capabilities. Typically Taiwanese firms work in partnership with non-Taiwanese technology companies with which they have formed licensing and production partnerships. Aggressive patenting around the margins of foreign-patented inventions, the strategy followed by Japanese companies, also enables Taiwanese firms to insist, as well, on licenses to technologies belonging to non-Taiwanese companies taking advantage of Taiwan's significant and highly competitive production capabilities.[67]

Sorting out the status of patents and other forms of intellectual property internationally within the framework of the World Trade Organization (WTO), established in 1995 and representing 125 countries and 90 percent of world trade, has been occasionally contentious. Members of the WTO in 1999 agreed to a comprehensive treaty on the Trade Related Aspects of International Property Rights (TRIPS). In their essence the TRIPS principles promote among the treaty's members uniform administrative procedures, domestic laws, and enforcement mechanisms governing intellectual property, as well as uniform civil and administrative remedies for patent infringement. The TRIPS negotiations became a battleground between poor countries—as well as countries afflicted with high incidences of AIDS-HIV—and western pharmaceutical companies seeking protection against unlicensed copying of their patented drugs. Pharmaceutical companies feared that if they gave free or low-cost licenses to poorer countries to enable them to produce copied drugs, those countries would re-market the drugs to third parties, including third parties in the United States. In August of 2003 the WTO agreed to a proposal—which the United States had previously opposed—to allow developing countries, as well as South Africa and Kenya, to import generic versions of more costly patented medicines from countries like India and Brazil, which produce generic drugs and do not normally patent medicines.

Given the importance of patents in a "knowledge-based" economy, efforts to bring some international harmony to patent laws and administrative procedures have followed the march of industrialization.[68] The oldest of these is the International Convention for the Protection of Industrial Property, adopted in 1883 and informally referred to as the Paris Convention. European countries then as now date the period of patent protection from the date of filing, and base priority claims on the filing date. Thus it was a signal benefit when the 140 Paris Convention signatories agreed to provide members' inventors the same patent protections they provide their own citizens—for example, a filing date of a patent application in one country would serve as the filing date for all other signatories to the Convention.

The Patent Cooperation Treaty of 1970 further simplified matters by providing for centralized filing procedures and a standard application among its ninety signatories—which did not include most of the countries in the Middle East, Southeast Asia, Africa, and Latin America, except Brazil. European inventors, at least, were further assisted by the European Patent Convention of 1977 which created a European Patent Office (EPO) and ensured that patents issued by the EPO would be recognized as a national patent in each of the member countries. More recently the 116 signatories to the General Agreement on Tariffs and Trade (GATT) of 1993 agreed to a twenty-year period of patent protection, and to allow developing countries ten years to repeal any compulsory licensing laws they may have instituted for manufacturing firms operating in their own borders. Further harmonization through a proposed Substantive Patent Law Treaty (SPLT) is the task of the World Intellectual Property Organization (WIPO). Established by an international convention in 1967, the WIPO became an agency of the United Nations in 1974.

The broad scope of patentability in the United States—one that entertains "anything made by man," including software, life forms, and business methods—is not supported in much of the developing world. Nor is it lacking in opponents even in the United States. The Consumer Project on Technology (CPT) has opposed any requirement that all countries adopt the broad definition of patentability used in the United States, noting the social cost of allowing some technologies to be patented. The CPT has argued that "nations should be free to exclude some areas of activity from the scope of patents. Certainly patents on business methods and software are highly controversial, and according to many experts, the costs far outweigh the meager benefits. There are also enormous controversies over the appropriate scope of patents in the life sciences area."[69] The National Association of Manufactures, meanwhile, advocates the broadest possible scope for patentable subject matter by eliminating technology-specific definitions of patentability, broadly defining patentable subject matter as things that "demonstrate some technical contribution above and beyond mere commercial impact."[70]

To those whose principal business is intellectual property, however, the most urgent harmonization issue must first be addressed at home: the adoption by the United States of a "first-to-file" date for the inception of the period of patent protection, the standard used by the European Patent Office, the Japan Patent Office, and most other countries offering patent protection. Adopting "first to file" would eliminate the difficulty of establishing novelty and resolve in a factually unambiguous way the philosophical issues surrounding the nature of novelty, which have kept court dockets full and intellectual property attorneys busily litigating infringement cases.

LURKING BENEATH QUESTIONS of patentability and ownership of biological life forms lies a no less serious political issue: free markets do not allocate their benefits equally, nor do they allocate their benefits equitably, if one's standard of social justice includes considerations other than market efficiency. Adam Smith's freely operating market place exists nowhere, nor has it ever existed anywhere, because there have always been other dynamics—political, sociological, ideological, ethnic, historical—to distort the theoretically perfect mechanism of supply and demand. Only well-capitalized companies that have already benefited from the open system of a relatively free market economy can pursue business strategies resting on technological priority and aggressive patenting.

How effective is U.S. patent policy at promoting technological innovation and the commercialization of the new products and services it yields? This is, of course, a compound question. Patent policy could indeed stimulate innovation, but successful commercialization—profitably bringing a new product to market—depends on many variables such as the availability of capital and the condition of a new product's market. Assuming, however, that those variables are propitious, how do we measure amounts of "technological innovation?" Systematically collected and complete empirical data is virtually impossible to obtain, in part because much of the critical information one might require is proprietary. Thus we are dependent on formulas developed by economists, formulas which must rely to some extent on estimates and gross aggregates.

A common solution to this problem is to simply assume that one patent represents one unit of innovation and to count patents, subjecting those counts to various arithmetic exercises to produce indicators (which should not be confused with reliable measurements, since the substance and consistency of what is being measured is debatable). This approach, however, is methodologically problematic from the start, for the reason that patents do not reflect uniform values of any kind other than the productivity of patent attorneys and examiners in the USPTO. We can have no idea what a patent is worth (beyond the cost of the application's preparation and submission) until it becomes a medium of exchange. And once it has become a medium of exchange, its value will have as much to do with market and structural conditions in its related industries than that partially subjective phenomenon we call innovativeness. The safest interpretation we can give to more or fewer patents is that they reflect more or less inventive activity in their corresponding field of endeavor.

WHILE THE CONTENTS of the U.S. science and technology policy toolkit are plentiful, the nature and direction of scientific research and technological innovation in the United States are rarely influenced by them one at a time. More often research institutions find themselves attempting to benefit from, and adapt to, the not always consistent opportunities and requirements of several government organizations, statutes, and administrative regulations at the same time. This is the price of pluralism in the open system of U.S. policy for science and technology. Transparency exacts a price as well, for the bargain embedded in Anglo-American patent policy is one that trades publication of an invention, in all its salient details for all to see and from which all can learn, for temporary monopoly ownership of the invention for the owner's exclusive use. As with much of public policy, we devise and use our science and technology policy tools as much because we *believe* they are important, as because we can *know* that they are important, and why. Pending proof to the contrary, the belief has prevailed, a natural companion to the U.S. propensity toward open systems in its public life and political economy.

5

Science, Technology, and Political Authority

In this chapter we turn to the tension between scientific and technical authority and political accountability in the United States. This tension is most typically acknowledged in media accounts of contests between "science" and "politics," such as the controversy over FDA approval of the "morning after" contraceptive pill. But the arena in which the authority of scientists and engineers competes with political accountability is vast, bounded on its sides by laws and administrative practices that govern citizen access to reliable information and its use in policy making. These boundaries are manifest by (1) the standing of scientific and engineering evidence in the federal judiciary, (2) the federal government's efforts to apply the scientific ideal of rational and systematic procedure to public administration, (3) the application of the Freedom of Information Act to scientific and engineering data used for policy making, (4) the controversy over the Information Quality Act of 2000 (also referred to as the Data Quality Act), and (5) the application of the Federal Advisory Committee Act to the use of scientific and engineering experts by executive branch agencies.

Science and Engineering in Federal Courts

The exercise of government power invariably affects the interests of someone by allocating, however indirectly, material or political benefits—or costs. Even the most abstract policy objective—preserving freedom, for example—entails costs. In a modern society such as the United States, scientific and engineering information is likely to be involved in determining the allocation of those benefits and costs. But policies are shaped not only in the Congress and the White House. In a government of laws, any policy is subject to challenge as contrary to federal law or the Constitution. And thus the federal judiciary, by interpreting and applying the Constitution and federal statutes to the facts in specific cases, sets the boundaries within which national policy is made—including policies with scientific or technological components.

Establishing the facts of a case is basic to all litigation, but the facts in any case can consist of complex or disputed scientific findings. Such was the case in *Brown v. Board of Education of Topeka* (1954), for example, in which the court reversed the "separate but equal" doctrine of racial segregation enunciated in *Plessy v. Ferguson* (1896).[1] But suppose both sides to a legal contest invoke contrary scientific or engineering evidence—as often happens in patent and tort litigation (which assigns legal responsibility for harm done by one person to another). How do courts determine the admissibility and validity of that evidence, and the weight it should carry in their rulings? Can science "trump" the law, which arises from a society's sense of what is just and fair? Since the early twentieth century the Supreme Court has grappled with these questions, most notably in four cases: *Frye v. United States* (D.C. Circuit Court, 1923), *Daubert v. Merrell Dow Pharmaceuticals* (92–102) 509 U.S. (1993), *General Electric Co. v. Joiner* (96–188) 524 U.S. Supreme Court (1997), and *Kumho Tire Co., Ltd. et al. v. Patrick Carmichael et al.* (97–1709) U.S. Supreme Court (1999). What these cases show is not only the centrality of reliable technical evidence in tort litigation, but the growing determination of the federal judiciary to retain both the authority and the responsibility for the evidentiary basis of its rulings, even when the evidence is predominantly scientific or technical.

Until the early 1990s, federal courts, when asked to determine the validity (as well as relevance) of scientific evidence, usually relied on the standard used in *Frye v. United States* (1923).[2] This case, which came before the District Circuit Court on appeal from a lower court, hinged on the question of the reliability of polygraph (lie detector) tests in establishing the guilt or innocence of a defendant convicted of second degree murder. In explaining its judgment that the defendant could not be exonerated by a polygraph test, the court introduced what was known for decades thereafter as the "general acceptance" test:

> Just when a scientific principle or discovery crosses the line between the experimental and demonstrable [the scientific community agrees on its validity] stages is difficult to define. Somewhere in this twilight zone the evidential force of the principle must be recognized, and while courts will go a long way in admitting expert testimony deduced from a well-recognized scientific principle or discovery, the thing from which the deduction is made must be sufficiently established to have gained general acceptance in the particular field in which it belongs.[3]

Frye's "general acceptance" test might have survived, had it not been for that "twilight zone" between a new idea and what Thomas Kuhn would later term "normal" science. As we have seen, the new disciplines of social studies of science and history of technology that gained acceptance on college and university campuses during the 1960s helped to illuminate that zone.

The 1960s also saw an effort, led by U.S. Supreme Court Justice Earl Warren, to codify standards courts should use in determining the admissibility of evidence, an effort that resulted in 1969 in the first draft of the *Federal Rules of Evidence*. The Congress, distracted in 1972 by the Watergate scandal, did not complete its own review of the draft until 1975, when, after various amendments, the rules became law. This was the context in which the federal judiciary's earlier deference to received scientific opinion began to give way before the principle that the courts, and only the courts, should determine the admissibility of evidence—including scientific evidence. Judges and juries remained "triers of fact," and courts would determine whether facts offered as evidence were relevant to the issues at hand. As for scientific evidence, the new *Federal Rules of Evidence* placed the burden of determining admissibility squarely on the court, acting as "gatekeeper." "If scientific, technical, or other specialized knowledge will assist the trier of fact to understand the evidence or to determine a fact in issue, a witness qualified as an expert by knowledge, skill, experience, training, or education, may testify thereto in the form of an opinion or otherwise, if (1) the testimony is based upon sufficient facts or data, (2) the testimony is the product of reliable principles and methods, and (3) the witness has applied the principles and methods reliably to the facts of the case."[4]

In 1990, the Federal Judicial Center published a guide for judges responsible for cases "involving complex scientific and technical evidence."[5] The result—the *Reference Manual on Scientific Evidence*—contained well crafted essays on the process of scientific validation commonly used in disciplines that most often bear on scientific or technical evidence in contemporary litigation: statistics, survey research, economics, epidemiology, toxicology, medicine, biotechnology, and engineering. As for the persuasiveness of scientific evidence introduced by litigants, three cases serve as sign posts on a path that leads to the Supreme Court's affirmation of the judiciary's gatekeeping role in *Kumho Tire Co., Ltd. et al. v. Patrick Carmichael et al.* (1999).

Frye's standard of validity for scientific evidence—that the "scientific principle or discovery . . . from which the deduction is made must be sufficiently established to have gained general acceptance in the particular field in which it belongs"—became an insufficient (though not in itself inadmissible) basis for evaluating evidence for the U.S. Supreme Court when it issued its ruling in *Daubert v. Merrell Dow Pharmaceuticals* (1993).[6] This case arose out of the suit filed by parents of children born with birth defects against Merrell Dow chemicals, maker of Bendectin, an anti-nausea drug taken by the children's mothers during pregnancy. Experts appearing on behalf of the Dauberts argued that

Bendectin *could* cause birth defects because test tube and live animal studies had shown this to be so; that the chemical structure of Bendectin was similar to other human teratogenic substances; and that their own "reanalysis" of previous epidemiological studies indicated a probable causal connection between Bendectin and birth defects. Merrell Dow's expert argued that in his review of thirty epidemiological studies involving 130,000 cases, *no* study found Bendectin to be a *human* teratogen. Rulings by the lower state (California) and federal circuit courts seemed uncertain over the grounds on which courts should treat some evidence as more reliable than others. Thus when the Dauberts' case made its way to the Supreme Court, the court recognized that part of its task would be to determine criteria for the admission of expert testimony.

Writing for the majority, Justice Harry A. Blackmun noted that the *Federal Rules of Evidence* made no mention of a "general acceptance" rule when deciding whether evidence, in addition to being relevant, was admissible. Moreover (drawing partly from *amicus curae* briefs filed by the National Academy of Sciences and the American Association for the Advancement of Science) Blackmun observed, scientific evidence could never be regarded as certain, for science is "not an encyclopedic body of knowledge about the universe" but a process or method for obtaining knowledge. While the general acceptance of a body of data or method should be considered, "general acceptance" alone was insufficient grounds for scientific evidence, wrote Blackmun. He then offered the court's own version of the "taxonomy of expertise" that judges should use in determining the reliability of scientific evidence: Courts should determine whether proposed evidence (a) entailed a theory or technique that could be tested (that is, verified), (b) had been subject to peer review, and (c) that the potential error rates governing the operation of the method on which the evidence was based were known and acknowledged.

Merrell Dow prevailed before the Supreme Court in 1993, and so also would the General Electric Company in 1997, when a case involving another chemical compound, polychlorinated biphenyls, or PCBs, came before the court. In *General Electric Co. v. Joiner* (1997), the issue was whether Robert Joiner's lung cancer had been caused by the PCBs contained in fluid with which he was regularly splashed during his work as an electrician for the town of Thomasville, Ga., or by the cigarettes which he had smoked for eight years, along with a family predisposition to lung cancer.[7] In its defense (as manufacturer of the fluid), General Electric argued that the evidence offered by Joiner consisted of little more than "subjective belief or unsupported speculation."

The Supreme Court agreed, with Chief Justice William H. Rehnquist writing that studies of mice directly injected with PCBs, as well as epidemiological studies of PCB exposure and incidences of cancer—the expert evidence offered by Joiner—did not prove a significant causal link between exposure to fluids containing PCBs and cancer of the kind contracted by Robert Joiner. Citing *Daubert*, Rehnquist affirmed the court's responsibility to review the expert scientific testimony offered in evidence, rather than to "admit opinion evidence which is connected to existing data only by the *ipse dixit* (or, "because *I say so*") of the expert." In short, the authenticity of a witness's expertise would remain subject to the judgment of the court.

Could legal issues involving engineering and manufacturing technology be subject to the same gatekeeping role described in *Daubert* and affirmed in *General Electric v. Joiner?* Justice Stephen Breyer, in *Kumho Tire Co., Ltd. et al. v. Patrick Carmichael et al.* (1999), ruled that it could, and in so doing introduced a subtle redefinition of the authority of scientific expertise in the eyes of the federal judiciary. At issue was the responsibility for the failure (blow-out) of a tire on a mini-van driven by Patrick Carmichael, which led to an accident involving loss of life and serious injuries to surviving passengers. Kumho Tire Co., maker of the tire, argued that the tire had been worn and under-inflated; Carmichael argued that, notwithstanding its wear and under-inflation, the tire's failure was due to faulty manufacture. The testimony of Carmichael's principal expert witness, Dennis Carlson, was based on a visual and tactile inspection of the tire made on the morning of his deposition. He was unable to testify accurately to the distance the tire had traveled, and admitted that he had examined few tires that had failed under comparable circumstances.

Not surprisingly, the court ruled in favor of Kumho Tire. But what is most interesting for our purposes can be found in the commentary embedded in Justice Breyer's written opinion. First, it contained numerous citations to the technical literature on tire construction, thus demonstrating the court's determination and ability to make its own judgment as to the reliability of various technical arguments introduced in evidence. Second, the opinion denies to science a unique standing as a qualitatively superior category of knowledge. "It would prove difficult," commented Breyer, "if not impossible, for judges to administer evidentiary rules under which a gate keeping obligation depended upon a distinction between 'scientific' knowledge and 'technical' or 'other specialized' knowledge. There is no clear line that divides the one from the others. Disciplines such as engineering rest upon scientific knowledge. Pure scientific theory itself may depend for its development upon observation and properly engineered machinery."[8]

Large corporations, thanks to their greater resources, are better able to sustain costly litigation for the years it may take for a product liability case to work its way through our state and federal legal system. But the outcome does not always or necessarily favor corporate defendants, nor will the best expertise "money can buy" necessarily exonerate them. What ordinary citizens do not hear about are the many cases that are settled out of court because a manufacturer recognizes that the court may, *on its own inspection of the evidence*, find that the facts support the allegedly injured party's claims. Science, in this context, is a servant of the law; it does not "trump" the law.

The Government Performance and Results Act:
Governing by Numbers

Scientific independence, or the principle that scientific integrity requires that scientists themselves judge the merit of each other's research, is a dearly held principle among academic researchers, many of whom rely at least to some extent on direct or indirect federal funding to sustain their work.[9] Thus the Government Performance and Results Act of 1993 (GPRA, or "the Results Act," as it also came to be called) has proven a substantial challenge to those who conduct federally funded research and development, even though the statute was not directed specifically at them. Championed by Senators William Roth (R-Del.) and Fred Thompson (R-Tenn.) of the Senate Governmental Affairs Committee (and, in the House, by Rep. Dick Armey [R-Tex.]), GPRA arose from the idea then in vogue that the holy grail of organizational effectiveness could be reached through strategic planning and management by quantitative performance measures.[10] "If you can't measure it, you can't manage it," was a favorite mantra of the Clinton administration's NASA administrator, Daniel Goldin. The belief that organizational discipline in public administration could be achieved with quantitative performance assessment linked to annual budgets bore a familial resemblance to the standards and testing movement working its way through state K–12 school systems during the 1980s and early 1990s. Both beliefs reflected the appeal of scientific positivism in politics, public administration, and public education—all enterprises that hover on the edge of chaos as their natural state.

Because "waste and inefficiency in Federal programs undermine the confidence of the American people in their Government," the GPRA begins, the Congress will "provide for the establishment of strategic planning and performance measurement."[11] The act acknowledges that perhaps not all federal activities could be fairly evaluated, or evaluated at all, "in an objective, quantifiable, and measurable form." Agencies might therefore make their case to the Office of Management and Budget for using an "alternative form" of assessment. But any such alternative form would have to describe the programs in question, and their success, "with sufficient precision and in such terms that would allow for an accurate, independent determination of whether [its] . . . performance meets the criteria of [its] . . . description." Thus program and budget examiners at the Office of Management and Budget, and not federally funded scientists themselves, would be deciding whether the money spent on the scientists' work was well spent, and, inevitably, how much more money should be budgeted for their future research projects and programs.

Shifting the authority for evaluating publicly funded academic research from scientists to budget examiners is not the only challenge posed by GPRA. Scientists, as well as every other political constituency, have historically judged the responsiveness of the Congress and the White House to their appeals for federal favor and funding by how much administrative machinery, and how many federal dollars, are allocated to their putatively urgent needs. However, new or growing budgets for constituency-specific agencies like the National Science Foundation are, in the jargon of management consultants, "input measures" rather than "outcome measures." If (as its funding advocates insist) basic research is known by its indifference to practical uses (that is, results), would GPRA not then impose an alien standard on the evaluation of scientific research?

Whatever impact GPRA may ultimately have on the effectiveness of government programs, its most immediate result has been to create a new occupation for administrative personnel throughout government. In every agency civil servants struggled to devise ways to meet GPRA's requirements that made sense within their own organizations and passed muster with program and budget examiners at the OMB. Their efforts were replicated among the staffs of the formal representatives of the scientific establishment in Washington— e.g., the National Academy of Sciences, the American Association for the Advancement of Science, and the Council on Government Relations.[12] Workshops, conferences, and reports flourished, all attempting to arrive at formulas for measuring the public benefit of publicly funded research in ways that would satisfy the Congress, the OMB, and the sponsoring agencies—and all without doing damage to the assumed prerogatives of the researchers and their institutions.

The National Academy of Sciences, through its blue-ribbon Committee on Science, Engineering, and Public Policy (COSEPUP), convened workshops and commissioned studies to find a way to respond to the new requirement without doing fatal damage to the existing system by which basic and applied research was supported by the federal government. Its first report on GPRA gamely accepted the notion of quantifiable performance goals and assessments, linked to budget items, for applied research. But basic research, by its nature, could not be measured in this fashion. Instead, urged COSEPUP, the performance of basic research should be measured for its quality, relevance, and leadership. Quality, relevance, and leadership, could, of course, only be measured by experts familiar with the work being done, and thus the COSEPUP recommended that expert panels (comparable to peer-review panels) of scientists *from outside the government* should be convened by federal science and technology agencies to provide the necessary assessments.[13] Granting the merit, in principle, of COSEPUP's recommendation, convening and staffing the recommended expert panels on the annual schedule by which GPRA had to be administered, would represent a considerable challenge to research adminis-trators within agencies such as NSF, NIH, and NASA.

To the historically minded, the products of agency efforts to comply with GPRA suggest how the federal government's activities might be portrayed by a later day devotee of Taylorism.[14] Numerous federal agencies produce glossy re-ports and colorful charts displaying their strategic and performance goals in terms of quantitative measures such as time spent, clients served, and actions completed, and then report their performance in terms of percentages of time spent, clients served, and reports filed. Bureaus have their designated places in graphic boxes that contain enumerated, quantifiable objectives; these then appear on their parent agency's series of charts or "vu-graphs" designed to show the connection between budget dollars requested or received, and per-formance goals achieved.

Prepared and submitted annually, first to the OMB and then to the Con-gress, federal agencies' performance reports are subject to scrutiny by the Gen-eral Accounting Office (after 2004 the Government Accountability Office, or GAO) and receive annual rankings on quantified scorecards issued by House and Senate government affairs committees. Agency performance reports may be criticized for failing to show how the many activities thus documented pro-duce objectively identifiable *outcomes*, rather than "outputs." To some extent, however, this criticism is unfair. Arguably half of all federal agencies came into being to serve as government patrons for influential constituencies. Those agencies' best—or most authentic—quantitative performance measure may be the number of votes they receive in Congress to reauthorize their budgets and appropriate the necessary (or perhaps increased) funds.

Meanwhile, NSF, NIH, and NASA's science programs have all attempted—sometimes quite well—to articulate the investigative goals of their respective research projects, and to disaggregate and enumerate the research steps to-ward those goals. In the process, they have produced descriptive accounts of their programs that can be more informative than the glossy, often hyperbolic, program reports intended primarily for public consumption. Meanwhile, how GPRA itself would fare if subjected to a rigorous cost-benefit analysis is open to question.

In view of the conviction with which the Congress has sought to evaluate the programs it has created with quantifiable performance metrics, a caution-ary note on relying on numbers to ascertain the success of public research ad-ministration is in order. Rarely will one hear or read a discussion of U.S. science policy in which federal budget numbers are not cited as proof of increasing or declining government support for basic research in this or that discipline, or the declining or growing share of federal research and development dollars being spent on basic research versus applied research, and so on. But what do these numbers mean?

Numbers associated with federal funding for scientific and engineering research must be interpreted with care because definitions of what is being funded vary among the principal accounting organizations in the federal government. Budgeted dollars will be classified by a numbered "budget function" as well as by the agency receiving the funding (as in funding data reported by the OMB). The NSF, however, classifies federal funding on the basis of whether the dollars are reported to have been spent on basic research, applied research, or research and development. Moreover, as the federal government has grown in the number of its departments, agencies, commissions, and other ad hoc organizations, neither the OMB nor the NSF has been able to provide data reduced to the smallest independent governmental unit. What trends, if any, are lurking in the "other" category in federal reports? And while both the OMB and NSF have attempted to harmonize their accounting categories, a definitive crosswalk from year to year, category to category, source and performer, is practically impossible.

Second, science and technology budget categories that attempt to capture what scientific and engineering research and development is getting done are largely abstractions relying for their meaning on one's interpretation of the data reported by their performers. The difficulty with the NSF's distinctions between basic and applied research, for example, is that they rely on a characterization of the researchers' *intent*—a characteristic that is one of the many imponderables of public policy. Basic research does not itself receive funding; it is people performing what they report to be basic research who receive federal dollars, and the funding they receive is available to them not as individuals, but as employees of institutions, who classify their work as part of the grants proposal and administration process. Since it is primarily faculty and research staff at universities and colleges who conduct basic research, federal dollars received by these institutions (dollars which will include such negotiated indirect costs as administration and research facilities maintenance), may be perceived as a form of public subsidy for those institutions.

While well intentioned, an OMB experiment initiated in FY 2001 with a different category—the Federal Science and Technology (FS&T) budget—may, by eliminating large systems development costs reflected in defense and energy R&D budgets, more closely approximate the government's investment in science and technology. However, the concept, which results in part from a recommendation by the National Academy of Sciences, could further complicate the challenge of determining convincingly the amount and ultimate disposition of the federal government's investment in science and technology.[15] Moreover, the technological systems development and their testing and evaluation—which constitutes the lion's share of defense R&D—are as much a part of engineering research as the initial phase of data collection and analysis.

As for the *development* half of the R&D pairing, federally funded research conducted by industrial firms is funded typically through contracts with specific demonstration and delivery dates, rather than research grants which normally do not oblige recipients to produce anything at the end other than an increase in our understanding of natural phenomena. Accordingly, most research *and development*—which anticipates the availability of a successful prototype technology at its conclusion—is funded with research and development dollars through mission agencies. But the substance of the work done in industrial facilities to support the programs of those agencies can contribute as well to our understanding of the behavior of natural phenomena. In short, the cerebral and sensory processes in which *all* researchers engage do not readily submit to accounting categories—indeed, cognition itself is a continuously evolving field of learning.

So it is that when we aggregate and manipulate series of numbers as indicators of a nation's scientific or engineering capabilities and productivity, we are committing the sin of aggregating, counting, dividing, and multiplying non-uniform things. All this said, can we deduce nothing from the innumerable volumes of science and technology accounting data collected by government agencies? Not at all; what we can deduce from such data, provided we consider it over extended periods of time (for example, in five-year increments at least) is the success with which research-performing institutions have achieved influence over the two most important things that any government has to offer: access to funding and decision making. The *precise* connections between that access and the patents researchers receive or peer-reviewed journal articles they publish—the two most common measures of research productivity—we can only surmise. Of one thing we can be sure, however: without resources, there would be little productivity to measure.

Scientific and Technical Data under the Freedom of Information Act

The United States' proclivity toward open systems is mirrored in the Freedom of Information Act (1966), which requires all federal officials to document their decisions with complete records and to maintain those records within their agencies or, after a certain of period of time, to transfer them to the National Archives. With exceptions for national security and protection of personal privacy and proprietary rights, the FOIA requires that the public be able to inspect the records—electronic as well as paper—thus generated.[16] A corresponding principle, asserted at least as early as 1834 in the Supreme Court's ruling in *Wheaton v. Peters*, is that written works prepared by federal officials belong in the public domain.[17] This principle is a fixture of U.S. copyright law, which denies copyright protection for any "work of the U.S. government."[18]

As federal information policy evolved after World War II, little if any distinction was made between scientific and technical and other kinds of information generated or collected by federal agencies. The basic principle of openness upon which the U.S. government's traditional information policy rests arises from the recognition that "government information . . . is a means to ensure the accountability of government . . . [while] the free flow of information between the government and the public is essential to a democratic society. . . . The open and efficient exchange of scientific and technical government information . . . fosters excellence in scientific research and effective use of Federal research and development funds."[19] "Information" is broadly defined as "information created, collected, processed, disseminated, or disposed of by *or for* the Federal Government," while an "information dissemination product means any book, paper, map, machine-readable material, audiovisual production, or other documentary material, *regardless of physical form or characteristic*, disseminated by an agency to the public [italics added]." Unlike proprietary rights—one of the three principal grounds for exceptions to the FOIA—the unfettered exchange of ideas and information is essential to the cosmopolitan culture of science, promoting discovery, and ensuring that scientific error is openly acknowledged and corrected.

Federal agencies that fund scientific research (e.g., NASA, NSF, NIH, the Environmental Protection Agency [EPA]) performed in university, non-profit, or industrial laboratories are authorized to request research data from externally funded researchers along with final contract or grant reports. (Once research data is accepted into a federal agency it becomes part of its official records, and thus possibly subject to FOIA.) Most federal agencies, however, decline to request research data, deferring to researchers' insistence that they could not release their data for a variety of reasons—protecting the identity of human research subjects, proprietary interest, or seclusion of data prior to its publication in peer-reviewed research journals. In 1980 the U.S. Supreme Court ruled (in *Forsham v. Harris*) that if a federal agency did not actually create, or obtain, research data itself, such data could not be considered a federal record subject to the FOIA. Thus the amendment that Senator Richard Shelby (R-Ala.) successfully attached to the Omnibus Appropriations Act for fiscal year 1999 created quite a stir—not only among federally funded scientists, but among those likely to be impacted by policies based on their research findings.[20]

The Shelby amendment required that federal agencies make available to the public "all data produced under a [federally funded research] award." What troubled many scientists about the measure was that it (a) required not only that information (that is, research results), but research *data*, be made public; (b) applied not only to information generated within federal research agencies by their own employees, but to information created by federal research contractors and grantees; and (c) required that such information be made publicly available under the authority and procedures of the Freedom of Information Act.

Senator Shelby's action had been provoked by the EPA's new National Air

Quality and Ambient Quality Standards released in 1997 during the Clinton administration. Shelby represented Alabama, a relatively poor state with some heavy smokestack industries, which complained of the high costs of complying with the new standards—estimates ranged from $7.5 billion to $16 billion.[21] Other critics pointed at several studies of toxic effects of air and water pollutants that had later proven faulty, and thus wanted to inspect the data behind the new EPA standards. Had those with a stake in EPA regulations known that the supporting research was faulty, they might have either lobbied successfully against the regulations, or for more favorable modifications.[22]

By law, OMB was required to publish revisions to its guidelines in the *Federal Register* and to invite and consider comments during a sixty-day comment period.[23] This it did in February of 1999. The comments OMB received were surprising for their intensity. Since both the norms of American policy making and the culture of science favor open systems, it was not readily apparent why Senator Shelby's amendment produced such controversy. After all, the National Research Council had recently stated that "data and information derived from publicly funded research [should be] made available with as few restrictions as possible, on a non-discriminatory basis, for not more than the cost of reproduction and distribution."[24] But, as it turned out, the phrase "as few restrictions as possible" was no trivial qualifier. In 1999 roughly 75 percent of all federally funded research was performed in non-federal facilities, or through grants and contracts to research organizations, about 60 percent of which were colleges and universities. Scientists maintained that their profession was governed by four basic principles: (1) peer review establishes the credibility of data and findings; (2) research results must be replicable; (3) peer-reviewed publication of research results establishes priority in discovery, and thus professional esteem and advancement; and (4) public discussion of preliminary findings must be allowed to proceed free of extraneous or political influence. It was the third and fourth of these principles that appeared to be threatened by the Shelby amendment.

Every White House, regardless of its incumbent's party, consults informally with its political constituencies to learn, or simply confirm, their views on the policy question at hand. The OMB's initial proposed revisions suggest that the Clinton White House's staff had indeed done its homework. While the OMB's initial proposal included the reference to FOIA now required by law, it included the qualifiers—"published," "used by the Federal Government in developing policy or rules," and "within a reasonable time." These effectively undermined the purpose of the amendment, inasmuch as the normal time required for research data to be published (several months at the very least), and the fact that regulations would have to have been enacted before anyone could determine whether research findings had influenced them, rendered the application of the FOIA meaningless. The intent of Shelby's measure was to enable those most likely to be negatively impacted by a pending regulation to examine—and possibly challenge—the supporting research before a regulation became law.

Senators Shelby, Trent Lott (R-Miss.), and Ben Nighthorse Campbell (R-Colo.) responded to Jacob Lew, director of the OMB, asking that the intent of Senator Shelby's amendment be adhered to with a requirement that agencies notify the public which studies will be used in the rule making or policy development, and process all timely requests for the studies' supporting data before the end of the public comment period. The U.S. Chamber of Commerce weighed in as well, estimating that compliance with federal regulations costs businesses $737 billion, while there was no public access to data used to formulate these regulations. Meanwhile, a study by the American Enterprise Institute-Brookings Institution Joint Center for Regulatory Studies concluded that data and analyses used for public policy making are rarely checked by anyone, and the scientific community's peer-review process was not adequate to detect and report inadequate or misused data or poor analysis.

The OMB responded to these objections in a second set of proposed revisions. These revisions, however, did not appear likely to accommodate its Republican critics. To the contrary, they were more likely to please the National Academy of Sciences' president Bruce Albert, who lobbied the White House to oppose allowing the FOIA to be used to obtain the research data at issue, partly because data from federally sponsored research might become part of a patent claim. Instead, the revisions added the qualifier "research" to the data that could be subject to FOIA, and raised the bar for eligible research data requests by specifying that the data had to have been used in making a regulation, rather than the vaguer phrase "developing policy or rules." It also added a clarifying definition of "research data" to satisfy the scientific community's fear that any and all things that might have served as data for a research project could be subject to release under the FOIA. The OMB explicitly excluded from research data "preliminary analyses, drafts of scientific papers, plans for future research, peer reviews, or communications with colleagues. . . . (A) trade secrets, commercial information, materials necessary to be held confidential by a researcher until publication of their results in a peer-reviewed journal, or information which may be copyrighted or patented; and (B) . . . files the disclosure of which would constitute a clearly unwarranted invasion of personal privacy." The OMB also proposed that the FOIA data release provision not apply to regulations having an estimated impact of less than $100 million.

The OMB's second revision of its guidelines revealed the extent to which the scientific community was willing to compromise its tradition of openness in order to protect its members' possible interest in "trade secrets, commercial information, materials necessary to be held confidential by a researcher until publication of their results in a peer-reviewed journal, or information which may be copyrighted or patented." It had come a long way from the ideal of scientists pursuing science for its own sake, "free to explore any natural phenomena without regard to possible economic applications"—Vannevar Bush's principal reason for the necessity of public support for basic research (see chapter 3).

But the controversy over the Shelby amendment involved more than the interests of federally funded researchers, on the one hand, and the industries most likely to be impacted by the EPA's new National Air Quality and Ambient Quality Standards. The citizen watchdog organization Public Citizen took an interest in the issue as well, pointing out that the OMB lacked the authority to revise a congressional statute or so qualify its application with "clarifying definitions" that it was emptied of its otherwise clear and unambiguous purpose. Citizens as well as industries had a stake in the timely availability of research data necessary to form "sound government policies on issues such as drug safety, automotive safety, nuclear energy, and workplace hazards."

The OMB's final revision (issued November 8, 1999) contained two minor changes to assuage Senator Shelby and the supporters of his amendment. First, the definition of research data that might have to be disclosed under the FOIA was expanded from data used in support of a regulation to research data used in support of "an agency action that has the *force and effect of law*." Second, the definition of exempt materials was narrowed from materials "necessary to be held confidential by a researcher until publication of their results in a peer-reviewed journal, or information which may be copyrighted or patented," to "materials necessary to be held confidential by a researcher until they are published, or similar information which is *protected under law*" (such as proprietary or national security information or data that could be connected to an identifiable individual). The restriction of the research data disclosure requirement to measures having an impact greater than $100 million had been removed.

Whatever the merits of Senator Shelby's amendment in principle (and Public Citizen considered those merits substantial), it had a serious operational weakness that was unaffected by OMB's efforts to interpret the measure administratively to serve the interests of the constituency represented by the National Academy of Sciences. Identifying the "research data" that supported research findings resulting from federally funded research was something that only the researchers themselves could do. The extent to which any of OMB's clarifying definitions applied to any item of research data could only be determined by those generating it—that is, the research scientists themselves. Federal sponsoring agencies might attempt to extract such information from the recipients of their research grants and contracts, but were unlikely to do so, given their predilections prior to the passage of the amendment. Thus, not withstanding efforts using the political process to subject scientific research used for public policy making to public scrutiny, scientists—in this instance—

retained their authority to shape policy without corresponding public accountability.

Reliable Information, Peer Review, and Federal Regulatory Policy

Tucked into the Treasury and General Government Appropriations Act for Fiscal Year 2001, which was passed by the Congress and signed by President Clinton near the end of his term, was a provision (Section 515) that seemed on its face to be eminently reasonable and in the interest of accountable government. The provision became known as the Information Quality Act of 2000, and turned out to be one of the most contentious measures of its kind. The legislation directed the OMB to issue guidelines ensuring "and maximizing the quality, objectivity, utility, and integrity of information (including statistical information) disseminated by federal agencies."[25] Agencies would have to establish "administrative mechanisms" allowing for corrections to information they had disseminated, and "report periodically" to OMB on any complaints they had received, and how they had disposed of the complaints.

While the act was signed under President Clinton, its implementation would be the work of an OMB director reporting to President George W. Bush, whose election in 2000 had been hotly disputed by Democrats. Democrats challenged the validity of the votes counted by the state of Florida, votes which gave Florida's critical electoral votes to Bush. The Supreme Court resolved the issue in favor of Bush, thus marking the beginning of what has proved to be one of the most divisive presidencies in recent American history. The broad political center that had formed during the presidencies of George H. W. Bush and Bill Clinton frayed rapidly in an atmosphere of political hostility familiar to those who recalled the contentious political environment of the Vietnam and Watergate years, which had driven two presidents, Lyndon B. Johnson and Richard M. Nixon, from the White House. Science found itself sucked into the cauldron along with religion, social policy, policies to combat terrorism, and economic policy.

Opposition centered on the Bush administration's application of Section 515 of the Information Quality Act, or the statutory direction to federal agencies that they use OMB-prescribed peer-review procedures to establish data quality. On the face of it, this seemed a reasonable and innocuous requirement. For over half-a-century academic science and its professional organizations had relied upon peer review as the principal means by which scientists ensure the merit of each other's research projects and resulting publications. The post–World War II federal science agencies—most notably the NSF and the NIH—had adopted agency-specific peer-review procedures to assist in making grant award decisions. That peer review in practice had its shortcomings was no secret; but this method of validating scientific research was nonetheless considered the best method available—analogous to the "general acceptance" test once relied upon by the federal judiciary to determine the reliability of scientific evidence.

The controversy over the Information Quality Act was fueled largely by Republican efforts to reduce the scope and number of environmental, consumer safety, and occupational health and safety regulations put in place during Democratic administrations. In this respect it was the progeny of Senator Shelby's effort to impose greater transparency on data used by the EPA and other regulatory agencies. Under Bush's OMB director, Joshua B. Bolton, the administration drafted guidelines that critics argued over-extended Bolton's authority in Section 515. OMB director Bolton would require federal agencies to create peer-review panels to determine the reliability and objectivity of scientific and technical data by ensuring that such panels excluded scientists whose current research was federally supported, and that the biases of some members of any panel be balanced by the opposing biases of others. Since those affected by the data's use in a regulation or policy would be allowed to challenge the data's quality, they could forestall regulatory action by time-consuming administrative procedures.

Just how far critics thought Bolton had reached was suggested by the responses of Democratic leaders of the House committees with jurisdiction over federal science policy (including representatives of states containing the University of California, the University of Texas, and the research intensive institutions surrounding Cambridge, Mass.). Democratic members of the House committees on science, government reform, and energy and commerce let go a salvo asserting that the administration, "under the guise of promoting sound science . . . is advancing a far-reaching policy that will impede efforts to protect health and the environment and open the door to conflicts of interest in the regulatory process." "For political reasons," continued the administration's congressional critics, "the Bush Administration has repeatedly distorted scientific data, manipulated scientific advisory committees, gagged scientists, and provided misleading information to Congress and the public. Yet the new OMB proposal [draft regulations] . . . erects new roadblocks to the use of high-quality science in agency decision making."[26]

The OMB received nearly 200 responses, largely from organizations, to the draft guidelines. These included comments from national organizations ranging from the National Academy of Sciences, American Association for the Advancement of Science, and Public Citizen, to the National Association of Manufacturers and the American Petroleum Institute.[27] While opinions were varied, they tended to converge on the guidelines' requirements that (1) the OMB exceeded its statutory authority in requiring agencies to establish new peer-review panels with scientists who were non-federally supported and whose biases were balanced, to determine the quality of their disseminated information—including information that had already undergone peer review, for example, journal articles; (2) individuals currently receiving or seeking "substantial" funding from agencies conducting the reviews be excluded from review panels; (3) individuals appointed to a review panel having one "bias" on the matter of hand be balanced by other appointees having alternate "biases"; (4) that review panel members and their evaluations be disclosed to the public; and (5) persons affected by the information be allowed to "seek and obtain correction of information maintained and disseminated by the agency."

While industry group comments generally approved of these provisions, academic, research, environmental, and public interest organizations pointed out that (1) the additional peer-review requirements were too prescriptive and would delay the release of information necessary for prompt agency action in the event of health or environmental emergencies; (2) the exclusion from panels of recipients of federal funding virtually ensured that panels would be populated by scientists funded by industry; (3) "bias" and "alternative bias" were vague and improper grounds for selecting expert members of peer-review panels; (4) expert members of peer-review panels would be less candid in their evaluations if their comments were disclosed publicly; and (5) that opportunities for public correction would enable regulated industries to stall indefinitely the implementation of federal regulations. In an effort to mollify its critics, the OMB shrewdly granted information from the National Academies (National Academy of Sciences, National Academy of Engineering, Institute of Medicine, and National Research Council) exemption from further peer review, and provided emergency exemptions for individual agencies which, however, could only be granted by the OMB.

Heads of federal agencies subsequently began to develop procedures to comply with the OMB guidelines. To what extent these procedures were actually new, or the re-packaging of practices already in place (peer review of grant proposals, for example), is debatable. Though often faulted for inefficiency, federal agencies have become adept at the survival strategies necessary to continue functioning when subjected to the erratic and expanded mandates (often accompanied with reduced budgets) that fall on them when the White House or the Congress changes party control. Moreover, the real significance of the new guidelines was not that they would be adhered to consistently, or even noticeably, but that they would legitimate efforts by those being regulated to frustrate the enforcement of the regulations they found unpalatable, and to do so in the name of better science.

Federal Advisory Committees: An Alternate Route to Power

As we saw in chapters 2 and 3, ideology has combined with historical practice in the United States to shape the contours of this country's political relationship with science and technology. That relationship is mirrored in a policy of using government resources to build into the nation's industry and academic institutions a capacity to produce, on call, world-class scientific research and a technologically advanced industrial production second to none. This same policy, however, has resulted in a mutual dependency between government and communities of expertise, a dependency that has provoked fears in a country that contains not only its share of populist sentiments, but critics of modern technological society in general. Experience during the decade since Eisenhower's farewell address (see chapter 3) proved that Eisenhower's fear was well-founded. It was shared by enough members of the Congress that in 1972 it passed the Federal Advisory Committee Act, designed to make the membership and activities of advisory committees more transparent by requiring public access to government policy deliberations and the records of those deliberations that were not otherwise protected for reasons of national security or personal privacy.[28]

Congressional efforts to gain some measure of control over advisory committees used by the Executive Branch—which some members of Congress feared had become a way for private citizens and interest groups to have an inordinate and unaccountable degree of influence over policy decisions—date back to 1942, when the Congress passed legislation prohibiting funding of any such groups without a special appropriation. A suspicion that industry groups were circumventing antitrust laws by meeting behind closed doors as putative advisors to the White House and federal agencies led the Department of Justice as early as 1950 to develop guidelines for advisory committees composed of private citizens. These guidelines have remained the basic components of all subsequent advisory committee legislation:

(1) There must be either statutory authority for the use of such a committee, or an administrative finding that use of such a committee is necessary in order to perform certain statutory duties.
(2) The committee's agenda must be initiated and formulated by the government.
(3) Meetings must be called and chaired by full-time government officials.
(4) Complete minutes must be kept of each meeting.
(5) The committee must be purely advisory, with government officials determining the actions to be taken on the committee's recommendations.[29]

Guidelines, however, are easily ignored. The head of the justice department's antitrust division testified before Congress in 1956 that there were 1,393 advisory committees in the executive branch not specifically established by statute, and that "less than 50 percent . . . were complying with the standards."[30] Responding to mounting congressional efforts to give the 1950 guidelines statutory force, the White House in 1962, probably as a preemptive measure, issued Executive Order 11007, which gave the earlier guidelines the near equivalent of statutory authority over federal agencies.

The use of advisory committees by federal mission agencies (see chapter 4), which receive the bulk of federal R&D dollars and then use them to award grants and contracts to university researchers and industrial firms, has been especially problematic because agencies rightly want to consult individuals with expertise in pertinent areas of science or engineering in the course of planning the programs that result in those awards. And indeed, most professionally established scientists, engineers, or university and industrial research administrators prefer to think of themselves as advisers whose opinions are sought on the basis of their individual expertise. However, under the U.S. criminal code, "special government employees" (a temporary and short-term federal employment category applied to government advisors) are forbidden to represent themselves or any organization in which they may have a financial interest. In order to serve, they must submit confidential financial disclosure reports—a requirement that some committee appointees, who consider providing their expert advice to the government a public service, find offensive.[31]

Thus, the dilemma: scientists, engineers, and research administrators are typically employees of institutions, and either they or their institutions (usually both) have a career or institutional interest in the same disciplines or technologies about which their advice is being sought. What's more, given the reliance of university-based research on federal funding, demonstrating the absence of a conflict of interest among academic scientists can prove challenging; avoiding even the *appearance* of a conflict of interest is the most challenging of all. Some agencies will waive a conflict-of-interest exclusion from advisory committee deliberations on the grounds that the nominee's value to the committee supercedes the risk of compromised judgment. In 2002 NASA eliminated the problem entirely for its highest-level advisory board, the NASA Advisory Council (NAC), by declaring that all of its members served as interest group "representatives"—that is, lobbyists—who would not have to submit financial disclosure reports because, as representatives of special interest groups, their financial interest in the advice being generated would be assumed.

As successive presidential administrations seek to shape the workings of executive branch agencies to reflect their own parties' ideologies and the platforms on which they were elected, appointments to advisory committees can become politically contentious. While neither of the major parties is immune from controversial scientific advisory committee appointments, the second Bush administration has attracted somewhat more than its share of media attention for allegedly subjecting nominees to advisory panels for agencies for

the Department of Health and Human Services to political or ideological lit-
mus tests. Given that most scientists will favor proceeding with research in
disciplines in which they have built their careers, the administration's efforts
to find panelists who share its generally negative views on stem cell research,
or the enforcement of strict environmental regulations on the energy industry,
are bound to raise alarms among those with contrary views.

Whenever a policy dispute erupts in the media and involves accusations of
improper influence over White House decisions, congressional (or political
opponents') demands for advisory committee records are likely to be met with
White House claims of "executive privilege" and complaints of improper legislative branch incursions into the executive branch's area of administrative discretion. Legitimate issues about the possible encroachment of the FACA on the
constitutional separation of powers were involved in two recent and politically
heated cases. The first was a by-product of the Clinton administration's effort
to develop a system of national health care that combined public with private
funding and management. Both the president and the first lady, Hillary
Rodham Clinton, were known to be earnest students of policy, and few were
surprised when Rodham Clinton was named head of a Health Care Task Force
to design a comprehensive plan for national health care.

In 1993 the Association of American Physicians and Surgeons (AAPS),
joined by other plaintiffs, sued in U.S. District Court (for the District of Columbia) that the FACA required that the task force's meetings be opened to the
public. The AAPS and its supporters charged that the task force was allowing
unacknowledged private interests special influence over the design of the plan.
Judge Royce C. Lamberth rejected the White House's argument that Rodham
Clinton was a "virtual" federal employee, and that therefore the FACA did not
apply to the Health Care Task Force. In considering the constitutional question
of whether the act encroached on presidential prerogatives, however,
Lamberth agreed that the president enjoys a constitutionally protected right
(Article II) to confidential deliberations in the process of developing legislative
proposals for the Congress, which may not regulate how, and from whom, the
president obtains advice. Thus requiring that meetings in which committee
members formulated advice and recommendations for the president to be
open to the public would impair the "candor" of the deliberations "impermis-
sibly."

The AAPS appealed Lamberth's ruling to the U.S. Court of Appeals (for the
D.C. Circuit), which ruled that Rodham Clinton was a "de facto [federal official]
or employee," and thus the White House did not have to open the task force's
meetings. Much—if not more—could be learned from the task force's records,
and most of these would have to be released, as required by the FACA (and the
FOIA). Delays and charges of destruction of documents ensued, until the administration decided to defuse the issue (which had not helped the political

prospects for health care reform) by releasing thousands of documents generated by the task force—a step that satisfied the AAPS.[32]

The political shoe shifted to the other foot when the administration of George W. Bush took office. The Bush administration's ties to large energy and defense/aerospace contractors had been the source of unusually sharp questioning of its appointments to senior policy advisory panels at the Department of Energy and Defense—appointments that appeared to include a large proportion of executives from those industries. Thus when, only a few days after taking office in 2001, the Bush administration formally established the National Energy Policy Development Group to be chaired by Vice President Richard Cheney, some observers were not persuaded by the administration's initial statement that the task force's membership would be limited to federal officials. If non-federal officials were to participate, then the task force might fall under the provisions of the FACA.

Two organizations—Judicial Watch, a citizen watchdog group, and the Sierra Club, an environmental organization—filed suit in U.S. District Court to ask the administration to release information about the task force. As with virtually all litigation, procedure entails pre-trial "discovery"; the discovery phase of this civil suit would necessarily reveal whether any non-federal officials consulted by the task force participated as *de facto* members. Early in the litigation both sides made important strategic decisions. Judicial Watch and the Sierra Club opted to cast their information request in the broadest of terms, as if the Cheney task force was presumptively subject to the provisions of the FACA. The White House, for its part, opted to withhold any and all information about the task force on the grounds of executive privilege and the constitutional separation of powers.

As a result, when the U.S. District Court issued a limited discovery order to the White House, the White House refused to comply and appealed to the U.S. Court of Appeals to direct the lower court to rule on the basis of information on record at that time. The Court of Appeals dismissed the White House's request, ruling that it must legally assert executive privilege and file objections to the lower court's discovery orders with "detailed precision." The White House then appealed to the Supreme Court, which sent the case back to the appeals court, ordering that the White House could not be held to ordinary discovery requirements.

As for the FACA, while the Supreme Court avoided ruling on its constitutionality as a congressional regulation of the executive branch, the principle of open government was never in jeopardy. The Administrative Procedures Act (5 U.S.C. Chapter 5) contains the provisions of the Freedom of Information Act, which are incorporated into the FACA, and provides for judicial review of federal agency actions (or actions "unlawfully withheld or unreasonably delayed"). The White House might be able to argue that executive privilege allowed it to withhold from public disclosure information about its internal workings and deliberations, but federal agencies could not, ordinarily, conceal their activities behind the same principle.[33]

But, how about the use of expert advisory committees by organizations such as the National Academy of Sciences, whose work is funded largely by federal agencies? In 1997 the Supreme Court let stand a ruling by the U.S. Court of Appeals for the District of Columbia circuit that the roughly 600 advisory committees operated by the National Academy of Sciences with funding from federal agencies were subject to the FACA. The ruling arose out of a 1994 suit filed by the Animal Legal Defense Fund to have access to the meetings and records of an academy committee preparing a manual of care for animals used in laboratory research. Given that the FACA requires that a federal official convene and be present at advisory committee meetings, the academy's ability to gather information in confidence would be diminished. Moreover, the academy argued (along with the eighty eminent scientists who filed a supporting brief) the effectiveness of its work depended on its independence from the government.

Faced with the Supreme Court's 1997 ruling, the academy threatened to halt convening further advisory committee meetings. Within weeks, however, the academy and the Clinton administration were able to persuade the Congress to pass legislation clarifying that it did not intend that the academy be subject to the FACA. The academy's victory was not complete, however. Legislative language provided by Rep. Henry A. Waxman (D-Calif.) specified that, while the academy could conduct meetings without regard to the presence or decisions of federal officials, it would still be expected to publish the names of committee members, avoid conflicts of interest among committee members, and ensure that a balance of interests are represented on their committees.

The concept that "candid" advice cannot be given openly, whether it is given to the chief executive or the clients of the National Academy of Sciences, is problematic at best. Its validity has been accepted in the case of the nation's chief executive, but whether it reflects well on eminent scientists when invoked by them is debatable. Part of the historic legacy of science is intellectual courage; its heroes—from Nicolas Copernicus (1473–1543) to Andrei Sakharov (1921–1989)—spoke what they thought to be true regardless of the consequences. The best in science has everything to lose, and nothing to gain, by the diminution of openness in society and politics.

Meanwhile, repeated skirmishing over the selection of members to advisory committees of federal agencies whose missions include programs with significant scientific or engineering content reflects the lack of unanimity within the larger public (including academic researchers in all disciplines) over whether scientists can be genuinely neutral when scientific questions intersect with policy choices. The belief that a career in basic or fundamental research releases intellectuals from ordinary material concerns—concerns which tend to be implicated in the distribution and use of political power—is essential to the ideology of science. But even if this innocence were possible, does it extend to all kinds of technological expertise? Populists would argue that decision making that is not entirely inclusive and transparent is a corruption of the democratic process.

CONTROVERSIES OVER THE Shelby amendment (1999) and the Information Quality Act (2000); evolving standards of reliability for scientific and technical evidence in the federal judiciary; the public accountability of scientists when helping to shape policy as members of federal advisory committees; and the effort to introduce scientific positivism into federal public administration with the GPRA of 1993, illustrate the different ways in which the immediate post–World War II aspirations of science have been challenged in the crucible of the nation's more recent political and economic history. They also illustrate the obsolescence today of the Bush paradigm as a cornerstone of federal science and technology policy.

The response of the scientific community (through its institutional advocates) to a statutory requirement subjecting all research data acquired with federal funds to disclosure under the FOIA revealed that the establishment's commitment to transparency in government *and* research was limited. The strength of that commitment had become contingent on preserving its members' careers (advanced by priority in publication) and proprietary interests in the possible commercial value of applications of its research. That there might be a legitimate public interest in prompt disclosure of research data obtained with public funds—quite apart from any policy applications of the data—appears to have been immaterial; public funding had become a virtual entitlement, and entailed no special obligations among its recipients.

This attitude can be seen as a natural outgrowth of the federal sponsors' concession to the community's preference for research funding with grants rather than contracts. Grants were and are regarded as subventions in order to increase a public good: the seed corn of basic research from which technological advance and economic prosperity could be harvested. Contracts, on the other hand, are purchase agreements for commercially available products or services. Scientists by definition, as their inherited ideology had taught, did not engage in commercial pursuits.

One of the arguments the scientific community used against the data disclosure requirement of the Shelby amendment was that it would interfere with the well-established process of internal (to the profession) peer review of research to validate scientific findings. When taken at its word, however, in the Information Quality Act of 2000, the community demurred. As implemented by a Republican administration, the requirement that data used in federal policy making be subject to peer review was suspect. The Bush administration (and Republicans in general) was opposed to federal environmental and safety regulations that might excessively burden smokestack and manufacturing industries; scientific data was often used to justify new or expanded regulatory standards. Delay in the public disclosure of federally acquired research data pending peer review was acceptable when required to protect a researcher's publication and proprietary interests. However, delay was not acceptable when it might retard the implementation of environmental or safety regulations.

The Information Quality Act also called for possibly additional "independent" peer reviews. The Bush administration defined "independent" as uninvolved in federal research; but vast numbers of scientists were funded by federal agencies, and just as many participated in their science advisory committees. That left largely industry-supported scientists to perform "independent" reviews. The assumption by the act's critics was that scientists funded by the federal government could conduct disinterested peer reviews, but scientists funded by industry could not. The act's supporters assumed just the reverse: scientists dependent upon federal agency funding would tend to support research used by agencies for regulatory purposes.

The scientific community's response to the Shelby amendment and Information Quality Act illustrates another aspect of the role of science and technology in federal policy development. Scientists have tended to shrink from congressional politics to achieve their aims, preferring to work through the executive branch policy making machinery—e.g., the White House Office of Science and Technology Policy, the OMB, the Defense Science Board, the NASA Advisory Council, and similar groups. Congressional efforts to accelerate disclosure of scientific and technical information used by federal agencies, and OMB efforts to verify it through additional reviews by scientists not dependent upon federal funds, can be seen as efforts to increase the public accountability of scientists' and other technical experts' influence on federal policy making.

The belief that the results of scientific research—or the study of nature, which rains on the just and unjust alike—could and should be politically neutral, was a critical corollary of the Bush paradigm. This was a belief, however, that did not survive public scrutiny of the workings of federal advisory committees. Whether rightly or wrongly (bias is difficult to prove), liberals and conservatives, Democrats and Republicans, suspected that scientific and technical experts selected to serve on federal agency advisory committees were selected to advance the current administration's political agenda, or at the very least ensure the continuation of particular federal research and development programs which provided funding to their own institutions.

The best response from the scientific community would have been to practice and otherwise affirm the transparency called for by the Federal Advisory Committee Act. This it chose not to do. The most that scientists on federal advisory committees could risk by speaking plainly might be a loss of favor among disgruntled colleagues, or the loss of new or renewed funding by a federal research administrator. It was a risk which the National Academy of Sciences would not take. Arguing that scientists could not give candid advice if that advice had to be given publicly, the NAS betrayed the principle of openness that has been critical to the emergence of modern science. The advance of science has been enabled not only by fine intellects, but by stout hearts—the courage of individuals who risked all that they had in order to preserve the integrity of their speech.

For the federal judiciary since the 1920s, evidence offered as scientific and therefore definitive was usually regarded as such if generally accepted by the majority of scientists in the pertinent discipline or area of research. By the 1990s, however, the Supreme Court acknowledged the tentativeness of scientific findings (a necessary condition of ongoing research) and reasserted the federal courts' responsibility to remain the triers of fact—including facts presented as scientific or engineering evidence—in order to fulfill their constitutional obligation to exercise the "judicial power of the United States."

The only element of the ideology of science to survive intact the experiences and scholarship of the post–World War II period was the belief that all human activity, including organizational behavior, could be understood and governed by the same scientific methods as natural phenomena—that is, through quantitative measurement and systematic management. The GPRA's design to make public administration in the federal government more accountable through strategic planning and the measurement of results contained echoes of the Congress's creation, in 1972, of an Office of Technology Assessment (OTA).[34] Testimony given during congressional hearings on the proposed OTA placed great emphasis on the importance and authority of scientific and engineering expertise in policy making. Government by experts who could formulate coherent, "rational," and comprehensive policies to remedy whatever ailed the country had its champions, but it also had its detractors. One of them was the economist Charles E. Lindblom, who argued that policy made through the messy process of democratic politics—"muddling through"—would "assure a more comprehensive regard for the values of the whole society than any attempt at intellectual comprehensiveness."[35]

The Bush paradigm itself was only a modern expression of an ancient political ideology that placed science at the pinnacle of intellectual and moral virtue. As such it enjoyed enormous success for nearly half-a-century as a justification for the generous public funding of scientific research, largely in academic institutions, with correspondingly minimal expectations of accountability. Thus the obsolescence of the paradigm in all aspects but celebratory rhetoric has serious and long-term ramifications for the public support for and regulation of scientific research and technology.

6

Open Systems in a Digital World

The open political system around which the United States' governing institutions were designed has combined with evolving information technologies to shape a revolution in the way we produce, manage, and communicate information. As a result most of us will be drawn into, and are already impacted by, the new and still expanding cultural and technological universe commonly called "cyberspace."[1] By virtue of the medium of which it is composed, this new universe transcends traditional geopolitical boundaries. As a consequence it has not only spawned its own set of policy challenges, but is forcing on us the need to reconsider the way we have framed and resolved older, yet still fundamental, issues of public and economic policy.

For example, open systems in the proceedings of government make it possible for ordinary citizens to be vigilant against abuses of power. But cyberspace has added a new dimension to the ancient tension between individuals and government power—the prospect of monopoly domination of the open communication networks important to a free civic society. While the economic dangers of business monopolies have been with us since the beginnings of commerce, whether U.S. antitrust legislation and jurisprudence are still applicable to the natural economic dynamics of cyberspace has become one of the more urgent questions before policy makers in this new age of the Internet.

In this chapter we will see how the approach taken by two agencies of the federal government—the Defense Advance Research Projects Agency (DARPA) and the National Science Foundation, which supplied federal funding and direction for the development of the Internet—shaped the emergence of the Internet as an open system. We will also explore how the open world of the Internet—in all its capacity to challenge prevailing political, cultural, and economic arrangements—has generated, or renewed, policy questions the resolutions for which are likely to have a profound influence on the global economic and political arrangements of the twenty-first century. The policy questions that have resulted from these developments tend to cluster around the applicability in cyberspace of (a) constitutional guarantees of freedom of speech and freedom from unreasonable search and seizure, (b) federal regulatory power over the networks over which Internet communications occur, and (c) the continuing usefulness of federal antitrust statutes in the electronic economy.

Cyberspace as an Open System

Three characteristics of the emergence of cyberspace ensured that it would become an open system: binary (digital) computing and the replacement of analog by digital communications, a cluster of innovations such as the graphical user interface that would foster the popularization of the computer, and the open architecture of the Internet. While most of us who use computers and travel the Internet will recognize the importance of these attributes, their relationship to some of the most pressing policy issues associated with cyberspace may not be so apparent.

The binary system is to the globalization of the world of the late twentieth century what classical Latin was to the emerging civilization of early modern Europe. Throughout the disintegration of the Roman Empire in the third century and the rise of Christianity—legalized throughout the empire by Constantine in 313—one thing above all else ensured the survival of the literature and learning of Greco-Roman and eastern Mediterranean antiquity: the use of Latin as a universal written language. Latin supplied a universal alphabet, it supplied a common vocabulary, and it ensured that the emerging Roman Catholic Church would have a cosmopolitan reach in learning, doctrine, and politics. Indeed, only in the twentieth century was Latin replaced (by English) as the basis of a nomenclature used in the learned professions of law, medicine, and science.

Unlike the decimal system, which relies on ten different digits and multiples of ten, the binary system relies on only two digits: 1 and 0. "1" and "0" can be arranged in infinite patterns to represent numbers of any size; but they can also be arranged in various patterns to represent an alphabet—*any alphabet*. Thus a binary code is as useful a tool for translating language as it is for numerical computations. Second, a binary system lends itself to use in mechanical and electronic computing devices with toggle or "on, off" switches. In short, the binary system and the modern digital computer are as mutually compatible as two hands clapping. They also provide a truly universal means of capturing, storing, managing, and communicating alpha-numeric and other digitized information of unlimited variety, including sound and images. The volume (rate as well as speed) at which they work their near-magic is theoretically limited only by the size and speed of electrons.

The digital computer, with its unprecedented capacity to store and manipulate alpha-numeric information, would have remained a sophisticated device useful to a limited population of scientists and engineers had it not been for a cluster of technical innovations and business decisions that enabled it to become as popular as the typewriter and telephone had become decades earlier.[2] Arguably the most important of these was "bit-mapping," first available in 1973 on Xerox Corporation's "Alto," which enabled the transformation of digital electronic signals on the computer screen into images. Combined with the "point and click" mouse, bit-mapping allowed for the development of the graphical user interface (GUI) which replaced the arcane text messages used to operate older electronic computers.

Colorful screens with images, icons, menus, and tool bars—once they had been adopted throughout the industry in the early 1990s—enabled ordinary people to operate computers without special training. The graphical user interface, in effect, standardized the operator interface for the personal computer, turning it—with its numerous business and personal uses—into a popularly accessible "open system" for information storage and management, document preparation, communications, and entertainment. While an image may only rarely be worth a thousand words, images can break down language barriers of all kinds, including technical language barriers. Internet publishers, like ordinary paper magazine publishers before them, were able to finance their ventures with colorful advertising, advertising that made ever more insistent appearances on potential customers' computer screens.

Some of these technical innovations were the by-product of grants and contracts from the Defense Advanced Research Projects Agency, which had funded extramural research and development in computers since World War II to support Cold War weapons operations and strategic systems development. For example, Douglas Engelbart, doing DARPA-funded work at the Stanford Research Institute, perfected the computer mouse, the multiple window display, and hypermedia (which enabled various texts, images, video, and sound files to be linked within a single document).

The popularization of the personal computer required not only technical innovation, but business decisions aimed at moving the computer from the laboratory or government installation into offices and homes throughout the country, expanding its role from large-scale "number cruncher" to text preparation, ordinary data management, communications, and video. Pioneer Apple Computer's best selling (for its day and market) Apple II, first introduced in 1977, was advertised in popular magazines rather than technical journals. It featured an open "bus" architecture using expansion slots to accommodate later increases in its capabilities, a spreadsheet program (VisiCalc) and a soon-to-be ubiquitous 5.3 inch "floppy disk" drive.[3] The more moderately priced and highly successful Macintosh (introduced in 1984) featured a graphical user interface and mouse.

IBM's immediately successful personal computer ("PC"), brought out in 1981, was the result of a deliberate open systems approach to product development. In developing the PC, which sold half-a-million units in its first two years on the market, IBM used an open architecture enabling it to buy materials and components from other companies and add on features. In doing so it promoted the success of two other future giants in the personal computer industry. Intel, which released the first micro-programmable computer chip in 1971, supplied the microprocessor for the PC's central processing unit. IBM then turned to the fledgling software company Microsoft, from which IBM obtained the PC's disk operating system, or "DOS." The popularization of the personal computer was the necessary first step for the creation of the Internet.[4] Once again the federal government supplied much of the impetus and funding for a new technological system. But just as important as the fact or extent of federal funding for inter-networking in the 1960s and 1970s, was the manner in which that support was distributed.

As we have seen, federal funding for research and development is largely spent extramurally, in university and industrial laboratories, for reasons that were both historical and ideological. There is also a pragmatic basis for this policy. Federal policy makers—whether heads of the government's mission agencies or their allies in the Congress—know that, at any given time, they can not predict what kind of technical program will be needed to respond to future exigencies such as war. But the federal government can promote the creation of technological *capabilities*—trained talent as well as ready facilities—necessary to develop and operate whatever technological system might be required. Some of this capability can be cultivated in federally owned and operated laboratories and test facilities; but typically 75 percent of federal R&D spending is in the form of contracts and grants to industrial firms, universities, and non-profit organizations.[5] For example, during World War II the Manhattan Project to build the atom bomb both benefited from, and helped to build, the remarkable research capabilities available at the Massachusetts Institute of Technology and the University of California.

The defense department's Advanced Research Projects Agency (ARPA, later DARPA), created like the National Aeronautics and Space Administration in 1958, was an exception, in that it maintained no research facilities of its own. Instead, the research it funded was carried out in numerous university and industrial laboratories scattered across the country. The agency's strategy was consistent with President Lyndon Johnson's (1963–1969) directive that federal R&D agencies foster the creation of extramural "centers of excellence" in basic research. The largest (but by no means only) beneficiaries of ARPA's Information Processing Techniques Office, established in 1962, were computing research centers at MIT, Carnegie Mellon, the University of California, and the Rand Corporation, created in 1946 as a non-profit research corporation focused primarily on military-related technologies. Meanwhile, the culture of scientific and engineering research—whether at universities, non-profit centers, or industrial laboratories—thrived on openness, or the unrestricted interchange of people and ideas. To the extent that any policy outcome can be considered inevitable, an open system was the likely outcome of ARPA's approach toward generating promising basic research in the relatively new field of computer science.

Computing researchers supported by ARPA, located in heterogeneous institutional settings, not only had to communicate with each other, but use geographically decentralized computing facilities. At the same time, their communications system or network needed to be extremely reliable and secure—or "robust." The solution to this two-pronged challenge devised by Paul Baran (at the Rand Corporation) and others would take full advantage of the possibilities of digital electronic communications.

First, they conceived of an inter-network architecture that used numerous connecting and transfer points (or "nodes"), allowing messages of varying length and type to reach their destinations by a variety of means. If, for example, a message encountered an obstruction in one path, it could proceed by another. Consider, by analogy, automobiles traveling on an interstate highway. Those going to the same destination must all travel together, using the same on-ramps and off-ramps. In contrast, automobiles traveling from upper Manhattan to lower Manhattan in New York City—thanks to the borough's grid street plan—can take an almost infinite number of routes to arrive at their destination. Similarly, electronic messages—such as electronic mail—traveling through the new inter-network system could be rerouted around obstacles such as non-functioning nodes.

Second, electronic digitized messages, unlike radio messages, are not broadcast to everyone tuned in to the same radio frequency. Digitization increased not only transmission speed, but relatively secure transmission. And third, but no less important than the others, was the development of "packet switching." Digitized messages from one computer ("server") could be broken up into small packets, electronically labeled to identify their source and destination, and sent in discrete packets by any number of different routes. Software in the receiving computer ("host") would reassemble the message. The "buffering" that occurs in computers receiving digital music sent over the Internet undergoes just this process of reassembling the digitized packets in which their music had traveled.

These innovations ensured a robust system of digital communications over computer networks that could not be readily shut down in their entirety. All that was required was the development of standardized format for electronic messages, packet labels, addresses for website and e-mail servers and hosts, and transfer or switching nodes to ensure that any and all kinds of computers using all kinds of software could communicate with one another. This was the function of the transmission control protocol (TCP) and inter-network protocol (ICP) developed by the International Network Working Group (INWG) in 1972—the same year ARPA demonstrated its newly created "ARPANET." Once again, standardization among interconnecting components helped to ensure the creation of an open system.

The Department of Defense, with its enormous appetite for data—capturing it, storing it, manipulating it, and sending it—necessary for logistics operations and weapons systems analysis, was not the only federal organization with an interest in promoting the development of computers and rapid as well as secure computer inter-networking. Whether modeling the globe's climate or its physical geography, or mapping the genome, scientists around the world had developed substantial data file storage, sharing, analysis, and transmission needs.[6] Thus, in the 1980s, it became the National Science Foundation's turn to fund the creation of computer science centers at universities throughout the country.

Computer science centers funded by the NSF were connected in thirty regional networks operated by university consortia. The NSF was helped in its task of connecting these regional networks into a national network by ARPA (which allowed NSF computer scientists to use the ARPANET as the NSFNET was being built); IBM, which built packet switches; and MCI, which leased its phone lines to the emerging NSF-funded and sponsored inter-network.[7] By 1990 ARPANET research sites were able to transfer their connections to the NSFNET, then rapidly expanding to include over 20,000 international research networks.

The emergence of the Internet as a thriving global and commercial phenomenon was promoted by federal policy in ways beyond the stimulus and financial support provided by the government's principal agencies for military and civilian research. As we have seen, since the earliest days of the republic the federal government has followed a policy of acquiring its goods and services from the private sector. Even intramural facilities such as weapons arsenals contracted out for ordinary equipment and services. Above all, the policy wisdom prevailing in Washington since at least World War II dictates that the government should not compete with the private sector. Once a technology advances beyond its R&D phase and becomes "operational," it should be transferred to the private sector, where its future will be determined by the market place. No market, no future.[8] And so, as NSF's regional centers began to "spin-off" commercial Internet ventures such as Mosaic software and regional

Internet service providers, or "ISPs," the time had come to turn the entire NSFNET over to the private sector.[9] By 1995 the NSF was able to disassemble its own NSFNET and rely on commercial ISPs to satisfy its Internet communication needs.

The final major innovation of the Internet was organizational: The creation, operation, and maintenance of an international communications system *not* normally subject to the control of a single government or group of governments. Because of the approach taken by ARPA and the NSF, the developers of the Internet were primarily contractors and university scientists who used the computer and networking systems they were developing. Such a process virtually defied state control. Instead, a decentralized community of computer scientists around the country and in Europe formed working groups that developed and managed the necessary rules to use and maintain the Internet. Chief among them are the Internet Activities Board (IAB), Internet Engineering Task Force, and the Internet Society. As a natural consequence of their origins, these organizations tended to agree on three general principles, all of which would ensure the continuation of the Internet as an open system: (a) that there should be choice and competition among Internet service providers; (b) that local networks should make their own operating decisions to the extent possible; and (c) that decision making should be inclusive and democratic.

THE EXPONENTIAL GLOBAL expansion of cyberspace has raised historically familiar policy issues. These issues are roughly three in kind. First, to what extent, and how, do the Constitution's First and Fourth Amendment guarantees affect federal and state policies governing "speech" and privacy in cyberspace?[10] Second, since the transmission of electronic information necessarily occurs over national and international telecommunication networks, do the regulatory precedents and approaches of the Federal Communications Commission best serve the public interest in cyberspace? And third, is monopoly power in the economy, which the Sherman and Clayton Antitrust Acts (1890, 1914) were intended to prevent, as great an evil in cyberspace as it has been judged to be in other industries?

The First and Fourth Amendments in Cyberspace

Freedom of Speech

Most controversies over the content of what one finds while exploring the "world wide web" focus on materials that some people find obscene, impugn the integrity of public officials, inveigh against one's political preferences, or serve as "fighting words" calling others to violence. None of these kinds of materials is peculiar to the Internet, and thus when federal courts have been asked to rule on the constitutionality of laws limiting the presentation of, or access to, such materials in cyberspace, they have been able to draw upon two centuries of jurisprudence applying the First Amendment guarantees of freedom of speech, the press, and assembly in particular situations. Moreover, while the Internet may be a relatively new phenomenon, it has analogous predecessors, enabling the Supreme Court to review cases involving efforts to regulate content on the Internet in a larger context than the technical capabilities of any individual generation of hardware, software, or associated communications capabilities.

Some of these foundations were more easily laid than others. For example, Justice Holmes's instrumental standard for judging the constitutionality of restraints on public speech, the metaphor of the "market place of ideas," has become solidly imbedded in Supreme Court decisions imposing severe tests on the constitutionality of laws abridging freedom of speech, regardless of the media in which the speech occurs. Among the earliest such questions to be resolved was whether First Amendment guarantees apply to state as well as federal laws. The Supreme Court had concluded by the end of the 1920s—partly in response to appeals arising from convictions under the Espionage Act of 1917—that they did, arguing that "It is no longer open to doubt that the liberty of the press, and of speech, is within the liberty safeguarded by the due process clause of the 14th Amendment from invasion by state action." The court added, however, "Liberty of speech and of the press is also not an absolute right, and the state may punish its abuse."[11] Just what constituted "abuse" and permissible restraints was, of course, the difficult part. Violations of the rights of others—as in libel—might justify abridgement of freedom of speech, as could legitimate needs for national security or public order.

Since protection of political speech has been unarguably the most important function of the First Amendment, federal courts have been most careful in deciding by which standard, and under what circumstances, freedom of political speech can be determined to have been "abused" and its suppression justified. In another classic formulation, Justice Holmes upheld the conviction under the Espionage Act (1917) of a man for circulating a paper urging resistance to the draft while the United States was at war with Germany. Arguing that the document at issue would not have been circulated except to result in the obstruction of the war (World War I) effort, Justice Holmes concluded that it presented a "clear and present danger," and was thus not protected by the First Amendment.[12] The court has also ruled that speech that serves no function except to incite others to violence—epithets or insulting and "fighting" words—can also be restricted by local authorities in the interest of maintaining public order.[13]

The civil rights struggle of the 1950s produced another important Supreme Court decision elaborating the circumstances under which political speech may be restrained or punished. During the desegregation demonstrations in

Montgomery, Alabama in March of 1960, *The New York Times* published an advertisement sympathizing with the protesting black school children and their supporters. The advertisement contained some factual errors which the local police commissioner, L. B. Sullivan, maintained would ruin his reputation. Sullivan sued *The New York Times* in state court for libel. The segregated proceedings and all-white jury at Sullivan's trial produced a guilty verdict, which the *Times* appealed to the U.S. Supreme Court. In its 1964 decision the court distinguished between ordinary citizens, who may sue for libel if statements about them can be shown to be false or damaging, and public officials. In his opinion Justice William J. Brennan, Jr. echoed Holmes's instrumental argument for free speech, saying that the press could not carry out its important role of criticizing public officials if it could be sued for libel in state courts over a simple factual error. As plaintiffs in libel cases, public officials would have to prove that the offending statements were made with "actual malice" and in "reckless disregard" of the truth. This additional and considerable burden—which came to be known as the "Sullivan rule"—vindicated the *Times*'s position that if the Alabama court's decision were allowed to stand, newspapers would be seriously handicapped in their constitutional responsibility.[14]

The presence on the world wide web of politically offensive opinions and images has created less controversy than the possibility that minors, "surfing" the Internet, will visit sites featuring obscene, pornographic, or otherwise unsuitable materials for children. Here, too, the Supreme Court, which has been called upon to review several cases involving attempted censorship of Internet sites, has proceeded upon a foundation well established in pre-Internet circumstances. While the court has ruled repeatedly that obscenity is not protected speech under the First Amendment, the challenge to federal judges has been determining just what constitutes obscenity.[15]

By the time of its decision in *Memoirs v. Massachusetts* (1966), the Supreme Court had evolved a definition of obscenity that relies less on the federal judiciary's prior notions of obscenity than on the ability of plaintiffs to convince a federal court that the material in question had these three characteristics: its "dominant theme" appealed to "a prurient interest in sex," it "affronts contemporary community standards relating to the description or representation of sexual matters," and it is established to be "utterly without redeeming social value."[16] This third standard—that obscene material must be "utterly without redeeming social value"—was challenged by Justice Warren Burger in 1973 when, in *Miller v. California* (1973), the Supreme Court let stand the conviction under a California law of an appellant for mailing brochures advertising "adult" material.[17]

In his opinion for the majority, Justice Burger took issue with the third part of the *Memoirs*' test, that material must be shown to be "*utterly* without redeeming social value," arguing that it was not possible to determine whether any book was *utterly* without social value. Instead, wrote Burger, states could constitutionally regulate material that "depicts or describes [sexual conduct] in a patently offensive way," in contrast to medical textbooks, for example.[18] However, a determination that material is obscene *is not, in and of itself*, a sufficient basis for abridging freedom of speech. The *circumstances* under which the public encounters the allegedly offending "speech" matters as well, and it has mattered especially in recent instances in which the Congress, responding to constituent pressures, has attempted to require the suppression (or "blocking") of Internet sites considered harmful to minors.

BECAUSE THE RADIO frequency spectrum is a limited resource, one that is used for emergency as well as commercial and military communications, the federal government has asserted a regulatory role over radio (and subsequently television) broadcasting since 1912. In the late evening hours of April 14 of that year, as the new British luxury liner *Titanic* sank into the ice cold waters of the North Atlantic, the ensuing frantic ship-to-shore radio communications were so chaotic that virtually everyone who relied on radio communications—journalists among them—demanded that the Congress institute federal regulation for wireless communications. The resulting Radio Licensing Act, passed the same year, required radio operators to be licensed by the Department of Commerce, which would allocate frequencies. Broadcasters would have to operate within their assigned frequencies and at specified times, and use identifying call numbers or letters.

By 1927 commercial radio broadcasting had become so commonplace and competition for spectrum allocations so keen that the Congress created a Federal Radio Commission (FRC) to ensure order on the airwaves by assigning broadcast frequencies and times, issuing licenses, and ensuring that stations operated in the "public interest, convenience, and necessity"—which it could do by withholding licenses from broadcasters not meeting that standard. While the newly created commission was expected to be a non-political group, the certainty that there would be controversy over what constituted "public interest" guaranteed that at least some of the commission's decisions would become politically controversial. Years later, because the backbone of the future Internet would be the nation's telecommunications network regulated by the FRC's successor—the Federal Communications Commission (FCC, est. 1934) —federal telecommunications policy would play a significant role in determining the government's regulatory role in cyberspace.[19]

But how did Internet communications differ from radio and television broadcasting? In 1978 the Supreme Court demonstrated when it upheld in 1978 an order issued by the FCC penalizing a radio station for broadcasting an "indecent" monologue. The fact that listeners were passively subjected to the controversial material was significant to the court's opinion. However, when exposure to obscene or indecent matter results from an active decision by an individual, the court has been less willing to uphold FCC bans, as it demonstrated in a 1989 decision which struck down an FCC order criminalizing telephone "dial-a-porn" services.[20] This variable standard—the extent to which individuals are exercising choice when they access material that some might regard as obscene—was proven critical to the Supreme Court's 1997 ruling in *Reno v. American Civil Liberties Union*, a case that arose from an early challenge to free speech over the Internet.

The availability of obscene or violent material to minors over our telecommunication networks has led to repeated congressional efforts to impose controls on the sources or purveyors of such materials. This was the case in 1996, when the Congress passed the Telecommunications Act of 1996, generally designed to promote advanced technologies and increased competition in broadcasting and telecommunications. Tucked away in the Act, however, was a section, Title V, the "Communications Decency Act" (CDA) of 1996, which was targeted at objectionable material that might be encountered by minors perusing the Internet. Two provisions in the CDA prohibited the knowing transmission or display to minors of messages that were "obscene," "indecent," or "patently offensive."

The Telecommunications Act was barely signed by President Clinton before the American Civil Liberties Union, joined by nineteen other plaintiffs, successfully sued in federal district court to prevent the enforcement of the CDA. The Department of Justice, then led by Attorney General Janet Reno, appealed. In upholding the lower court's decision the Supreme Court demonstrated that it was not convinced that new technologies required new readings of the Constitution. In fact, in his majority opinion Justice John Paul Stevens III likened the Internet to that oldest of information management systems, a "vast library including millions of readily available and indexed publications." Justice Stevens distinguished between the words and images found on the Internet and those that might be heard or seen on television. One selects what one will read and see in a library, but is a passive recipient of televised words and images. Precisely because individuals can choose whether or not to visit certain websites, the burden is on them, rather than the government, to discriminate among the possibilities for what they might see and hear.

As for the availability of objectionable material to minors, the court put the responsibility squarely on the shoulders of parents and guardians; it is they, and not the government, that society holds responsible for the rearing of their children. Parents and guardians could use software for blocking Internet sites they found objectionable. Stevens echoed Holmes's metaphor for public life as an open system as Stevens wrote, "The Government asserts that . . . the unregulated availability of 'indecent' and 'patently offensive' material on the Internet is driving countless citizens away from the medium because of the risk of exposing themselves or their children to harmful material. . . . The dramatic expansion of *this new marketplace of ideas* contradicts . . . this contention. . . . As a matter of constitutional tradition, in the absence of evidence to the contrary, we presume that governmental regulation of the content of speech is more likely to interfere with the free exchange of ideas than to encourage it [italics added]."[21]

Undeterred, congressional supporters of a clean Internet continue to write and marshal support for legislation to keep obscene or pornographic sights and sounds from children exploring the Internet that can pass constitutional muster. A second effort, the Child Online Protection Act of 1998, was struck down by a panel of the U.S. Court of Appeals for the Third Circuit (Philadelphia). Congress tried again with the Children's Internet Protection Act (CIPA) of 2001, which would have required libraries to use blocking technologies "to prevent access to child pornography and materials considered obscene or 'harmful to minors.'" The American Civil Liberties Union, this time joined by the American Library Association and others, successfully challenged the law. If filters are to be used to restrict access to some Internet sites in order to protect minors, ruled Chief Judge Edward R. Becker of the U.S. Court of Appeals for the Third Circuit in May of 2002, let them be used by parents. Filters are "blunt instruments . . . it is currently impossible, given the Internet's size, rate of growth, rate of change and architecture, and given the state of the art of automated classification systems, to develop a filter that neither underblocks nor overblocks a substantial amount of speech.'"[22]

But Judge Becker's ruling was not to be the last word. The Department of Justice appealed his decision to the Supreme Court, which upheld the use of site blocking software by libraries receiving federal funding. The essence of the majority's opinion (written by Chief Justice William H. Rehnquist) was that the Congress was not exceeding constitutional boundaries in its desire to ensure that public libraries continue to "fulfill their traditional missions of facilitating learning and cultural enrichment . . . [by having] broad discretion to decide what material to provide their patrons. . . . [The public forum analogy] is incompatible with the broad discretion that public libraries must have to consider content in making collection decisions."

In their dissenting opinions Justices Souter and Stevens conceded that the Congress had a legitimate interest in protecting children from pornography. However, the CIPA offered no assurance that adult library patrons could "obtain an unblocked terminal simply for the asking." Indeed, the CIPA even invites restrictions on *adult* Internet access by "allowing unblocking only for a 'bona fide research or other lawful purpose.'" Moreover, arguing that libraries exist for "cultural enrichment" as opposed to providing a "public forum" is an outdated nineteenth-century view. In an earlier (1975) decision the court had already ruled that *"the policy of the First Amendment favors dissemination of infor-*

mation and opinion, and the guarantees of freedom of speech and press were not designed to prevent censorship . . . but *any* [government] action . . . [that] might prevent" the "general discussion of public matters [italics added]."[23]

The global nature of the Internet has sharpened distinctions among national attitudes toward the indiscriminate worldwide availability of infinite varieties of public "speech." Absent a universal "free speech" regime in international law, national boundaries may entail varieties of censorship ranging from outright blocking of particular sites—such as neo-Nazi websites, which are unlawful in France and Germany—to limits on the proportion of English language or American-originated websites available in particular locations. However, national Internet barriers are the exceptions that prove the rule in U.S. federal courts' instrumental rationale for protecting free speech in order to preserve political liberty. The fall of the Iron Curtain at the end of the 1980s reflects the ability of ordinary people to communicate rapidly across borders, while the vigilance of human rights organizations has come to depend substantially on Internet-based reporting from citizens struggling under politically oppressive regimes.

The Fourth Amendment

The growth of the Internet has challenged not only those tempted to use the power of U.S. state and federal governments to limit speech that they might find politically, socially, or morally offensive. It has also created an inviting opportunity for the federal government to expand domestic surveillance of criminal suspects to domestic surveillance of presumably innocent individuals in the interest of national security in order to pursue the current U.S. "war on terror."[24] To civil libertarians this is an ominous development. The procedural safeguards for protecting citizens from the "unreasonable searches and seizures" of their "persons, houses, papers, and effects" guaranteed in the Fourth Amendment are well established in U.S. criminal law. Courts have greater difficulty enjoining possible violations of the Fourth Amendment when they must "second guess" executive branch claims that national security justifies suspension of certain freedoms. If Abraham Lincoln could suspend *habeas corpus* during the Civil War, and if American citizens of Japanese origin could be interned in California during World War II, then certainly, some argue, the Fourth Amendment can be suspended justifiably during the "war on terror."

What does it mean to be anonymous? Individuals standing in a crowd listening to a speaker may be in a public place where they can be observed and recognized. Travelers through the networks of the world wide web may have "cookies" placed on their computers allowing operators of websites to recognize individual users. Often, however, organizations and Internet businesses will alert computer users that cookies are being placed on their computers, or even request permission to do so. In both instances individuals have some choice; they may exchange a degree of anonymity for the benefits they seek from being at a certain place at a certain time—whether on a busy street corner or the Internet. When government is the observer, however, the possibilities resulting from a loss of anonymity change profoundly.

In January 2002 the Department of Defense created a special Information Awareness office and launched a project to create "Total Information Awareness," a surveillance system that would enable federal intelligence agents to hunt for terrorists by scanning all manner of electronic transactions conducted by millions of individuals on the Internet every day. Public alarm over the potential for abuse and probable Fourth Amendment violations in the plan was strong enough to lead the Congress to terminate it a year later. Undeterred, Bush administration officials in the Pentagon's Defense Advanced Research Projects Agency floated a new project, "eDNA" (or electronic marking of unique individuals on the Internet) to reconfigure the Internet into "public network highways" and "private network alleyways." Highway users would have to be identified, while alleyway users could proceed anonymously. The unique personal identifiers individuals would be required to use to navigate the "highways" would enable electronic surveillance personnel to trace every site visit or transaction on the Internet. Preliminary discussions even among participants in the project were so contentious—privacy issues being the chief concern—that eDNA, too, has been abandoned.

Nonetheless, the notion of "data mining"—the computerized analysis of electronic data bases containing information about identifiable individuals—has proven irresistible to official Washington, especially to those agencies, like the defense and homeland security departments, involved in classified programs to pursue the second Bush administration's "war on terror." Passed by Congress in the wake of the September 11, 2001 destruction of the twin World Trade Center towers in Manhattan and a portion of the Pentagon in Washington, the USA Patriot Act, among other things, authorizes the federal government to conduct broad personal electronic surveillance and personal data gathering activities in secret. Hence the ability of individuals and organizations to challenge implementation of the act is handicapped by the difficulty of learning how the act's provisions are being used, or abused.[25]

Three decades earlier DARPA's predecessor organization first demonstrated the highly decentralized open system architecture of the Internet, so designed to ensure the completion of digital electronic communications through enumerable paths and nodes in order to frustrate hostile efforts at interference or destruction. In a political irony worthy of Shakespeare, today that same open system architecture, now expanded many times over, would in all probability serve as the critical obstacle to the federal government's designs for centralized Internet surveillance or control. Any such government system for user identification and scrutiny, or routine mining of transmissions for advance warnings of terrorist attacks, would require the most improbable well-orchestrated cooperation of thousands of local Internet service providers, various search engine operators, thousands of local telephone and cable companies, and thousands of website operators. The volume of digital data to be watched as it passed over computer screens defies imagination. And so, this most technical of open systems could stand not only as one of its own best defenses, but as an important ally of the federal judiciary in its efforts to preserve the constitutional protections of the First and Fourth Amendments.

Federal Regulation of Internet Highways

Throughout much of our history Americans have looked upon government power as the largest potential threat to freedom of speech and assembly. We have relied on constitutional law as interpreted by federal courts to protect us from efforts of federal, state, and local government agencies to regulate the content of the many ways in which we express ourselves. At the same time we have willingly accepted government regulation of the means—or media—by which we communicate. We have done so believing that public telephone and broadcast networks are comparable to public utilities, to be regulated in the public interest. Technically this belief has been well founded, since chaotic noise would result from the failure of broadcasters to operate within their assigned frequencies. Thus, for example, the Federal Communications Commission, created in 1934, was expected to function as a non-partisan agency regulating radio broadcasting and ensuring the availability of efficient telegraph and telephone services at reasonable cost. It was authorized to set and enforce technical standards for broadcasting equipment and emissions, to allocate increasingly scarce radio spectrum, and to issue licenses, which could be withheld from non-compliant broadcasters.

Defining and applying a standard of "public interest" has not been a simple or uncontroversial process. Today it may seem quaint, but commercial product or broadcast station advertising on the airwaves was not always considered consistent with the "public interest," and broadcasting talk and music recorded on phonograph records, rather than original programming, was once thought deceptive.[26] Commercialization of the airwaves was probably inevitable, however. The owners and operators of telephone wires and wireless transmissions have always been largely privately owned companies with a natural interest in finding ways of covering their costs and making a profit. Charging for use of equipment has been one way to do that, and buying and selling spectrum rights, allocated by the FCC and its predecessor, the Federal Radio Commission, another.

The young entertainment industry of the 1920s was able to expand to developing programming—whether original or recorded—to sell to broadcast networks. In turn, an advertising industry built itself by marketing the products whose sales from radio and television commercials generated the revenues that the networks use to purchase their programming. As a result, government regulation of telecommunications, exercised through the decisions of the FCC, has had a secondary consequence of considerable importance for most citizens: the shaping of a broadcast and telecommunications industry of great economic power and political influence.

The FCC consists of five commissioners, all appointed by the president and approved by the Senate, who serve for a period of five years unless completing the unexpired term of another commissioner. No more than three of the five commissioners may be appointed by the same party in the White House. While the commissioners are forbidden to have any financial interest in a matter under their review, the nature of their appointments ensures a political cast to the appointees' judgments. And indeed, absent a consistent procedure for awarding radio and broadcast licenses, the FCC has been accused periodically of personal or political favoritism in its decisions. With the growth of television broadcasting, which involved ever greater revenues, the power of the FCC has become even more important.[27] In 1993 the Congress authorized the FCC to use competitive bidding (auctions) as a fairer and more efficient way to assign licenses.

With the exception of the stations of the National Corporation for Public Broadcasting, which are supported partly by government subsidy and partly by direct listener and viewer contributions, a large commercial broadcast industry provides Americans with all the news and entertainment they receive over radio and television. In this context, speech is "free" to the extent that it is not government controlled. But "speech," unlike the ordinary commodities and services that we exchange in the commercial economy, is not infinitely interchangeable. Its importance is qualitative, and its content highly liable to selection or manipulation by the media that controls its dissemination. In an ideal laissez faire world, the "market place of ideas" may be the best judge of the value of our speech. But in the actual commercial market place in which the resources necessary to disseminate our ideas are accumulated, bought and sold, perfect competition may not prevail. When it doesn't, the open systems by which we form and communicate our ideas and values are threatened not by government power, but by market power.

The Internet relies on the same communications technologies for its "backbone" as do our telephones—whether wireless or land wired—and our televisions and radios.[28] Thus the policies and decisions of the FCC can and do effect the way we access and use the Internet, most especially its benefits as an "open system." This is so because, as Justice Stevens observed in *Reno v. American Civil Liberties Union* (1998), unlike radio or television, which we receive as passive listeners or viewers, our use of the Internet is active (or, "interactive"). Individuals become actively engaged in shaping the content of the ideas that are communicated over the Internet in the same way that they are actively engaged in any truly open political forum. Thus policies that affect public access and use of the Internet can have serious ramifications for the quality of our political, social, and cultural life.[29] In the area of telecommunications, "technical" policy decisions rarely remain simply that.

The rapid growth of Internet use has intensified two kinds of policy issues that have accompanied the growth in the telecommunications industry since its inception. With the exception of local telephone companies, most telecommunications businesses cross state lines and are thus subject to federal regulation by the FCC. Beyond setting technical standards to minimize signal interference and ensure equipment reliability, the FCC's decisions have their largest influence on (a) the availability and most efficient use of radio spectrum, and (b) the structure of the telecommunications industry. Both policy areas—spectrum use and industry structure—can, for good or ill, influence the extent to which the Internet remains the open system for which it was originally designed, and has accounted for its considerable social and political impact.

Today well over two-thirds of the U.S. population uses computers, and the largest percentage of U.S. computer users is children between the ages of five and seventeen. Over half of all U.S. computer users are also "online," using the Internet for the many capabilities it now offers: communications, research, information management, and entertainment. Internet use is largest among young people as well, with 65 percent of ten–thirteen year olds and 75 percent of fourteen–seventeen year olds online.[30] All of this activity, which is magnified exponentially on a global scale, makes great demands on our land-line and wireless communications networks.

All telecommunications entail the transmission of electromagnetic impulses through a medium—wires, fibers, or cables—or the atmosphere (radio or wireless communications). Wireless communications occur in some portion of the radio-frequency spectrum, which is measured out in kilohertz, megahertz, and gigahertz.[31] Radio frequencies can also be expressed in wavelengths, which vary in direct proportion to radio frequencies, that is, the lower the frequency, the longer the wavelength.[32] Many individual transmissions may travel at the same time, and at the same radio frequency bands, *or* they may travel concurrently in the same geographic area, but they cannot travel at the same wavelength, in the same geographic area, and at the same time—any more than two automobiles can drive down the same road at the same time, at the same speed, and in the same location. For this reason, the radio spectrum is one of the world's most precious natural resources, comparable in importance in the modern economy to oil and natural gas. Unlike oil and natural gas, however, radio spectrum is a truly global resource, the use of which is regulated internationally by the International Telecommunication Union and within the United States by the FCC.

When the FCC assigns (or licenses) spectrum use, it does so in terms of patterns of radio transmission, transmitter power, and the range of frequencies in which radio operators can broadcast. Assuming a growing market and sufficient capital, a telecommunications company's growth will be limited only by the amount of bandwidth it is licensed to use and the volume of data it can transmit within its bandwidth. Analog transmission, the technique used in radio and television (prior to High Definition television [HDTV]), transmits sound and images in signals that correspond directly to continuous changes in the sound or image being relayed, and is thus a less efficient broadcast technique than digital broadcasting, which translates sound and images into discrete bits of digital data. Digital information can be broadcast in much higher volume in a given bandwidth, and, as we have seen, can be fragmented and reassembled. Resistance to signal distortion and dissipation—or normal signal strength—is another attribute coveted by the commercial telecommunications industry. Generally, radio transmission over short distances, like radio-controlled garage door openers, can occur at very low frequencies while transmissions that must span over very long distances (such as satellite communications and radio astronomy) require the superhigh or extremely high frequencies (SHF-EHF) in the neighborhood of 300 Ghz.[33]

Is the spectrum most akin to a public utility, or is it a commodity? For those who see the availability of efficient and reliable broadcasting and Internet use as a public necessity, the continuing regulation of and telecommunications along the "public interest" model of the mid-twentieth century may seem good public policy. But it would be difficult to overstate the influence on policy deliberations in Washington of the political climate that resulted in the elections of four Republican and one Democratic president since 1980. Moreover, the centrist Clinton administration espoused the more typically Republican view that the public interest is best served when market place competition is allowed to produce the presumed efficiencies and customer service promised by the ideology of laissez faire economics. (Commercialization, privatization, and deregulation have been the answer offered in the halls of Congress and the White House for virtually every issue associated with traditional government programs.)[34]

This trend is reflected in the George W. Bush administration's policy statement on spectrum management, released in June of 2003. Consistent with the conservative Heritage Foundation's recommendation that all federal spectrum management be assigned to the National Telecommunications and Information Administration of the Department of Commerce, the statement announced the creation of a Federal Spectrum Task Force to develop recommendations for improving federal spectrum management and for the Department of Commerce to hold public meetings toward the same end. Noticeably absent from this announcement of a relatively routine executive branch policy process was mention of the Federal Communications Commission.

The statement then goes on to praise the agreement between the Department of Defense, the Department of Commerce, and the FCC to make an additional 90 MHz of spectrum available for the wireless communications industry to use in developing its third generation ("3G") wireless services "while accommodating critically important spectrum requirements for national security." In exchange for releasing some spectrum for commercial use, the Department of Defense was able to set detailed technical standards to minimize interference. The Bush administration's statement also announced release of an additional spectrum at 5 Ghz for new Wireless Fidelity ("WiFi") services allowing direct wireless Internet access for anyone within a few hundred feet of a special transmitter, and "in conjunction with the FCC," approved use of ultra-wideband (UWB) technologies.[35]

These initiatives, singly and as a group, reflect intensive lobbying by the commercial telecommunications industry for a federal spectrum policy driven by its needs, as well as the view that the public is served best when it is continuously offered innovative products by a thriving industry. In all probability federal spectrum policy under a Republican administration would have been similar under a Democratic administration. Both parties are considerate of the information technology, telecommunications, and media industries. The public is generally enthusiastic over technological progress in communications (as reflected in the Internet stock bubble of the 1990s), while the telecommunications industry lobbies hard, helps pay for political campaigns, and could (members of Congress might fear) sabotage the career of a noncompliant politician.[36]

Audiophiles dispute whether analog or digital reproduction offers the greatest fidelity to original sounds, while photographers debate the relative merits of chemical and digital image processing. Nonetheless, the FCC, citing "ten years of research and public comment," in November of 1999 directed that the "public interest" required that by 2006 all of the country's approximately 1,600 television stations broadcast in digital format (DTV). The public would gain several advantages: Television viewers would be able to enjoy a "much clearer and brilliant picture," while television broadcasters would be able to transmit as many as four channels, both digital and analog, as well as offer Internet connections for downloading data and other interactive services.[37] The FCC's decision was undoubtedly welcome by television equipment manufacturers as well as media production interests which had converted to digital audio with the production of CDs in the 1980s. The costs to local networks of conversion to DTV are well into the millions, though the National Telecommunications and Information Administration has softened the impact by awarding grants to help meet conversion costs to numerous public television broadcasters.

The number of consumers who can take part in the wonders promised by the new era of high-speed, high-volume digital telecommunications depends today upon whether they have access to "broadband" land lines or direct satellite communications.[38] Access, in turn, depends partly on geography. Some locations are too obstructed for satellite communications, while broadband lines or cable requires the laying of new lines by the local telephone or TV cable company. This mixing of commercial telecommunications carriers—telephone, satellite, and cable—in order to supply broadband service has unsettled previously established regulatory arrangements, with uncertain consequences for Internet users. Internet access over the local telephone companies lines may be slow, but it is relatively certain and uncomplicated because the local telephone company, while a monopoly, is a "common carrier" and must, therefore, allow any Internet service provider to operate over its lines.

If one lives in a fairly well populated area, where the local or regional phone company has invested in the installation of digital subscriber lines (DSL), one can opt for high-speed Internet access, usually for an additional base and subscription fee. Or, one might live in an area where the local cable company, a monopoly by virtue of its franchise, will offer not only digital cable for TV viewing, but digital high-speed (broadband) cable for Internet access. In that case, the user has no choice of Internet service providers, and must settle for the service provided by the cable company. A third option, if one's geographic location permits it, is to subscribe to a direct broadcast satellite service, again for a considerable base and subscription fee.

Policy issues have arisen largely out of the different rules under which regional phone companies, satellite operators, and TV cable franchises all operate. The FCC, committed to stimulating innovation and increasing competition (which now prevails as the standard for "public interest"), is experimenting with new rules in ways likely to impact the open systems legacy of the Internet's development. While telecommunications satellite operators do not have to pay for spectrum licenses, commercial communications satellite operation entails both high cost and risk, especially in an uncertain market, with the result that two of the most ambitious early ventures, Iridium and Globalstar, have gone bankrupt. Thus in early 2003 satellite operators successfully appealed to the FCC to allow them to branch into land-based cellular (wireless) communications networks to enable them to operate in urban and other areas where satellite signals are obstructed. Established wireless phone companies such as AT&T, Verizon, and Cingular, citing million dollar investments in spectrum licenses, are pressing the FCC to set conditions to stabilize what has already become a chaotic market for wireless communications services. Conditions might include requiring the customers of combined satellite and land wireless services to buy special equipment enabling them to reach satellites directly.

To the extent that Internet access depends on networks other than the lines of the local telephone company, users' access to an open market of Internet services and entertainment is at risk of becoming constrained by the emergence of integrated monopolies resembling the freight rail carriers of the late nineteenth century. Telephone companies are regulated by the FCC as common carriers (comparable to public utilities) which must allow other telephone services to use their lines. The FCC, however, in an effort to boost the prospects of the growing broadband service providers, has designated cable Internet access an information service, thus lifting the requirement placed on common carriers of allowing competitors to use their lines. Then, in a three–two decision on February 20, 2003, the commission agreed to relieve regional telephone companies from the requirement to share their lines with competitors in the case of broadband, or DSL lines.[39] Defenders of the FCC decision argued that if local phone companies were able to rely on broadband service developed by the regional companies, they would have no incentive to invest in broadband connections themselves, to the long-term detriment of Internet users.[40]

As the availability of broadband Internet access spreads, increased speed and capacity will become not only welcome advances for private users, but necessary to the survival of businesses, large and small, throughout the country. (For example, retailers who cannot maintain inventory at the same rate as their competitors could be at a fatal disadvantage in both local and mail order markets.) However, if broadband Internet access is allowed to become a monopoly of the local cable operator or regional telephone company, businesses no less than individuals may find themselves with no choice of Internet service providers. Internet service providers, in turn, could deny use of their networks or charge discriminatory rates to some websites and online personal and business services as they begin to integrate compatible services in order to monopolize segments of the Internet market place.

Further complicating the prospects for universal broadband access is the FCC's decision in the fall of 2004 to allow electrical utility companies to join telephone and cable companies in offering broadband communications services (BPL, or broadband over power lines). Power lines are virtually ubiquitous, and broadband access through them can be obtained through a special modem. The prospect concerns some amateur radio operators and emergency services, who fear electromagnetic interference with their own radio communications networks. Whether the use of the electric power grid for broadband communications would increase or decrease the danger of monopoly controls over Internet content is not yet clear.[41]

As a market place of ideas, the Internet and its content are at greater risk of content controls by economic monopolies than as a result of government censorship. That such fears are not imaginary is suggested by the opposition to the FCC's decisions, rendered in the name of improved "competition," being voiced by groups such as the Coalition of Broadband Users and Innovators and Media Access Project. Both are groups which include corporate interests otherwise in competition themselves, companies such as Microsoft, Disney, Yahoo, eBay, Apple Computer, and Radio Shack. Shortly after the FCC lifted the requirement on cable companies to share their broadband networks with competing Internet service providers on the grounds that Internet access was an information service, Earthlink and other Internet service providers filed suit in federal court. In October 2003 the U.S. Court of Appeals for the Ninth Circuit ruled that, to the contrary, broadband (Internet) service providers provide, at least in part, telecommunication services, and were thus obliged to allow competing ISPs to use their networks.[42]

The "strange bedfellows" that controversial policies can assemble were never so much in evidence as in the opposition to the FCC's June 2, 2003 revision of rules governing the extent of media ownership in single markets. Among other things, the new rules increase from 35 percent to 45 percent the proportion of the national audience that could be reached by a single television owner's stations, increase from two to three the number of television stations that can be owned by a single company in the larger markets, and lift a ban on a single company owning both a newspaper and a TV or radio station in the same locality.

Opponents of the FCC's rule changes cited the growing monopoly power being accumulated by a handful of media giants, such as Rupert Murdoch's News Corporation, Clear Channel Radio Communications, Viacom, Disney, and AOL/Time Warner (all of whom advocated the changes). That such concerns were not trivial was illustrated by the decision of Sinclair Broadcast Group during the 2004 presidential election campaigns to require the sixty-two television stations it owned across the country to broadcast—without charge to those stations—a film highly critical of the Democratic party's candidate, John F. Kerry, without offering to broadcast a comparable film, under the same conditions, critical of the Republican candidate, George W. Bush. The FCC, dominated by Bush administration commissioners, refused to intervene.

Critics of the FCC's new media concentration rules also argued that media conglomerates would soon own major cable systems and Internet service providers. Constituent opposition—not to mention the unlikely combination of the National Rifle Association, the National Organization for Women, the Consumers Union, the United States Conference of Catholic Bishops, and media mogul Ted Turner—spurred a bipartisan effort in the U.S. Senate to roll back the changes.[43]

An indication of the direction the FCC would continue to pursue under a Republican administration (which would ensure that three of the five commissioners were Republicans) was the Commission's approval in December 2003 of News Corporation's purchase of Hughes Electronics Corporation's DirecTV home satellite system. This substantially increased share in the television broadcasting market meant that News Corporation, which owns the Fox television network of several successful cable channels, could raise prices charged its cable rivals to drive customers to DirecTV. To mitigate this possible result, FCC's approval was conditional on News Corporation's providing local channel service to all of its markets by 2008.[44]

The emergence of Internet telephone service only underscores the gradual dissolution of technological and regulatory distinctions between older telecommunications networks, historically treated as a "public service," and Internet-based digital communications, now treated as a commercial service in which public benefits are thought most likely to accrue from unfettered market competition. Pressure on the distinct regulatory regimes of each is likely to continue to result in contentious decisions from a Federal Communications Commission which is as subject to partisan influences as most other federal regulatory agencies.

Monopoly Power in the Software Industry

In the early years of computers, during the 1940s and 1950s, designing and building effective computer hardware was the preoccupation of a small industry which, to the extent it had customers other than the U.S. government, served the computational needs of American industry. Today, other than the memory and speed of the chip that sits inside the computer's box, software is the engine that drives the computer industry.[45] Probably nothing better exemplifies the growing policy challenge of market power in the industry than *United States vs. Microsoft Corporation*, the antitrust case initiated by the U.S. Department of Justice against Microsoft in 1997 and resolved in a federal court-approved settlement in 2002. Focusing on a single case, however, may be somewhat misleading. Microsoft, given its enormous size by any corporate measure except physical plant, has been forced to plan for continuous antitrust litigation as an ordinary cost of doing business. It has earned this distinction by virtue of the power it wields over the market place for software.

Microsoft was not the first successful corporation in the computer industry to run afoul of U.S. antitrust laws. Since the 1930s IBM has had to tailor its own business arrangements to hold the Antitrust Division of the U.S. justice department at bay. The experience may have helped IBM recognize the merit of developing and marketing its computer hardware and software as separate product lines, as well as following an open systems approach with its PC (introduced in 1981) in designing its architecture and acquiring "off the shelf" software and components from other sources.

By the time Microsoft founder William Henry ("Bill") Gates III reached the age of thirty-one, the company he created in 1980 at the age of twenty-five to market his new disk operating system (MS-DOS) was earning $2 million annually and Gates himself had become a billionaire. Setting aside any question of technical or business genius in Microsoft's rise, Gates learned the value of leveraging market power right from the start. First, in 1980, Microsoft licensed MS-DOS to IBM for use in its first personal computer.[46] The next year, Microsoft and IBM joined forces with Intel, pioneer of the programmable microchip, and IBM introduced its hugely successful "PC" (personal computer).[47] By the end of 1983 Microsoft was ready with its new Windows operating system and a business plan that emphasized continual upgrades, similar to "planned obsolescence," the strategy taken by the automobile industry to promote frequent trade-ins of still well-functioning cars. As the decade came to its close, Microsoft had become a software giant, supplying an estimated 90 percent of the *world's* computers and earning a billion dollars in annual revenues.

The essential legal question in the case was whether Microsoft violated federal antitrust law. But there was a larger policy question embedded in the volumes of documents disclosed and hours of testimony given during the case. Supposing the court found that Microsoft *had* engaged in "restraint of trade," had consumers been demonstrably harmed by it? Might they not have benefited indirectly by Microsoft's practices? Indeed, were U.S. antitrust laws— originally written to curb monopolistic practices in such older industries as sugar, petroleum, and freight rail transportation—meaningful in the software industry and cyberspace?

The cornerstones of federal antitrust law are the Sherman Antitrust Act of 1890 and the Clayton Antitrust Act of 1914, congressional efforts to respond to constituent outrage over the abuses of "big business"—most notably the "trusts"—formed by vertically and horizontally integrating firms in the post– Civil War era. Processors and distributors of raw goods such as petroleum—for example, Standard Oil—integrated horizontally to ensure control over their product and market. The largest companies either bought out their competitors, or formed trust agreements with them to apportion their markets. Vertically integrating firms were those that produced finished consumer goods, such as cigarettes and sewing machines. To enhance their market position they either bought or formed exclusive agreements with suppliers, manufacturers, sales, and product maintenance businesses. Extraordinary fortunes combined with non-competitive pricing and monopoly control of regional and local markets to provoke rebellions in state legislatures and ultimately the Congress.

The Sherman Antitrust Act attempted to address this controversial aspect of the rise of big business by declaring unlawful "every contract, combination in the form of trust or otherwise, or conspiracy, in restraint of trade or commerce among the several States, or with foreign nations." The Act authorized the federal government to sue in federal court to obtain the dissolution of trusts. The language was emphatic enough, but not clear enough. What, precisely, was a "trust?" A "combination?" How does one demonstrate "restraint of trade?" And did the Act apply to common carriers such as the railroads, which too often charged higher, discriminatory rates for shorter intra-state hauls than the more remunerative long hauls between states? Did the Act apply to labor unions? (To these questions the Supreme Court eventually answered "yes.")[48]

The Clayton Antitrust Act attempted to clarify some of the uncertainties of its predecessor. It made clear that labor organizations were not subject to antitrust prosecution, and specifically prohibited price discrimination tending to create a monopoly, "tying" contracts, or contracts conditioned on purchasers' refusal to buy or handle competitors' products, and interlocking directorates of firms capitalized at $1 million or more. Individuals could sue in federal court for injury and damages, courts could issue injunctions and cease and desist orders, and corporate officials would be held *personally* liable for judgments. Three weeks earlier the Congress had passed legislation creating the Federal Trade Commission, authorized to investigate allegations of unfair competition in interstate commerce, trade boycotts, mislabeling and adulteration of commodities, and false claims to patents. This was the edifice of federal antitrust law with which Bill Gates and Microsoft were destined to collide.

The story begins not with Bill Gates, but with Apple Computer, which in 1984 introduced its "Macintosh" to the world in a commercial during the January Super Bowl football extravaganza. The "Mac," which had a faster speed and memory use than its Apple predecessors, featured the first graphical user interface (GUI). But it was Microsoft's Windows (the Apple Macintosh ran on Apple's own operating system) that would popularize the GUI on the IBM PC. In 1988 Apple sued Microsoft for copyright infringement, initiating nearly six years of litigation which concluded in 1993 with the court dismissing Apple's suit. That same year, the FTC, which had by now begun to receive complaints about Microsoft's business practices, deadlocked over whether to bring an antitrust action, while the U.S. Department of Justice's antitrust division began its own probe. As a result of that probe, Microsoft and the justice department in 1994 signed a consent decree. At issue were (a) Microsoft's tying (or "bundling") with Windows its own applications, thus undercutting competing software producers, and (b) its practice of requiring purchasers of Windows to buy other Microsoft applications as well. Under the consent decree Microsoft agreed to limit its bundling practices and contract restrictions, while insisting on being able to "integrate" features and applications into Windows, provided it did so without anti-competitive intent.

That same year Netscape Communications released its browser, which soon captured 85 percent of the Internet browser market. Microsoft responded in 1995 by tying to its updated Windows 95 its own Internet Explorer browser, at no additional cost.[49] Evidence later produced in federal court showed that Microsoft had already threatened to (and for three months did) deny Netscape its applications programming interfaces (APIs, which Netscape needed to program its browser to work with Windows) if Netscape did not agree to sell its browser for use only with older Windows programs and on non-Windows computers. The justice department also charged Microsoft with including in its contracts with computer manufacturers and other software producers like America Online provisions restricting their ability to do business with Netscape.[50] In August of 1996 Netscape submitted an extensive brief to the justice department, complaining of Microsoft's anti-competitive "strong-arm" practices. In October 1997 the justice department filed a motion in Federal District Court, asserting that Microsoft had violated the 1994 consent decree by tying its Internet Explorer to Windows.[51]

Moving with unusual speed, Judge Pennfield Jackson two months later ordered Microsoft to decouple its browser from Windows, and, since Jackson was not himself a computer user, appointed a "special master" to advise the court on the technological aspects of the legal issues. (Nor was Attorney General Janet Reno, then head of the justice department in the Clinton administration, a computer user.) After initial settlement discussions failed, the justice department filed a general antitrust lawsuit in May 1998. It was joined by the attorneys general of twenty states. Trial proceedings began in October of that year. By then 90 percent of the personal computers to be found in nearly 50 percent of all American households used Windows as their operating system, and Microsoft was earning annual profits of $4.5 billion, or twice those of General Motors.

Whether or not Microsoft violated federal antitrust laws as flagrantly as the justice department charged, Microsoft and the justice department settled the case in November 2002, and the populist view of monopolies that inspired the Sherman and Clayton Antitrust Acts was severely put to the test. That view held that bigness was by nature bad, because it led to market monopolies, which would naturally be used to enrich undeserving businessmen and subject consumers to diminished choices among alternatives. Monopolies controlled access to markets, access to capital, and access to the information and patents necessary to sustain innovation, went the classic antitrust argument.

But how could any software company, even the biggest, monopolize the market when anyone could order or download competitors' software from the Internet? And what constitutes "excessive" size and profits, which to the critics of Standard Oil, the bete noir of the populist movement, seemed on the face of it evidence of a "bad" monopoly? Had consumers been harmed by Microsoft's market power? How? And if no harm to consumers could be shown, did it matter how big Microsoft was? What's more, some economists were arguing that monopolistic arrangements could be to the public's benefit if they introduced efficiencies—for example, the combination of Microsoft operating systems, Cisco Systems' Internet "plumbing," and the telephone wires of local telephone companies serving as common carriers, had all contributed to the enormous growth of the Internet in the United States. The best test of a monopoly's value in the economy, critics argued, was whether it survived. Inefficient monopolies, failing to respond to the demands of the market place, die of their own weight—or so went the counter argument at the end of the twentieth century.

One argument Microsoft clearly could not win. Its lawyers argued before Judge Jackson that Microsoft simply could not control the software market, because software, being merely lines of programming code, is "infinitely malleable"; any and all competitors could write software in a variety of ways to perform any variety of tasks. That being so, observed Judge Jackson, then Microsoft's defense that Internet Explorer was "inextricably integrated" into its Windows operating system was simply disingenuous, and Microsoft could be ordered to unbundle its browser from Windows.[52]

By June 2000 Judge Jackson had heard enough. He had ordered settlement discussions between Microsoft, the justice department, and the now nineteen states that had joined the suit, and the discussions had broken down. Jackson found Microsoft to be a monopoly guilty of anti-competitive practices, most especially in tying its web browser to its operating system. As a remedy, he ordered the corporation to be broken into two companies, one to produce operating systems, and the other to produce what had been Microsoft's principal software applications—Microsoft Office, the Excel spreadsheet program, its browser, and other Internet ventures. The two companies would function under court supervision for a period of seven years. Within a week Microsoft appealed to the U.S. Court of Appeals, prompting the justice department to appeal directly to the U.S. Supreme Court, by-passing the lower court. The Supreme Court declined. In June of 2001 the Court of Appeals ruled that Microsoft had indeed behaved like a "predatory monopolist," but also accepted Microsoft's argument that Jackson was biased; thus the court rejected Jackson's order to break up the corporation. Instead, it ordered that a new district court judge oversee the negotiations to arrive at a remedy. By 2001 events unfolding external to the Microsoft case would also influence its outcome. Bill Gates discovered the necessity of bipartisan political lobbying and gift giving, and in January 2001 a Democratic White House was transformed into a Republican White House. Political winds that had carried anti-business sentiments in the 1990s now blew with more business-friendly breezes. Two years after George W. Bush was elected, Judge Kollar-Kotelly, selected by lot, announced the conclusion of settlement discussions between Microsoft, the justice department, and the attorneys general of the eighteen states that had stayed the course.

The settlement provisions prohibited Microsoft from writing contracts designed to exclude competitors, directed it to use the same contract for all computer ("hardware") manufacturers, allow manufacturers and users to remove some of its icons from their computers' desktops, and furnish programming codes to other software developers enabling them to develop products that would work as well with Windows as Microsoft's own applications. Finally, Judge Kollar-Kotelly affirmed that the court would hold Microsoft executives "individually responsible" for their compliance.

Judge Kollar-Kotelly's ruling turned out to produce only a brief hiatus in an ongoing war. A year later, in November of 2003, the U.S. Court of Appeals for the D.C. Circuit heard appeals from the attorney general of Massachusetts and technology trade organizations, alleging that Microsoft had failed to separate its Windows operating system code from the code for its Internet Explorer browser, thus contributing to an 80 percent loss of market share on the part of Explorer's chief rival, Netscape Navigator. Massachusetts also faulted the settlement's failure to require Microsoft to release its browser code under an "open source" license, enabling Microsoft competitors (such as Sun Microsystems) to distribute software applications through the Internet.[53]

Moreover, the settlement of 2002 addressed only Microsoft's practices in the browser and operating system market. But America Online and Sun Microsystems, which were among the principal aggrieved parties during the preceding five years of litigation, were not reassured that they could succeed in a rapidly changing Internet-based market without another imbroglio with Microsoft. For example, Sun's Java software programming language that could run on virtually any operating system, threatens the market dominance of Windows. As a result, Sun has found itself tangled in its own antitrust and copyright violations litigation with Microsoft to protect Java.[54]

America Online meanwhile, is only one company eager to exploit electronic commerce over the Internet by serving as a "gateway," where consumers can store personal data; Microsoft, with its ".NET Passport," is another. And still another is "Liberty Alliance," led by Sun Microsystems. Based on an open source system that anyone can use (rather than a proprietary system like Microsoft's), the Liberty Alliance enables groups of about 150 companies to set up shared entry points for consumers (with shared data), but would not itself gather personal information in a central database. As a reminder that the Internet is a global space, Microsoft recently had to yield to pressure from the European Union's oversight group on Internet privacy protection to amend .NET Passport to allow consumers to review personal data maintained by the service as well as a summary of their privacy rights under the law.[55] Market dominance may prove ultimately elusive for Microsoft as the entire computer industry—including online commercial services—becomes increasingly global. The largest PC maker at the dawn of the twenty-first century, Hewlett Packard, chose in early 2004 to equip the computers it markets in Asia not with Microsoft, but with the Linux open source operating system.[56]

The high cost of continuous litigation may have become an ongoing tariff that Microsoft is prepared to pay to continue its practice of "bundling" applications to its operating system. In December 2003 RealNetworks, maker of software for playing computer-based and Internet-distributed digital audio and video content, filed a $1 billion antitrust lawsuit against Microsoft for bundling its competing Windows Media Player with its Windows operating system. Microsoft and AOL Time Warner have also tangled over software necessary to distribute digital entertainment over the Internet, with Microsoft finally agreeing to license to AOL Time Warner its Windows Media software.[57]

Meanwhile, as if to affirm Judge Jackson's skepticism about Microsoft's claim that the nature of software necessitated "inextricably" integrating Internet Explorer into Windows, in June 2003 Apple Computers introduced its own browser, "Safari," described by one software reviewer as "an elegant piece of work. . . . Instead of waiting for Microsoft to update the aging Internet Explorer, [Apple] wrote its own software."

> In other words, bundling a Web browser with an operating system makes sense—just as Microsoft said! But unlike Microsoft's version of browser integration, Apple's preserves choice. Uninstalling Safari is as simple as dragging its icon to the Trash. Furthermore, Safari's underlying Web-rendering code is free for everyone else to use—even Windows developers. That is because Apple used open-source code . . . which both allowed it to build on other people's work and required it to release its own improvements under the same "free software" terms. The collaborative effort enables Safari to function on a Web in which many sites are written to work only in Internet Explorer.[58]

JUSTICE HOLMES'S APHORISM that the value of ideas is best left to the judgment of the market place, which in turn requires an open sphere in which public opinion can form and change with circumstances, was emblematic of the American pragmatist's aversion to idealism of any sort. Holmes, wounded in battle during the Civil War, had seen idealism turn into carnage, and believed he had found a formula to protect the state from capture by a single idea to be imposed—perhaps with war—on the rest. That his formula was not entirely original matters less than its long-standing appeal to American jurists arguing for the vigilant application of the Constitution's First Amendment. Louis Menand, in his parting reflections of his study of Holmes's circle of pragmatists, observed:[59] "once the Cold War ended, the ideas of Holmes, James, Peirce, and Dewey reemerged as suddenly as they had been eclipsed. . . . For in the post–Cold War world, where there are many competing belief systems, not just two, skepticism about the finality of any particular set of beliefs has begun to seem to some people an important value again. And so has the political theory this skepticism helps to underwrite: the theory that democracy is the value that validates all other values. Democratic participation isn't the means to an end, in this way of thinking; it is the end."[60]

Since the 1980s, which saw another congenial Republican president serve two terms, democracy's analog in economic policy—preserving the competition of a minimally regulated market place—has acquired a similar aura: for Democrats and Republicans alike, competition is the surest means to achieving and maintaining a free market economy, itself the most certain path to national and international prosperity. Still, here and there skeptics have grown so bold as to question the mantra. Partisans of social justice in the new global economy openly doubt whether unalloyed competition in the market place is in the best interest of everyone in every nation, while even some members of the U.S. business elite have joined consumer advocates in challenging the "efficiency" of unfettered competition in building and maintaining the U.S. communications infrastructure.[61] Nor does the ever-expanding content of the Internet oblige us to amend the First and Fourth Amendments. Designed and built as an open system, the Internet and the policy issues raised by its far-reaching adoption have only served to demonstrate, once again, the resilience of the fundamental principles upon which the original, open American political system rests.

7

Open Systems in
Outer Space

Breathtaking photographs of inter-stellar space from the Hubble Space Telescope; images of the Earth's changing climate, oceans, and land formations; astronauts, an orbiting space station's lights blinking overhead on a clear night, and Space Shuttles going up into space and twice catastrophically failing to return: All provide the personalities and visual interest much desired by our television news and feature networks, and hence a broad public awareness of the National Aeronautics and Space Administration, which is arguably the best known public R&D institution in the United States. The agency's programs in space science, Earth science, microgravity and biological sciences, and aeronautics research are so varied that it would be impossible to do them justice with a summary in this chapter. But NASA's considerable visibility in the constellation of federal R&D programs is not the only reason the agency's past and probable future merit a close examination in any study of U.S. science and technology policy.

The U.S. space program that sent the first men to the moon is sometimes compared with the Manhattan Project, but the two differ in important ways. Begun in 1941 under the management of the U.S. Army, the Manhattan Project to develop the first atomic bomb (successfully deployed over Hiroshima and Nagasaki in August of 1945, thus bringing to an end the war in the Pacific) was necessarily conducted in secret and thus shielded from the ordinary political processes that otherwise create and sustain federal R&D programs. Second, its objective was singular and unambiguous, and thus success could be clearly recognized and the project brought to an administrative and budgetary end—though, of course, the harnessing of atomic energy itself would continue. The U.S. space program, by contrast, was a campaign launched with a vague set of ideological and uncertain military objectives, pursued largely in the public eye by a civilian agency whose fortunes—technological, administrative, and budgetary—were and remain entirely at the mercy of congressional and presidential politics. As such, for both good and ill, the space program illustrates to an unusual degree the conditions and dynamics set in play by the nation's proclivity toward open systems in its policy making and management of large-scale research and development.

Eisenhower, "Open Skies," and the Making of U.S. Space Policy

Within a year after the Soviet Union successfully launched the first man-made orbiting satellite, Sputnik I, in October 1957, the administration of Dwight D. Eisenhower and Congress created the National Aeronautics and Space Administration, initially out of the cluster of federal aeronautical laboratories operated by the National Advisory Committee for Aeronautics (est. 1915). The NACA was thus transformed in 1958 into the federal civilian space establishment with a renewed and much enlarged mission. Within a few years it had doubled in size, thanks to the addition of the Army Ballistic Missile Agency (later renamed the George C. Marshall Space Flight Center) and the new Goddard Space Flight Center in Beltsville, Md. and Wallops Station on Wallops Island, Va. The Jet Propulsion Laboratory of the California Institute of Technology, a contractor-owned and -operated facility involved in rocket research since 1936, was transferred from the U.S. Army to NASA in 1958. The Manned Spacecraft Center (renamed Lyndon B. Johnson Space Center in 1973) in Houston, Texas was added in 1961 and the John F. Kennedy Space Center at Cape Canaveral, Florida in 1962.

The bill creating NASA that the Eisenhower administration sent to Congress in April 1958, and which by July 29 the president was able to sign into law, contained a broad policy statement of the measure's general and benign purposes. The act begins with a congressional declaration that "it is the policy of the United States that activities in space should be devoted to peaceful purposes for the benefit of all mankind." It then places upon the agency responsibility for a substantial menu of open-ended aerospace R&D programs, directing NASA to

—expand "human knowledge of phenomena in the atmosphere and space";
—improve "the usefulness, performance, speed, safety, and efficiency of aeronautical and space vehicles";
—develop and operate "vehicles capable of carrying instruments, equipment, supplies, and living organisms through space";
—conduct long-range studies of "the potential benefits" of "aeronautical and space activities for peaceful and scientific purposes";

—preserve "the role of the United States as a leader in aeronautical and space science and technology and in the application thereof to the conduct of peaceful activities within and outside the atmosphere";

—share with U.S. defense agencies "discoveries that have military value or significance";

—promote international cooperation in the peaceful applications of aerospace science and technology; and

—ensure "the most effective utilization of the scientific and engineering resources of the United States . . . to avoid unnecessary duplication of effort, facilities, and equipment."

A newcomer to Washington in 1958, reading the Space Act for the first time, might also have been forgiven for not recognizing the policy import of the first few pages of the statute. But beneath the innocent sounding declaration of peaceful scientific intent and innocuous references to cooperative information exchange with the military, lay a national Cold War strategy of great importance in its own time, and for the decades following. That strategy was designed by President Eisenhower who, as supreme commander of the Allied Expeditionary Force in western Europe that launched the successful invasion of Europe in June 1944, had an exceptional appreciation of the high cost of large-scale military mobilization.

During the early years of the Cold War, the dangers that worried President Eisenhower included not only the Soviet Union, but fear and greed—demands, in the name of national security, by the military's supporters and its industrial contractors, for more bombers, more missiles, and more nuclear warheads. Eisenhower feared that if such demands were met, the United States would become a "garrison state," emptied of its ability to provide the very "way of life . . . we are striving to defend." As the highest ranking general in the European theater during World War II, Eisenhower had opposed the use of the atomic bomb against Japan as unwarranted overkill.[1] For Eisenhower the president, avoiding an arms race involving nuclear weapons and the missiles to deliver them was a principal preoccupation. The warning delivered in his Farewell Address is worth revisiting in some detail, for it mirrored the salient context in which U.S. space policy initially took form:

> [The] conjunction of an immense military establishment and a large arms industry is new in the American experience. . . . We recognize the imperative need for this development. Yet we must not fail to comprehend its grave implications.
>
> In the councils of government, we must guard against the acquisition of unwarranted influence, whether sought or unsought, by the military-industrial complex. The potential for the disastrous rise of misplaced power exists and will persist. . . .

The prospect of domination of the nation's scholars by federal employment, project allocations, and the power of money is ever present—and is gravely to be regarded.

Yet in holding scientific research and discovery in respect, as we should, we must also be alert to the equal and opposite danger that public policy could itself become the captive of a scientific-technological elite.[2]

Preventing catastrophic war between the two great powers of the Cold War would require, most of all, reliable intelligence, whether to confirm compliance with any moratorium on existing weapons, or to prevent military mobilization in response to exaggerated or non-existing military threats. For example, in August 1956 the CIA had estimated that by mid-1958 the Russians would have 470 Bison and Bear bombers and 100 intercontinental ballistic missiles (ICBMs). But in June of 1958, the estimate was that the Soviets actually had 135 bombers and no ICBMs. Eisenhower commented that "the Soviets have done much better than have we in this matter. They stopped their Bison and Bear production, but we have kept on going, on the basis of incorrect estimates and at a tremendous expense in a mistaken effort to be 100 percent secure." Later he remarked that he "just didn't know how many times you could kill the same man!"[3] And so it was that for Eisenhower the principal reason for the U.S. to operate in space was to be able to collect routine and reliable intelligence on the military activities of the Soviet Union.

This objective was consistent with Eisenhower's broader Cold War policy of containment, recommended in 1946 by George Kennan, then U.S. minister in the Soviet Union. The policy was to limit (rather than attempt to appease or overwhelm) Soviet expansionism at a military level that would be economically sustainable for the foreseeable future.[4] It was consistent with the outlook of a president who was prepared to take the initiative in 1958 in seeking a test-ban treaty with the Soviet Union lest the United States find itself in "moral isolation" in the international community. And it was consistent with the judgment of the Eisenhower who, supreme commander of the allied forces that brought about the military defeat of Germany in 1945, would not put more American lives at risk for the political objective of seizing Berlin before the Russian army reached the city.

The Rand Corporation, principal U.S. Air Force contractor for operations research, had been studying the possible national security uses of Earth-orbiting satellites since 1947. Rand reported as early as 1951 that the use of orbiting satellites for weather and military reconnaissance was technically feasible at reasonable cost using technologies then available or under development. Having used air reconnaissance effectively as a military commander during World War II, Eisenhower required no persuading. By May 1955 his administration was ready with the United States' first statement of its space policy, which the president announced in general terms to the American people in a radio address reporting on his summit meeting at Geneva with Soviet and other European leaders. A shrewd use of the principle of open systems in shaping national security strategy was central to that policy, which continues to this day to enable the U.S. to gather space-based intelligence over the entire globe.

At Geneva Eisenhower had proposed to Nikita Khrushchev an "open skies" regime by which both the United States and the Soviet Union would open their airspace and airfields to each other, allowing continuous air reconnaissance. Eisenhower's intent had been to build the confidence necessary for an effective moratorium on arms or nuclear weapons testing. But Khrushchev had rebuffed his initiative, dismissing it as a "bald espionage plot."[5] Thus the policy laid out on May 20, 1955 in National Security Council (NSC) document NSC 5520, "Draft Statement of Policy on U.S. Scientific Program," was all the more important.

When NSC 5520 was issued in 1955, the military had already proposed two satellite programs for intelligence gathering, while the National Academy of Sciences and National Science Foundation had appealed to the federal government for support for a third satellite—a scientific satellite to be orbited as a major contribution of the United States to the International Geophysical Year (IGY).[6] The NSC knew that Russian scientists were at work on a satellite program and recommended that the defense department "develop the capability" (that is, develop the launch vehicle and system) necessary for the U.S. to "launch a small scientific satellite under international auspices, such as the International Geophysical Year." Information about the satellite and its orbit would be public; information about its launch system (largely indistinguishable from a ballistic missile) would not.

The NSC's endorsement of U.S. participation in the IGY was not, in and of itself, the significant policy declaration. Rather, it was the *conditions* and *circumstances* under which the satellite would be flown that would set important precedents, and thus policy. First, while contributions to scientific research are almost always a good thing, in this case the international scientific community would provide a moral high ground—"peaceful purposes"—from which U.S. space-based reconnaissance could be made acceptable internationally, whether the information being gathered served scientific or national security purposes. Moreover, no country need fear that any military use would be other than defensive—that is, that a bomb might be dropped from a satellite— because anything released from a satellite would simply follow in its own orbit.

Second, the satellite's flight would test the willingness of the international community to extend to space the fundamental principles of the international law of the sea, namely that none may assert sovereignty or ownership over any portion of the high seas, and that all should assist mariners in difficulty. In other words, "freedom of the seas" would extend to "freedom of space." The U.S. IGY satellite should be flown, wrote the NSC, "in a manner which . . . does not involve actions which imply a requirement for prior consent by any nation over which the satellite might pass in its orbit, and thereby does not jeopardize the concept of 'Freedom of Space.'"[7] Woven into this legal point was the notion that the law of the open sea—rather than national airspace—would provide the legal foundation for satellites flying over the territories of sovereign nations. John F. Kennedy's later reference to space as "this new ocean" was more than a nice turn of phrase.

Meanwhile, U.S. representatives in the United Nations (UN) managed to embed the principle of "open skies" into evolving international space law by successfully negotiating UN acceptance of four cardinal U.S. positions: (1) all nations could traverse space in the same way that all nations could traverse the high seas; (2) any UN space-related agreements would apply to space craft, but not to launch vehicles; (3) peaceful purposes did not preclude defensive military purposes (for example, intelligence gathering), which could serve peace by ensuring compliance with disarmament treaties; and (4) the international space regime would be adequately served by an *ad hoc* UN planning committee, rather than control by a permanent international space agency. The U.S. representatives' efforts resulted in the creation of the UN Committee on the Peaceful Uses of Outer Space, or UNCOPUOS, established in 1958. UNCOPUOS was able to reach agreement on four treaties, the most important of which was the first, informally referred to as the "Outer Space Treaty," in which the signatories agreed that:[8]

> The exploration and use of outer space, including the moon and other celestial bodies, shall be carried out for the benefit and in the interests of all countries, irrespective of their degree of economic or scientific development, and shall be the province of all mankind.
>
> Outer space, including the moon and other celestial bodies, shall be *free for exploration and use by all States without discrimination of any kind, on a basis of equality and in accordance with international law, and there shall be free access to all areas of celestial bodies.* There shall be freedom of scientific investigation in outer space, including the moon and other celestial bodies, and States shall facilitate and encourage international co-operation in such investigation [italics added].[9]

Thus by the spring of 1958 the United States had crafted a policy by which it could proceed with the technological capability and diplomatic latitude to operate in space. The policy was reinforced by the decision to lodge the nation's space program in a new civilian agency, rather than the defense department. Assuming the success of the U.S. entry into the IGY, the federal government could acquire important information about the physical characteristics and dynamics of the upper atmosphere, ionosphere, geodesy, and orbital mechanics and the Earth's rotation necessary to predict and track satellite orbits. This information would yield important military as well as scientific returns.

National Transcendence in Space as Eisenhower's "Complex" Takes Hold

More problematic, however, was the NSC's assertion that "considerable prestige and psychological benefits will accrue to the nation which first is successful in launching a satellite."[10] While it was and remains certainly true that whatever any country launches into space is a demonstration of its space launch capabilities, a demonstration any government is duty bound to evaluate seriously, claiming "prestige and psychological benefit" for a national policy of such serious moment introduced a structural weakness into the foundations of U.S. space policy.

Just how prestige or psychological impact were to be assessed or measured to the extent necessary to support *responsible* presidential decision making could not be known then, no more than it can be known today. Nonetheless this assertion was inflated and elevated to the opening pages of the NSC's subsequent policy statement, "U.S. Policy in Outer Space," issued June 20, 1958. The statement begins with a sweeping claim for the significance of human activities in space—"it is probable that the long-term results of exploration and exploitation will basically affect international and national political and social institutions"—and proceeds to list three of "the starkest facts" impinging upon the "security of the United States." Only the third refers to tangible military capabilities that might "create an imbalance of power in favor of the Sino-Soviet Bloc" and pose a direct military threat to U.S. security. The first two of "the starkest facts" were not facts at all, but unsustainable assertions of the relative capacity of the US and USSR to exploit space, and the importance of space to the people of the world, ergo the United States: "The USSR has surpassed the United States and the Free World in scientific and technological accomplishments in outer space, *which have captured the imagination and admiration of the world*; (2) the USSR, if it maintains its present superiority in the exploitation of outer space, *will be able to use that superiority as a means of undermining the prestige and leadership of the United States* [italics added]."[11] The remarkable assertion being made here by the NSC was that the political and religious freedoms, constitutional government, social mobility, and wealth generating economic opportunities which, for well over a century, had sustained the"prestige and leadership of the United States," were no longer sufficient; all that mattered now was how the United States performed in the new arena of space technology.[12]

This presumptuous claim, made by unelected government officials on behalf of a policy fashioned by themselves, ventured beyond the realms of "prestige" and "psychological benefit" to soar into the intangible and emotive realm of transcendence and "great values to be discovered." This was surely a new realm for the NSC, one that ordinary citizens normally navigated with the help of clergy or other private means of personal reflection. Notwithstanding the political and media uproar that followed the Soviet Union's first launch of a satellite, or its first successful flight of a human (Yuri Gagarin) on April 12, 1961, it did not necessarily follow that as a matter of good policy the United States need follow in kind, with a massive crash program to "beat the Russians."

Eisenhower biographer Stephen E. Ambrose has argued that only Eisenhower, with his unassailable credibility as a military leader, could resist the pressure to increase military spending led by such influential institutions and individuals as the Ford Foundation, the Rockefeller brothers, the joint chiefs of staff, and much of the U.S. Congress.[13] But Eisenhower's reluctance to lead the nation into a space race was not simply the conservatism of a budget-conscious Republican. The only argument Eisenhower would accept for engaging in an ambitious space program was that it would bring clear scientific or military benefits. A program that might send men into space promised neither, as far as Eisenhower was concerned, an opinion shared by his President's Science Advisory Committee as well as the heads of the National Academy of Sciences and the National Science Foundation.[14]

However, the U.S. Air Force had been engaged in exploratory work on a manned space flight capability since 1956 and had begun to seek approval for such a program in 1958, at the same time the administration prepared its proposal for the creation of a new civilian agency to manage the nation's space program. That such an agency should be civilian was consistent with Eisenhower's concern over the growing defense budget. By the time he signed the Space Act creating NASA, he had determined that there was no *military* need for "man in space," and assigned responsibility for any such program to NASA rather than to the Air Force. NASA immediately set to work on what became Project Mercury, which successfully launched six American astronauts into space during 1961–1963, including John Glenn, who was the first U.S. astronaut to orbit the Earth (February 20, 1962).

Undeterred by the lack of congressional or White House authorization or direction to do so, NASA proceeded to work on plans for an ambitious manned space program to follow Project Mercury. While the concept of a "follow on" program might not have been a NASA invention, the agency used it as a strategy to ensure that the technological and human capital marshaled to carry out one program would not be dissipated once the program was completed. Convincing the Congress and White House to approve a follow-on program thus became a driver for continuing tax-payer funded R&D in those areas for which the bureaucracy had been given statutory authority. While such programs were always given high-minded objectives, it also was and remains the case that the continuation of jobs, future of careers, and untold millions of contract dollars were at stake.

Indeed, "the [Eisenhower] administration's attempt to keep the space budget at a low level meant that the government space agencies [principally NASA] were not able to win significant support from an industrial constituency, especially in comparison to the industrial support for the Air Force and Navy strategic missile programs."[15] Ensuring the continuation of jobs, careers, and large procurements to the military and aircraft industries—Eisenhower's "military industrial complex"—was translatable into constituent political pressure on members of Congress to approve and fund the next big program. Regard for the wishes of space lobbies such as the National Space Society and the American Institute of Aeronautics and Astronautics could also result in increased political support.

For two years before the agency finally received authorization from President John F. Kennedy in May of 1961, NASA planning groups put together a plan to send a man to the moon (which the agency would name Project Apollo). The plan anticipated being able to proceed from manned circumlunar reconnaissance flights to a moon landing within ten years, and projected an annual budget requirement not exceeding $1.6 billion (NASA's actual budget by 1967 had grown by three times that amount). The agency likened its plan to Columbus's first voyage to North America and claimed that it would "win [for the U.S.] more gold medals in the space Olympics than any other nation." Eisenhower, when he first learned of it from a PSAC report in December 1960, was not pleased, asserting that he was "not about to hock his jewels' to send men to the Moon."

But six months before Eisenhower was told of NASA's plans, the agency had assembled probable industrial contractors for the Apollo project for a conference to alert them to what might be in the offing; that the industry could then work its quiet influence on the Congress and the next administration needed not be said. NASA had already briefed its congressional committees on its plans.[16] Lest industry representatives fail to appreciate their stake in what NASA was proposing, the agency's officials advised them that "they would be 'invited to participate, by contract, in a program of system design studies'" and "a systems contract for the design, engineering, and fabrication of the manned spacecraft and its components will probably be initiated in fiscal year 1962." Notwithstanding the White House's and Bureau of the Budget's continuing resistance to a costly follow-on program after Project Mercury, NASA on October of 1960 awarded "contracts for six month Apollo spacecraft feasibility studies . . . to Convair/Astronautics, General Electric, and the Martin Company."[17]

And still President Eisenhower would not approve an ambitious manned space project. He left office in January of 1961 without doing so. Only he had stood in the way of the creation of a government program that would expand and further empower the "military-industrial complex"—today's aerospace/ defense industry—in order to realize the mix of vague and mystical ambitions laid out in the NSC's early space policy directives and national security strategy.

Kennedy, Johnson, and Webb: The Campaign Without End

Kennedy's approval of the Apollo project, with his promise to the nation on May 25, 1961 to land a man on the moon and return him safely before the end of the decade, introduced a major aberration into the space policy developed during the Eisenhower administration.[18] Eisenhower had fashioned a policy that authorized publicly funded space activities for scientific and/or military objectives having inherent merit, while avoiding further growth of a military-industrial complex fueled by the pursuit of ill-defined political objectives such as "prestige." Given the political noise of the early days of the Cold War and the Soviet Union's initial achievements in space, shaping a policy of restraint in the use of power (for example, containment v. conflict in our Cold War foreign policy) required considerable leadership.

Since the 1950s a space policy emphasizing the exploration of space using instruments and robotic technologies, the use of satellites for communications, navigation, and Earth science, and a separate national-security space program—all unarguably "public goods"—have shared bipartisan political support. What has not enjoyed the sustained political support necessary to convince the Congress to fund it at NASA-desired levels has been manned space flight. Indeed, well before Neil Armstrong stepped onto the moon in July of 1969, NASA administrator James Webb resigned in frustration over diminishing congressional interest in the Apollo project and resulting NASA budget cuts.

Within months of becoming president in January 1961, John F. Kennedy found himself in the center of a political uproar. On April 12 the Soviet Union successfully launched Yuri Gagarin into orbit around the Earth and recovered him safely. Five days later a CIA-trained and financed group of Cuban exiles attempted to invade Cuba at the Bay of Pigs, only to be forced to retreat in three days. These events were a near overwhelming challenge to a man whom some observers believed saw Soviet-American relations as a "'contest of wills . . . in which the prize was pride at least as much as any substantive outcome.'"[19] The Democratic party had chosen to balance Kennedy on the 1960 presidential ticket with Lyndon B. Johnson of Texas, since 1937 a member of Congress (save a year's World War II service in the Navy) and since 1949 one of the smoothest operators in the Senate.[20] As Senate majority leader, Kennedy's vice president had advocated a strong response to the Soviet challenge in space, and he worked closely with the Eisenhower administration to pass legislation creating NASA. Kennedy, perhaps looking for a way to keep Johnson busy, asked Congress for approval to name the vice president chair of the National Aeronautics and Space Council created by the Space Act, thus positioning Johnson for the role he would play in providing Kennedy with a solution to the depressing events of mid-April 1961. Successfully sending a man to the moon would surely recapture the political advantage for the United States—and Kennedy.

Announcing such an ambitious undertaking would be going out on a considerable political and budgetary limb, and so Kennedy did what political leaders often do in such cases: reinforce the limb with recommendations from a panel of "experts" and political heavy-weights. This approach held promise not only because the Congress might be swayed by such a panel's views, but because members of a distinguished panel could be counted upon to use their personal influence to lobby for whatever program they recommended. Johnson obliged his president, and was careful to add to just such a special panel: Democratic Senator Robert S. Kerr, the millionaire Oklahoman who had been head of Kerr-McGee Oil and a strong Johnson ally (Kerr replaced Johnson as chair of the Senate Space Committee); George Brown of Brown and Root, a major Texas construction company headquartered in Houston; and Wernher von Braun of NASA, the German rocket engineer whose visions of orbiting space stations found wide circulation in the pages of *Colliers Magazine* in 1952. Thus it was that Kennedy put before the world the American challenge to the Soviet Union. Saving the best for last, toward the end of a speech on "urgent national needs" before a joint session of Congress on May 25, 1961, he made his bold promise, finally adding White House endorsement to NASA's efforts to proceed on the course it had already begun.

The genie that Eisenhower so feared was released from its lamp, the lamp's sides having been rubbed assiduously by the recently appointed NASA administrator, James E. Webb. A large man of restless ambition, Webb's ability to combine his genuine administrative skills with a gift for making good use of powerful personal and political connections enabled him to parlay each of his jobs into another, better one. His work for fellow North Carolinian R. G. Lassiter (owner of a large construction firm) led to a job as law clerk for Lassiter's brother's law firm. After a brief interval in aviation for the U.S. Marines to escape the worst ravages of the depression, Webb took a job with Lassiter's friend Edward W. Pou (D-N.C.), then chairman of the House rules committee.

Hard work and a growing circle of connections in Washington—which included Lyndon B. Johnson—propelled Webb forward. Before he arrived at NASA he had worked as vice president for Sperry Gyroscope, a regular contractor to the aircraft industry; served as President Truman's budget director, where he honed his bureaucratic battle skills during the creation of the

Economic Cooperation Administration (which administered the Marshall Plan) and the defense reorganization of 1947, creating the Department of Defense. Webb also served as an undersecretary of the Department of State, where he chafed under the patrician leadership of Dean Acheson; member of the board at McDonnell Aircraft; and vice president of Kerr-McGee Oil of Oklahoma, where he salvaged Kerr-McGee's subsidiary Republic Supply from near financial ruin, and earned the admiration of the politically powerful Senator Kerr.

When Webb took the helm at NASA he was well schooled in the ways of Washington and understood that if the Apollo program was to succeed, it would require not only great technical and managerial skill, but a broad base of political support for the substantial increases in the NASA budget that would be required. One of the best ways to achieve this was to make sure that members of Congress had voters employed in their districts by NASA contractors, and thus Webb saw to it that NASA spread its procurements throughout the country, and not just among those districts that supplied congressional leadership. In so doing he made NASA's procurement strategy as open—in current parlance, as "inclusive"—as possible to ensure that the congressional political support base could be called upon as needed.

When deciding to relocate NASA's Space Task Group (which would oversee the Apollo Project) from Langley Research Center in Virginia to Houston, Texas, Webb told the Task Group's head Robert Gilruth, "'What did Harry Byrd [senator from Virginia] ever do for you? . . . We've got to get the power. We've got to get the money, or we can't do this program. And the first thing, we got to move to Texas. Texas is a good place for you to operate. It's in the center of the country. You're on salt water. And it happens to be the home of [Albert Thomas] the man who is the controller of the money.'" The NASA map soon traced the political topography of the U.S. Congress, with NASA's Florida launch center (renamed for the assassinated president in 1963) fortuitously located in Donald Fuqua's (D-Fla.) district; while a rocket test annex to the Marshall Space Flight Center in Huntsville, Ala. was located in John Stennis's (D-Miss.) district (and renamed Stennis Space Center in 1988).[21]

As long as the director of the Bureau of the Budget was David E. Bell (who had been Webb's subordinate when he, Webb, was Truman's budget director), Webb could and did appeal over Bell's head to the president for the initial budget increases Webb wanted to expand the agency to support an accelerated lunar mission.[22] But by the end of 1962 a budget battle between Webb and his Apollo program head, Brainard Holmes, over reallocating additional funds to keep the Apollo program on schedule laid bare the hazard of politically driven commitments of government resources to open-ended objectives.

The dispute finally erupted in front of Kennedy, who argued that he had committed only to landing a man on the moon and bringing him back safely, not to a post-Apollo manned project for which Webb was already seeking more funding. But Webb boldly corrected the president: The objective was "preeminence in space," which required preeminence in "all important aspects of this endeavor and to conduct the program in such a manner that our emerging scientific, technological, and operational competence in space is clearly evident." And when would that be? Given that "scientific, technological, and operational competence" is a continuously evolving thing, how would the country know when it had reached its goal? When was "enough?" Neither Webb nor Kennedy nor anyone else could say. In the meantime, the race for preeminence would require a FY 1964 budget request of $6.2 billion, a nearly 60 percent increase over the agency's FY 1963 appropriation. Kennedy finally gave in, agreeing to a budget request for $5.1 billion.[23]

When Johnson succeeded the assassinated Kennedy in November 1963, the rising costs of the war in Vietnam and Johnson's Great Society program threatened to create a significant budget deficit. Johnson's economic advisers, who by 1965 included a new budget director, Charles L. Schultze, began to urge that Johnson request a tax increase. Schultze warned the president that the administration was in danger of running up an $8 billion deficit in FY 1966. The space program for that fiscal year would spend $788 million more than appropriated, "costing more than any of the domestic reform programs." Cuts would have to be taken; Schultz recommended a $300 million cut for NASA. Webb protested, defiantly asking the president to tell him how to manage such a budget cut at NASA, "since it would mean releasing 'some 20,000 people . . . from NASA operations, plus 60,000 from research and development and an additional five to ten thousand from construction by July 1, 1967.'"[24] The vast majority of these 85,000–90,000 people would be employees of contractors.[25] Webb was holding before Johnson the threat of congressional opposition fueled by constituents' fears for their jobs. Schultze, in reply, "questioned the assumption that the United States, for fear of falling behind the Russians, should do everything in space that was technically feasible. Above all, he challenged the idea that it was necessary to strive to keep 'the *peak level of industrial manpower*' that had been achieved during the Apollo build-up. 'The space program,' he reminded LBJ, 'is not a WPA.'"[26] Johnson reluctantly yielded to his budget director and agreed to a FY 1967 NASA budget $1.2 billion smaller than he had requested. The prospect for FY 1968 looked no better, while in Congress sentiment was growing to trim NASA's budget by $400 million.[27] Frustrated, and no doubt wearied by the grueling political and bureaucratic aftermath of a launch pad fire on January 27 in an Apollo capsule, which took the lives of three astronauts, James Webb resigned on September 1968.

Webb's struggle with the White House over NASA's budget was exacerbated by the agency's request for funds for an Apollo "follow-on" program to maintain the "level of industrial capability" that Schultze understood to be a program to keep NASA's aerospace industry contractors engaged with (and thus actively supporting) the program. Thus began a pattern which plagues the agency to this day: as it struggles to maintain existing programs, NASA must also struggle to define and obtain funding for new programs. Failing to receive full funding in FY 1970 for an ambitious Apollo applications program, the agency terminated production of the mighty Saturn launch vehicle.[28] NASA had already modified the spent Saturn IVB rocket stage so that it could be used as an Earth-orbiting laboratory or workshop, named "Skylab," which was flown on three separate and successful missions during 1973. However, the agency's 1970 decision to terminate Saturn precluded development of the Skylab as a proto-type space station, and thus ensured that any realization of the long-held dream of space enthusiasts to orbit a space station around the Earth would require another ambitious new "manned" space flight program.

The burden on NASA to continuously devise and fund new human space-flight projects was symptomatic of the larger policy issue that encompassed the Kennedy-Johnson administration's travails in Vietnam as well as the U.S. space program. Eisenhower and his generation learned the responsibilities of leader-ship during the country's previous, virtually all-consuming public policy chal-lenges—the Great Depression and World War II. Those responsibilities included committing lives and public resources to clearly defined tactical and strategic objectives. Simple numbers could tell when the country's economic health had been restored, and the official surrenders of Germany and Japan verified when allied forces had achieved victory. In contrast, winning an ideo-logical war in space was an open-ended campaign for vague and subjective victories like "preeminence" and "prestige," luring larger-than-life personali-ties like John F. Kennedy, Lyndon Johnson, and Jim Webb into short-sighted decisions rich in immediate political gratification.

Thus the military mobilization necessary to defeat the Axis powers in World War II could be followed by a popularly welcome demobilization, while popular insistence on demobilization of U.S. troops from Vietnam helped to bring about the end of this country's role in that conflict. But the national mobilization necessary to engage the Soviet Union in the Cold War knew no bounds, and was thus easily exploited for political and economic gain. More than 80 percent of the $17.3 billion the Congress appropriated to NASA be-tween 1960 and 1965 flowed back out of the agency in contracts and grants to the country's universities and the aerospace industry. In 1960 NASA's procure-ment offices managed 44,000 procurements; by 1965 NASA officials were pro-cessing and monitoring almost 300,000 contracts spread geographically throughout the country, in keeping with Webb's broadly distributed procure-ment strategy.

The fact that congressional appropriations to NASA continued to flow back into numerous congressional districts in the form of NASA contracts and grants (normally 80 to 90 percent of any given year's annual appropriation) does not, in itself, negate the genuine fascination with space travel and exploration reflected in attendance figures at special expositions and space museums around the country. Even so, polling data tell us that more Americans saw the Apollo program as another effort to "beat the Russians" than as an essential goal of U.S. space activities. As the sequence of Apollo missions unfolded from the first lunar landing in July of 1969, public support for the space program did not increase. Instead, the proportion of Americans opposed to more government expenditures in space increased between 1965 and 1975 from one-third to one-half of all adult Americans. A survey conducted for NASA in the late 1980s by the International Center for the Advancement of Scientific Literacy found only a little under two-thirds of the adult population in the U.S. expressing any support for space exploration, while respondents might have been even less enthusiastic if they had been asked to choose between federally financed space exploration and other government programs of more immediate benefit to them.

Another aspect of public interest in space exploration revealed by opinion surveys is the general demographic distribution of the civil space program's enthusiasts and its likely detractors. Americans who say they support the space program are more likely to be male, Catholic, white, college-educated (but not holders of graduate or professional degrees), not yet in their forties, Republican, and receiving annual incomes well over the median average annual household income. Those who express a lack of interest in the space program tend to be women, minorities, the less-educated in non-salaried occupations—people who may have a more intimate experience of the daily challenges of putting food on the table, raising children, and caring for the elderly. This less interested grouping is also more likely to include Democrats and generally politically liberal holders of advanced academic degrees—presumably not including those employed in university science and engineering departments that benefit from NASA contracts and research grants. Had the senator from Texas not become Kennedy's running mate and been elected vice president in 1960, Eisenhower's restrained approach to a federally financed campaign for U.S. "preeminence" in space might have prevailed.

The Space Shuttle and Commercialization

After the 1969 moon landing, NASA's budgets continued to be trimmed by the Bureau of the Budget (reorganized into the Office of Management and Budget by President Nixon in 1970), forcing Webb's successor Thomas O. Paine (1969–1970) to choose between further production of the Saturn launch vehicle and the Viking project (which landed two unmanned spacecraft on Mars in 1976), in order to pay for continued Apollo missions to explore the moon.[29] Viking survived, as did the proto-space station "Skylab," fashioned from Apollo-Saturn hardware and flown three times during 1973.

Between 1965 and 1968 over 150,000 NASA-supported jobs evaporated, while during the 1970s an additional 236,671 persons once employed by NASA or its contractors had to find other jobs. Some of those jobs were undoubtedly absorbed by the Department of Defense, where civilian employment increased by 160,000 between 1965 and 1970. But after the end of U.S. military involvement in Vietnam in 1973, most of those positions were lost as well.[30] The organization that built America's civil space program in the high-noon of the Cold War had to find a new, politically marketable human space flight mission in order to maintain the flow of funds into NASA and the aerospace industry. Webb's procurement strategy would be put to the test.

In 1971 NASA's deputy administrator George Low contemplated recasting NASA as a national technology agency, responsible not only for aeronautics and space research and development, but for a wide range of "technological solutions" for national problems such as alternative power and energy sources, environmental pollution, improved transportation systems, health care systems, improved productivity of services, education, and housing.[31] Others in Washington were thinking in the same vein. In 1973 the White House's National Aeronautics and Space Council, which could have served as a vehicle by which the executive branch crafted an interagency consensus around a well-defined space program, was abolished. From 1974 onward Congress added to NASA's authorizing statute numerous additional responsibilities (for example, to develop an electrical or hybrid fuel system ground vehicle), most having only tangential relation to the agency's original purpose.[32]

If NASA did not define another new and substantial "follow-on" program and persuade the White House and Congress of its necessity, the agency might not survive. Deputy administrator George M. Low and administrator James C. Fletcher tried successfully to turn misfortune into opportunity by persuading the Nixon administration of the need to approve a new program to develop a reusable space launch vehicle to "take the astronomical costs out of astronautics."[33] Low promised that the agency would abandon the strategy of developing "individually tailored technologies" and, instead, "focus on *multiple-use, standardized* systems" (emphasis author's)—an argument readily appreciated by budget-minded officials.[34] NASA, in short, would replace its sophisticated, one-of-a-kind space systems design with an open systems approach to engineering originally developed by early nineteenth-century U.S. arms manufacturers (see chapter 2). To shore up this argument NASA pointed to a study it had contracted from the economics research firm Mathematica, Inc. that concluded (on the basis of figures and formulas that had to have been somewhat speculative) that NASA's proposed Space Transportation System would be an economical solution to the high cost of space launches.[35] This notion assumed, however, a flight rate of "between 300 and 360 Shuttle flights in the 1979–1990 period, or about 25 to 30 Space Shuttle flights per year."[36]

Low and his successors were unable to deliver on their promise. The most critical technologies in what would become the "STS," or Space Transportation System—the Shuttle main engine and the Shuttle's ultra-high temperature resistant tile skin—had to be developed *de novo*. The engine and tile skin, as well as the entire STS, required such complex and sophisticated development, testing, and manufacture that the feasibility of their multiple use, standardized production was nil—even if the wildly optimistic flight rate projections of Mathematica had been realized, which has yet to be the case. Whether in politics, social arrangements, economics, or engineering, the productive dynamism of open systems is predicated on a critical mass of players (e.g., citizens, competitors) or standardized components functioning in analogous ways.

President Nixon announced his decision to proceed with the Shuttle program on January 5, 1972. Continuing to promote the Shuttle as a cost-saving alternative to expendable launch vehicles, NASA was able to obtain and maintain congressional authorization for sufficient funding to undertake the program, authorizations that continued even as cost overruns began to mount and completion dates were postponed. Nixon's successor, following the brief term of Nixon's vice president, Gerald R. Ford, after Nixon's resignation, was Jimmy Carter. A Democrat trained in physics, Carter supported space science, space applications such as environmental Earth systems monitoring, and cooperative activities with international partners. But the Carter administration issued no ringing summons for further U.S. human *in situ* exploration of space. Instead, space activities beyond space science, applications, and technology would be pursued in order to advance "United States' domestic and foreign policy objectives."[37] An analysis of congressional votes for and against Shuttle funding from 1971 to 1983 found that whether or not a congressman's district contained a primary Shuttle contract—not party membership or ideology—was the single most important variable in predicting a pro-Space Shuttle funding vote. Committee members tended to vote in a block for the program. Only after the Shuttle's initial flights during 1981–1985 did congressional support waver, a reflection of the shallowness of any general public commitment to a federally financed and managed human space exploration.[38]

The arrival in the White House in January 1981 of Ronald Reagan was a promising turn of events for NASA. Reagan was probably the most enthusiastic about space of any president since Lyndon Johnson. The time was propitious for the agency to attempt to persuade the White House to announce the "logical next step" after the Shuttle program, the building of a permanently occupied manned orbiting space station. Throughout its twenty-five-year history prior to 1983 NASA engineers had been working on various designs to ensure the agency would be ready when and if political approval for an Earth-orbiting space station came—indeed, the Shuttle had originally been proposed in tandem with a space station, which it would help to build as the principal means of transportation to and from orbit. Following what had become a familiar pattern, NASA had already issued contracts to its leading contractors for studies of "Space Station Needs, Attributes, and Architectural Options." The study reports included well-crafted presentations on the various needs a space station could serve—in other words, provided NASA with help in developing justifications for its next human space flight project.[39]

Space station advocates' hope was realized when Reagan in January 1984 announced that the United States would build "Space Station Freedom." (During the Clinton administration the space station was renamed the "International Space Station," to acknowledge that the European Space Agency, Canada, Japan, and Russia would participate in building and operating it.) But Reagan's announcement turned out not to be the good omen human space flight enthusiasts hoped it would be. Three circumstances complicated the nature of White House support for a NASA-managed human space flight program: (1) the Republican party's renewed commitment under Reagan to reducing taxes and the size of government by replacing non-defense government enterprises with commercial market-based activities, (2) the Space Shuttle *Challenger* accident in 1986, and (3) the collapse of the Soviet Union and the Iron Curtain during 1989–1990. Moreover, while President Reagan demonstrated his interest in human space exploration with his endorsement of NASA's space station program, it was the possibility of the development of a space-based anti-ballistic missile system—or Strategic Defense Initiative (SDI) referred to by its critics as "Star Wars"—that most excited him. As a result, 1982 marked the first year that the Department of Defense's space budget began to overtake NASA's, becoming roughly twice as large in as little as three years.[40]

Meanwhile, the Reagan administration and its allies in the Congress were able to institutionalize "commercialization" as federal policy to the extent that it would remain in place even after the return of the Democrats to the White House in 1993. Commercialization consisted of the application to publicly funded programs of the open systems ideology of free-market economics. Federal agencies were directed by the Office of Management and Budget to obtain all products and services they required that could not be considered "inherently governmental" (that is, program oversight and executive decision making) from the private sector.[41] If the country truly supported space exploration, as NASA and aerospace industry polls attempted to show, the government should step aside and allow a commercial space industry to emerge and flourish.

In May of 1983 the National Security Council issued a directive announcing the government's policy to promote the growth of a viable expendable launch vehicle industry, which it would do principally by encouraging "free market competition among the various systems and concepts within the U.S. private sector" and providing "equitable treatment for all commercial launch operators for the sale or lease of government equipment and facilities consistent with its economic, foreign policy, and national security interests." In late 1984 Congress passed the Commercial Space Launch Act, committing the U.S. government to encouraging the growth of a launch vehicle and launch services industry, and authorizing the president to establish an office in the Department of Transportation to license and otherwise oversee and promote the development of a commercial expendable launch vehicle industry. Language was added to the Space Act directing the agency to "encourage to the maximum extent possible the fullest commercial use of space."[42] Thus NASA was reminded that since it had the Space Shuttle—which it had sold as an all-purpose vehicle for access to space—there was no longer any justification for the agency to build or fly expendable launch vehicles. Any further ambitious human space flight program(s), including the "logical next step" of a manned orbiting space station, would be tied, for good or ill, to the fortunes of the Space Shuttle.

Then, in the morning of January 28, 1986, the Space Shuttle *Challenger* exploded shortly after lift-off. It was carrying, in addition to its crew of seven, all of whom perished, a tracking and data satellite/inertial upper stage, several science payloads associated with the visit to our solar system of the Comet Halley, and student/teacher payloads accompanying the first "teacher in space," Christa McAuliffe. It was also carrying a commercial payload, a fluid dynamics experiment for Hughes Aircraft Company. Setting aside the predictable post-accident vows to honor the lost astronauts with a renewed commitment to human space exploration, NASA's manned spaceflight program would never afterwards command the political support in U.S. space policy that it had enjoyed in the 1960s.

A commission headed by former secretary of state William P. Rogers issued its report on the *Challenger* accident, attributing it largely to NASA management failures.[43] And yet within a few months Reagan, a former governor of California, authorized NASA to build a replacement for the lost *Challenger*. The reason was not hard to fathom. The Shuttle was built in California by Rockwell International, NASA's top contractor during 1979–1988 and already a recipient of $15 billion in NASA's procurement dollars from 1973 (when the STS program was approved) to 1986. This decision also ensured continuation of the major program managed by NASA's Johnson Space Center in Houston, Texas—home state of Reagan's vice president, George H. W. Bush.

At the same time, Reagan announced that henceforth the market place would determine the extent and cost of launching payloads into space—all payloads, that is, that did not require astronaut tending. The August 1986 White House statement read that NASA "will no longer be in the business of launching private satellites. . . . Free enterprise corporations will become a highly competitive method of launching commercial satellites and doing those things which do not require a manned presence in space."[44] Reagan subsequently signed a new national security directive prohibiting NASA from launching commercial or foreign payloads on the Shuttle unless those payloads had unique requirements that could only by met by a Shuttle launch, and it reiterated that NASA was not to maintain or use its own expendable launch vehicles (ELVs), contracting instead to the industry when it wanted an ELV launch.

This change in policy was a deft maneuver best characterized as "having it both ways." The most costly portion of the civil space program—manned space flight—would continue to be financed and managed by the federal government, while everything else would be turned over to the private sector. The policy of "space commercialization" was continued by the Clinton administration, which issued in 1994 a "National Space Transportation Policy," declaring that the Space Shuttle "will be used only for missions that require human presence or other unique shuttle capabilities, or where use of the shuttle is determined to be important for national security, foreign policy or other compelling purposes."

Reagan's successor, George H. W. Bush, like Kennedy, decided to keep his vice president occupied with responsibility for White House space policy. By March of 1992, a presidential election year, Vice President Dan Quayle was able to announce a new policy directing NASA to begin a "Space Exploration Initiative," the goal of which was to "return to the moon—this time to stay—and human expeditions to Mars."[45] This election-year announcement was undoubtedly most welcome to the people of Texas and California, which together had eighty-five representatives in the House and over a quarter of the electoral votes (seventy-six) necessary to elect a president in 1984 and 1988. (The gambit would be repeated by President George W. Bush, who successfully ran for reelection during the 2004 presidential election year.)

The elder President Bush was not reelected, however, and a presidential summons for a return to the moon and a human mission to Mars was not repeated during the Clinton administration. As for the commercialization of hitherto federally financed and managed space expeditions: Would this ideologically simple notion—sure to appeal to a nation schooled in the economics of Adam Smith—suffice to maintain the complex of aerospace industry contractors that had grown up around NASA's "manned space flight" programs? Reagan's 1986 decision to replace the lost Challenger suggested not.

The transfer of the White House from the Reagan and Bush administrations to the Clinton administration in early 1993 meant that NASA continued to struggle with budget austerity. Space was not a Clinton administration priority, and NASA's survival would depend on its ability to hew to the new administration's priorities. The Clinton administration was like the Kennedy administration in that Vice President Al Gore, like Lyndon Johnson, had served as chair of the Senate committee overseeing the civil space program, and Clinton, like Kennedy, gladly handed the space portfolio to his vice president. In addition to a background and interest in space policy, Gore was especially interested in environmental issues such as global climate change—issues that could benefit from space-based monitoring of the Earth's climate, atmosphere, and natural resources. And, as happens to every presidential administration, events in remote and foreign places forced the White House to take an interest in international affairs.

Gore was eager to support space-based Earth observations to advance his own political future as a champion of environmental protection. Second, he could create and burnish his foreign relations image by brokering a U.S.-Russian cooperative space regime centered on Space Station Freedom (renamed the International Space Station, or ISS). Russian participation would provide employment under U.S. eyes for hundreds of former Soviet aerospace engineers and space scientists, who might otherwise become employees of countries unfriendly to the United States. The European Space Agency, Canada, and Japan joined the U.S. as partners in the ISS program, thus further enlarging the circle of those with a politically significant interest in the ISS's survival. It turned out to be an effective strategy, as the ISS program has survived, notwithstanding numerous "downsizings" and cost overruns.[46] However, since the loss of the Shuttle *Columbia* in February 2003, the U.S. has had to rely on its Russian ISS partner to help ferry crew and supplies back and forth to the space station.

Meanwhile, promises to shrink the federal government made by candidates from both major parties would have some unadvertised consequences. While Democrats have emphasized improving government efficiency as a way to maintain or increase public services with fewer employees, Republicans have emphasized "contracting out," "outsourcing," "commercialization," or "privatization" as ways to reduce civil servants without reducing federal services. But for both parties reducing the aggregate numbers of federal civil servants has been a common cause. No federal agency has been immune, NASA least of all. During the two Clinton administrations (1992–2000) the number of civil servants employed by NASA declined steadily from its high point during the Apollo program—36,000 in 1967—to half that number today. Did the agency's programs then shrink by half? Not at all; NASA procured more of its work from contractors, whereupon the numbers that mattered politically were the numbers of jobs created (or preserved) in various congressional districts, rather than the number of civil servants employed by NASA.

In 1995, arguing that the agency's internal resources were better spent on genuine research and development, NASA declared its space and Shuttle operations "routine," and thus more appropriately performed by contractors.[47] Accordingly, in the space of two years NASA awarded billions of dollars worth of contracts to Boeing North American, Inc. Space Systems Division (located in Houston, Tex.), Lockheed Martin Space Mission Systems and Services, Inc. (located in Houston, Tex.), Rockwell Space Operations Company (located in Houston, Tex.), and Lockheed Space Operations Company (located at Kennedy Space Center in Florida) to consolidate and carry out (1) NASA's satellite communications, command and control, and operate deep space tracking systems, and (2) all Space Shuttle operations.

NASA's two principal Shuttle operations contractors—Lockheed and Rockwell—soon created United Space Alliance ("USA"), a new joint organization solely for Shuttle operations. NASA's own senior-level responsibility for the Shuttle program also shifted from NASA Headquarters in Washington, D.C. to Johnson Space Center, in Houston, Texas. NASA referred to this change in Shuttle management policy as privatization. But critics charged that this transfer to aerospace industry contractors of responsibilities for both safe and cost-effective Shuttle operations would come at the expense of Shuttle safety—a concern that seemed well-founded in the aftermath of the loss of the Shuttle *Columbia.*[48]

National economic policy that sought to transfer as many government functions as possible to the private sector was not designed initially as a form of demobilization for the federally financed and operated human space program. But to the extent that commercialization has placed the future of that program in the hands of private firms, rather than government agencies, that is the direction in which the policy is tending—but only if public dollars are replaced by private investors. This, however, is unlikely. At the end of the century the U.S. aerospace industry was receiving 63 percent of federal R&D funding, or 30 percent more in federal procurements than the next best funded industry (see Table 7.1).

In 1981 the Council on Economic Priorities published a study of the politics of U.S. defense and NASA (that is, aerospace) contracting that concluded "a powerful flow of people and money moves between the defense contractors, the Executive branch [DoD and NASA], and Congress, creating an 'iron triangle' on defense policy and procurement that excludes outsiders and alternative perspectives."[49] Campaign contributions, seats on federal advisory committees, and a "revolving door" through which executives passed back and forth between positions in industry firms and DoD and NASA—all ensured the perpetuation of programs to be "outsourced" to the aerospace industry.

Federal election and government integrity laws intended to weaken the grip of the iron triangle on defense and space (as well as other) Washington policy choices had only marginal effect. In FY 2002—a typical recent year—including their respective shares of the $1.79 billion USA contract, components of Lockheed Martin Corporation held $1.32 billion in NASA contracts, while components of the Boeing Co. had won $1.43 billion in NASA contracts. Thus nearly

TABLE 7.1

Industrial R&D Receiving Federal Funding
(v. Private Funding) in Excess of 10% in 1999

Industry	Percent R&D Federally Funded
Aerospace products and parts	63.2%
Navigational, measuring, electro-medical, and control instruments	39.8%
Architectural, engineering and related services	32.9%
Transportation equipment	29.6%
Scientific R&D services	29.2%
Professional, scientific, technical services	24.3%
Computer and electronic products	16.7%
Utilities	12.0%

* Source: National Science Board, *Science and Engineering Indicators, 2002,* Vol. 1 (Washington, DC: National Science Board, 2002), Text Table 4-7.

half of the $9.57 billion NASA awarded in contracts to business firms in FY 2002 went to Lockheed Martin, Boeing, and Rockwell corporations for manned space flight research, development, and operations.

At the same time that NASA participated in the concentration of firms in the aerospace industry, Webb's open systems approach to NASA procurement ensured a broadly scattered accumulation of economic—and thus political—interest in the preservation of NASA's programs and associated contract dollars. In 2003 NASA could boast that fifty states and the District of Columbia received NASA contracts and/or grants worth more than $25,000; these awards went to 3,422 different organizations in 1,083 different cities, ensuring a broad dispersion of NASA dollars throughout hundreds of congressional districts. At the same time, 80 percent of the total $11.2 billion in that year's procurement awards went to the nine states and District of Columbia where NASA installations are located; 50 percent went to California and Texas.[50] The open systems principle of federal procurement policy—that awards should be made as a result of "full and open competition"—failed to sustain a full and open market place of competing aerospace firms, but did ensure a critical mass of political support in congress.

Concentration in the defense-aerospace industry has only increased since the 1980s, with the Lockheed Corporation—both DOD's and NASA's top contractor—receiving almost $22 billion from defense contracts alone in 2003. Because R&D "know how" is cumulative, if not always progressive, early winners acquired a cumulative advantage in successive competitions. Only part of that advantage is greater experience in conducting and managing high-technology programs; what remains is attributable to the personal contacts, the "revolving door" of executives back and forth between private and federal jobs, political balance sheets, and the informal channels of influence that characterize the "iron triangle."

U.S. Space Policy at the Dawn of the Twenty-first Century

In the early days of commercialization policy during the Reagan administration, proponents had assumed that allowing the market to provide all necessary space products and services (including launches) would ensure greater efficiency and sufficient availability of routine and low-cost access to space. A thriving commercial launch and space industry would satisfy both private sector and government needs for space transportation. But the market where launch vehicle manufacturers might find profits sufficient to develop new launch vehicles (ones representing significant technological advances over the current ICBM-based U.S. launchers) has become a complex place.[51] It is an international market for launch services, one in which not only the U.S. but France and Europe, with their Ariane system, Japan, China, Russia, and the Ukraine are all competitors. And in each case some mixture of government and private investment has been essential for survival.

The market place for space launch vehicles has proven to be a very uncertain one, for it depends upon the market for space-based products and services. Unfortunately, the hyperbole that inflated the technology investment sector in the 1990s had wafted through the space products and services industry as well. Thus the optimistic forecasts of launch needs of the 1980s became the excess capacity of the late 1990s. While communications satellites have been the basis of roughly 60 percent of the commercial space launch market, fiber optic and other land based systems may, in the end, prove less costly, more reliable, and entail lower risk. Similarly, although space-based remote sensing can provide continuous coverage over large areas, airborne remote sensing can be less costly and just as useful for local or regional observational needs.[52]

In August of 1999 two of the three principal new mobile satellite telephone companies—Iridium L.L.C. and ICO Global Communications Ltd.—found themselves in bankruptcy court, leaving the third, Globalstar Telecommunications, owned by Loral Space and Communications, to serve as the bellwether for the industry. By the summer of 2003, Loral too had filed for bankruptcy and announced plans to sell its fleet of communication satellites to the international consortium, Intelsat. Meanwhile, Boeing announced losses in excess of $1 billion from its satellite and rocket-launching business. As the aerospace giant launched its new Delta IV rocket (competing with Lockheed Martin's Atlas V) in late 2002, it had to concede that its anticipated launch rate of fifteen to twenty annually had dwindled to four to six. Meanwhile, the European consortium Arianespace was facing deficits that threatened bankruptcy, leading European Space Agency members to consider a policy requiring all European satellites built with public funds to be lunched on Ariane launch vehicles.

The only portion of the U.S. civil space program that has seemed generally effective and untroubled is NASA's space science program, conducted largely by robotic vehicles and devices and remotely controlled instruments. In August 2003 NASA launched the last of its great Earth-orbiting space observatories, the Space Infrared Telescope Facility, or SIRTF. Designed to peer back to the earliest moments of cosmic time, SIRTF (renamed the Spitzer Space Telescope) joined the Hubble Space Telescope (launched 1990), Compton Gamma Ray Observatory (launched 1991), and the Chandra X-Ray Observatory (launched 1999) to bring ever greater human illumination to the darkness of the night sky.[53] A few months later, late in the evening of January 3, 2004, excited NASA-funded scientists at the Jet Propulsion Laboratory received the first signals from "Spirit," a roving robot explorer that had landed successfully on Mars to spend the next three months searching for evidence of liquid water on the planet during its history. An identical companion craft, "Opportunity," landed a few weeks later to serve as a back-up.

At the urging of President George W. Bush's national security advisor Condoleeza Rice, the White House in early 2002 launched a comprehensive review of national space policy under the joint leadership of the National Security Council and the White House Office of Science and Technology Policy.[54] The suite of issues before the executive branch committee that undertook the review was impressive, reflecting the complexity of policy choices created by the now twenty-year-old effort to commercialize space activity. For example, now that the federal government must rely on commercial firms for its remote imagery needs, what are the best options for ensuring the availability, reliability, and security of this critical data? How can the United States remain at least a generation ahead of other countries in remote imaging technology, while allowing non-U.S. customers to purchase products containing that technology as well? What manner of coordinated commercial "sponsorship" of the ISS program—e.g., advertising, entertainment, tourism—might all of the international ISS partners be willing to accept? (Russia already ventured down the tourism path with its sale to U.S. businessman Dennis A. Tito of a multi-billion dollar seat on a Soyuz rocket trip to the ISS.) In managing radio spectrum allocations, how should the federal government accommodate the growing appetite of the new broadband communications industry and third generation cell-phone transmissions without adversely impacting government (for example, national security) operations? How should—or perhaps how could—the United States respond to the growing determination, and apparent ability, of an economically competitive China to become a major space power? What measures (if any) should the federal government use to improve the global competitiveness of the U.S. expendable launch vehicle industry? And, a not unrelated question—just how critical was "assured access to space" to the U.S. government? The fact that this last question was even being asked marked these new policy deliberations in historical time.

In 1996 the Ansari family of Dallas, Texas—following the example of the $25,000 Orteig Prize that challenged Charles Lindbergh to make his successful solo trans-Atlantic flight in 1927—offered a $10 million "X Prize" to anyone who could fly into space (generally accepted to begin at sixty-two miles high) a pilot and two passengers or their equivalent weight, twice within two weeks. With a funding assist of over $20 million from Microsoft co-founder Paul Allen, a team led by Burt Rutan (designer and pilot of the first aircraft to complete a non-stop circumnavigation of the world) won the prize in October of 2004 with an airplane they designed, built, and successfully flew, once the plane had been lifted nine miles up by a larger airplane (which is analogous to the Shuttle's dependence on booster rockets). The impressive feat appeared to signal the beginnings of a new era of space tourism, as the founder of Virgin Airlines, Richard Branson, announced his intension to launch a space tourism company based on Rutan's and Allen's victory. Given the clear indication that the commercial sector is likely to be able on its own to generate and sustain a human space travel industry, might the nation's publicly financed manned space ventures be retired? After all, a robotic program of space exploration had repeatedly been shown to be effective at a fraction of the cost. This was not to be, however.

During October 2004 the Congress considered a FY 2005 federal budget proposal from the White House containing a significant budget increase (from $15.5 billion to $16.2 billion, or 4.1 percent of the FY 2005 federal budget) for NASA. The increase would fund President Bush's vision of astronauts returning to the moon and landing on Mars, which he announced at the beginning of his 2004 successful reelection bid. House Majority Leader Republican Tom DeLay, whose Texas home district includes NASA's Johnson Space Center, ensured that the NASA increase was included in the FY 2005 budget without debate. To a cheering Johnson Space Center audience DeLay explained, "NASA helps Americans fulfill the dreams of the human heart," echoing the same transcending justification for sending humans into space embedded into U.S. space policy by the National Security Council in 1958, the summer that NASA was created.[55]

THE PROCLIVITY TOWARD open systems characteristic of U.S. policy, law, public administration, science, and engineering has shaped every important facet of our past and current civil space program. The Eisenhower administration's decision to lodge those space activities not dedicated to national security in a civilian agency was a deliberate effort to prevent their captivity by the military-industrial complex. Public dissemination of information about NASA's activities was and has remained a statutory requirement, as well as an organizational imperative. An "open" space program was matched by the "open skies" policy shrewdly designed to ensure (as it did) lawful freedom of movement for U.S. observation satellites over the entire globe. The policy had a no less important second function: It enabled the U.S. to compete with the Russians in space without engaging in provocative gestures that might ignite Cold War hostilities.

In one important respect however, Eisenhower did not prevail in the early design of the civil space program. His insistence on specific military or scientific objectives for federally financed space activities was defeated by the open-ended nature of the space policy goals that were articulated by the National Security Council in the 1950s. The amorphous quality of these goals, with their vague promise of a quasi-mystical transcendence (to persist well into the twenty-first century with Tom Delay's "NASA helps Americans fulfill the dreams of the human heart"), would remain a constant feature of justifications for the continuation of taxpayer financed human space flight adventures after the moon landing.

Meanwhile, NASA's Apollo program administrator, James Webb, was no less shrewd than Eisenhower when Webb understood that federal procurement policy, with its principles of openness in administrative and selection procedures (see chapter 4), could be used to create in congressional districts across the country an economic and political interest in the preservation of NASA. Because human space flight programs have required the highest funding levels in NASA's budgets, the size of those programs' procurements have been larger, and the degree of economic and political interest vested in them have been correspondingly large as well. Had Webb's objective been otherwise, that is, to favor a handful of powerful defense/aerospace contractors, the procurement process would have been an obstacle, rather than a convenient and legitimate tool.

Ironically, however, NASA's human space flight program contributed to concentration in the aerospace industry that would mirror the concentration of firms in the defense industry. Both NASA and the Department of Defense, consistent with bipartisan commercialization policy (itself an effort to impose on the government the discipline of the open-system free market of Adam Smith), relied on industrial contractors for much of their technical systems research, development and testing, and production. Because of the nature of the work these contractors did, they benefited more than ordinary contractors from the cumulative advantages of incumbency.

Competition for procurements for relatively undifferentiated services such as housekeeping and facilities management are likely to be won on the basis of effectiveness and cost, qualities that small or new firms can offer as well as older or larger firms. In the case of competition for R&D procurements, however, early winners acquire a cumulative degree of technical know-how and program management experience that new, less experienced contenders cannot match. Along with that cumulative technological and management advantage comes increased familiarity with the federal officials and political insiders who define and lobby for the projects to be contracted out. Thus what begins, in principle, as an inclusive open system—federal R&D procurement—contains within itself the inexorable tendency toward exclusion. The result is federally promoted industrial concentration.

As we have seen, this concentration of economic power in the aerospace industry has been accompanied by a concentration of political power over civil space policy decisions. (That power also extends over defense policy decisions.) One of the most critical years for NASA may have been 1974, when the Federal Election Campaign Act (FECA) of 1974 lifted its predecessor's (FECA of 1971) ban on government contractors making campaign contributions. Opposition to the ban had originally come from trade and labor unions, whose members might be contractors. Today political action committees affiliated with such major defense/aerospace contractors as the Lockheed Corporation are regular contributors to federal political campaigns, thus adding to a member of Congress's legitimate concern for the economic well-being of his or her constituents, a no less compelling concern for the well-being of his or her campaign purse.[56]

The question remains, to what extent, if any, have NASA's science programs been shaped by similar forces? Most of NASA's science (and education) programs are outsourced as well, but agency funding for these purposes flows largely into universities and non-profit research institutes, and it flows more typically in the form of grants, which fund research and educational activities rather than purchase products or services. To be sure, a successful scientific career in academia constitutes as much of a material interest in funding for one's research as an outright financial interest, but the amount of funding involved in any given grant is substantially smaller than in manned space systems' R&D.

NASA's practices in funding space science also tend to mimic the research grant competitions used elsewhere in the federal government (for example, the National Science Foundation), which aspire to be based on merit and normally rely on peer review. This, and NASA's heavy use of both its own advisory committees composed of external scientists in designing its programs and the National Academies' space science and aerospace engineering committees, would suggest a counter-tendency to the concentration apparent in industry procurements for other NASA programs. However, a trend toward concentration is apparent here as well, recalling Robert Merton's "Matthew Effect" (see chapter 3).

The two largest NASA awards to educational institutions in 2003, California Institute of Technology ($260 million) and The Johns Hopkins University ($187 million), funded the operation of the NASA's Jet Propulsion Laboratory and Space Telescope Institute, respectively. Together these awards represent 25 percent of the $1.8 billion NASA awarded that year to more than 100 educational and non-profit institutions. Setting aside these two large awards in NASA's space science and education programs, as well as the agency's $1.7 billion contract with United Space Alliance to operate the Shuttle (or 18 percent of the agency's $9.5 billion awards to over 100 business firms), the next ten recipients among educational and non-profit institutions received 28 percent, while the next ten recipients among industry contractors received 41 percent. Thus, in each category of awards, the top ten recipients out of more than 100 contractors and grantees, along with the recipients of large institutional management contracts, received more than 50 percent of the total awards in their category.[57] The cumulative advantages of incumbency operate in NASA funding for the space sciences and education, just as they do in NASA's space R&D—thus defeating the broad distributive intent of "full and open competition" in U.S. procurement policy.

8

The Crisis in American
Health Care

Our dependence on the private automobile, with its complex technological network of industrial, energy, and infrastructure systems, has generated great demands on our capacity for public policy that effectively coordinates the desires of individuals with the shared desires of the entire society. Yet the content of the national policy choices posed by the ubiquitous private automobile are relatively accessible to most informed citizens. For example, few people question that automobile engines can and must burn cleaner and more efficiently; debates occur over how fast and by what policy mechanisms (including reliance on the market place) that objective can be met. Or, citizens may dispute whether communities should build more roads, and whether roads should be paid for with taxes or with highway tolls, or whether improved public transportation would reduce our reliance on personal vehicles.[1] However one argues these issues, as a society we need not and do not defer to automotive engineers to make such policy choices for us.

By contrast, when we discuss the many interconnected policy issues posed by our probable need for medical attention at some point in our lives, we habitually defer to the authority of experts—e.g., physicians, radiologists, pharmacists, biologists, bioethicists, medical researchers, hospital administrators, epidemiologists, and health economists—while we debate largely over how to manage the ever increasing costs of health care. Aside from acknowledged administrative inefficiencies, the widely perceived high quality of American medicine, due to its scientific and technological components, is the generally accepted trade-off for escalating costs.

Unfortunately the more we defer to scientific and technical expertise for authority over our public policy choices, the less open our political culture becomes. A political culture that is relatively immune from the influence of ordinary citizens becomes an eventual danger to science as well, for science is as much a creature of society as our politics. Similarly, insofar as access to health care is restricted, and opportunities for informed choices among consumers and patients reduced, health care takes on the characteristics of closed systems: the dominance of received wisdom discourages curiosity, criticism, and the spread of knowledge and understanding—with its promise of improved well-being—throughout our society.

This reliance on a complex network of experts in making health care policies mirrors the extent to which all that is entailed by "modern medicine" has become arguably the single largest and most complex research-based technological system in the United States today.[2] Thus no survey of today's science and technology policy choices can fail to examine the issues posed by the various aspects of the way we as a nation deliver and receive medical care. At the same time the subject is so nearly overwhelming in its complexity that no single chapter in a single book can do it justice. Instead, this chapter will examine a selection of issues which, taken together, illustrate the depth and range of considerations that challenge the political boundaries of biomedical expertise.

We begin with observations on the role of scientific and technical expertise in mediating contemporary policy issues arising from biotechnology-based medicine. We examine as well the politics of the United States' increasingly costly health care system and the challenges of health care reform, finding that the single most stubborn failing of U.S. health care policy is the extent to which it tolerates a largely closed system. Patient access to the system is rationed on an ad hoc basis of socioeconomic factors, while widespread reliance on medical experts for information that remains largely opaque deprives the system's users—patients and consumers—adequate knowledge to make informed decisions about their own care. We then consider how well the federal regulatory framework that oversees the pharmaceutical industry ensures public benefit and safety from pharmaceuticals—a critical policy issue today, for this framework will determine who actually benefits (and how) from the promised wonders of sub-cellular biological research.

Expert Authority

As we saw in chapter 3, the notion that science properly pursued is the disinterested discovery of objective truth—with its corollary that science cannot posit or resolve questions of right and wrong, or good and evil—lay at the heart of the establishment of federal patronage of basic science after World War II. But the ideal of "pure" science as a pursuit that could isolate itself from public moral concerns had already been swallowed up in the moral wreckage of World War II. The dropping of atomic bombs by the U.S. over Hiroshima and Nagasaki brought an end to the moral innocence of modern physical scientists, but medical researchers had to confront their complicity in Nazi war crimes.[3] Three weeks after the end of the war in the Pacific an international military tribunal in the German town of Nuremberg convicted twenty-two of twenty-four defendants for one or more of four war crimes:[4] (1) military aggression in violation of international agreements, (2) crimes against humanity, (3) war crimes, or acts in violation of the laws of war, and (4) conspiracy to commit the previous three crimes.

Among the disclosures in the courtroom were the accounts of medical experiments performed by German physicians on women, Jews, Gypsies, prisoners of war, children, elderly, mentally retarded, and non-Germans—all categories of persons experiencing either some form of incarceration or social isolation. These experiments, conducted by an estimated 200 doctors, were known to their physician colleagues—none of whom is recorded as having overtly protested. The defendants justified their activities as required by science, inasmuch as the eastward movement of Allied forces in late 1944 threatened the continuing supply of "test persons" needed for their medical experiments.[5]

Lest the German medical profession be singled out as the only physicians capable of inhumane experiments in the name of science, the Japanese army began human experiments in bacterial warfare research in 1936 using thousands of unwilling human subjects in Japanese-occupied Ping fan, Northern Manchuria. Like their counterparts in Germany, the hundreds of Japanese medical experimenters sought to identify differential reactions to morbid conditions among various races. However, Japan's medical experimenters fared better at the hands of the Americans than did the German doctors at Nuremburg. In an unprincipled quest for research data on biological weapons, the U.S. authorities declined to prosecute them in exchange for their research findings.[6]

The German researchers had abandoned the physician's first moral duty—to do no harm—in the interests of scientific research. That they had done so was not lost on the doctors assisting the prosecution during the trial. After having heard defendants argue that no legal code distinguished between lawful and unlawful human experiments, Dr. Leo Alexander in 1947 drafted a memorandum detailing ten points of permissible medical experiments. What would become known as the "Nuremberg Code" served as the basis for ethical codes developed by medical societies during the next two decades. Its key principles are familiar today: human research subjects must provide voluntary and informed consent; every effort should be made by experimenters to avoid unnecessary physical and mental suffering and injury or death; the human subject should be free at any time to terminate an experiment if indicated by his physical or mental state; and the scientist-researcher also must be prepared to terminate an experiment that appears "likely to result in injury, disability, or death to the experimental subject."[7]

The Nuremberg Code represents the first effort within a modern research community to grapple with the moral ramifications of scientific research when it intrudes upon the physical and mental integrity of individual persons. In so doing, it acknowledges that there *is* a moral dimension that must be reckoned with when the aspirations of science can only be met by viewing individual humans instrumentally as test subjects.

Throughout the next two decades an international body of physicians led the preparation of codes detailing the ethical treatment of human research subjects not only in their own countries, but to serve as the needed international guidance alluded to at Nuremberg. The result of their efforts was the Declaration of Helsinki, adopted during the 1964 Helsinki (Finland) meeting of the World Medical Assembly (first convened in 1947), which has been reviewed and periodically revised since 1964. Helsinki's guidance is structured around a distinction between clinical research and non-clinical (or non-therapeutic) research, a distinction that reflects the priority it gives to the physician's prior moral responsibility: "The health of my patient will be my first consideration."[8] While the Helsinki Declaration includes a provision for "informed consent" from research subjects, this provision is not preeminent, and the fact of having obtained such consent does not absolve the researcher of absolute moral responsibility for the consequences of the research for the subject: "The responsibility for the human subject must always rest with a medically qualified person and never rest on the subject of the research, *even though the subject has given his or her consent.*" Of no less importance, "*concern for the interests of the subject must always prevail over the interests of science and society.*"[9] There is no delicate balancing of interests here, no broad gray space for philosophical speculation or parsing of relative risk and benefit, individual and societal interests.

The post–World War II decades saw a rapid acceleration in the institutional and social forces that would produce today's large and technology-intensive medical infrastructure in which only about 6 percent of all medical personnel are practicing physicians. Much of this growth was due to the harvesting of medical research that expanded rapidly along with advances in organic chemistry, physiology, and biology. The corresponding shift in the nature of healing from doctors caring for persons to medical staff treating cases, so well described in the late Lewis R. Thomas's autobiographical *The Youngest Science: Notes of a Medicine Watcher*, was both cause and effect of the growth of hospitals as the seats of scientific medicine:[10] "The transformation of the hospital from a poorhouse to the nerve centre and headquarters of the new medicine had profound implications. . . . One consequence was that, by mid century, hospitals were absorbing about two thirds of the resources spent on health care in the United States, and the percentage continued to rise. These hospitals became key centres [sic] of medical research, in the conviction that this would generate health improvements. Medical research and medical education grew inseparable, and bigger, costlier and more prestigious hospitals were their status symbols."[11]

As medical research, medical education, and medical treatment of patients became intermingled in hospitals and clinics, there would emerge an increased risk of blurring the distinction between patients, beneficiaries of medical science, and research subjects. Compounding this gradual replacement of the older model of physician-patient care with one of many cases in a hospital ward was the intellectual deconstruction of the patient's physical person into body parts or physiological functions treated by medical specialists. In time the growing dominance over the treatment of human illness and mortality by medical science and technology provoked the kinds of resistance represented by Elisabeth Kubler-Ross's book, *On Death and Dying*, first published in 1972, which urged compassionate care for the dying, rather than subjecting them to various technologies to artificially prolong life.

The professional codification of ethical practice represented by the Nuremburg Code and Helsinki Declaration, codes that asserted the primacy of physicians' therapeutic obligation to their patients over medical research interests, did not suffice to protect human research subjects in the new era of hospital and research-based medicine. During the summer of 1972, the same year that saw the publication of *On Death and Dying*, *The Washington Star* newspaper and *The New York Times* broke the story of the Tuskegee Syphilis Study.[12] Begun in 1932, the study was a cooperative undertaking between The Tuskegee Institute (a vocational education school in Alabama for Negroes) and the U.S. Public Health Service (USPHS). One account attributes the inspiration behind the study to a reduction in funds to the Tuskegee Institute from one of its principal benefactors, the Rosenwald philanthropy. To demonstrate the need for a syphilis treatment program that would end if funding stopped, the institute and the USPHS enlisted six hundred black men for the study.

The men were led to believe that the study was intended to evaluate various treatments for syphilis. In fact, only 201, or one-third of the men received treatment; the remaining 399 received only clinical evaluations enabling the USPHS to study the natural history of the disease. The 399 black men continued to go untreated, even after 1947 when the USPHS began offering treatment using penicillin at "rapid treatment centers." Uninformed about the study's real purpose and possible consequences to them, the subjects continued to participate, thanks to such inducements as free medicines, hot meals, and transportation on the days of clinic visits. After the United States entered World War II, local selective service officials were alerted to the identity of the men not receiving treatment, lest they be drafted, their condition detected, and treatment given. The study violated Alabama state law, which since 1927 required that all persons with venereal disease obtain treatment until their condition was no longer communicable. However, even as late as 1969 the Public Health Service, with the agreement of local chapters of the American Medical Association (AMA) and the National Medical Association (NMA), reaffirmed the need for the study.

It took four months after wide public disclosure for the study to be stopped (on November 15, 1972), by which time its results were: twenty-eight men dead of syphilis, 100 dead of syphilis-related illnesses, no fewer than forty wives infected, and nineteen children infected since birth. A year later (July 1973) attorneys for the study's victims filed a $1.8 billion class action suit against the PHS and the Tuskegee Institute, which was settled out of court, each survivor receiving $37,500 in damages and heirs of those who had died receiving $15,000. Much can be said about the deplorable factors of race and socioeconomic status implicated in the Tuskegee syphilis study. Beyond this, the episode illustrates the impotence of professional ethical codes, absent substantial enforceable sanctions, when confronted by the power of institutional interests—in this case the Tuskegee Institute's and local USPHS's desire to maintain its funding, and the AMA's and NMA's unwillingness to oppose the interests of their local affiliates.

Meanwhile, as federal research funding for the USPHS and National Institutes of Health increased eightfold from $68 million in 1955 to $523 million in 1965, allegations mounted of health care abuse during medical research. Elderly people, children, and prison inmates reportedly were being used as test subjects for studies of novel therapies or to enable researchers to examine the natural etiology of various diseases. In 1966 the Senate Committee on Labor and Human Resources began to hold hearings on the issue. The hearings led to passage of the National Research Act of 1974, which required the NIH to codify the policy for the protection of human research subjects that it had issued intramurally in 1966. The result was an expansion of NIH's internal regulations in 1974 into Title 45, Part 46 of the Code of Federal Regulations, "Protection of Human Subjects" (46 CFR 45). These regulations, periodically amended but substantially unchanged since 1974, require that any research using federal funds and involving human subjects proceed only after a research protocol has been reviewed "taking into account the rights and welfare of the subjects involved, the appropriateness of the methods used to secure informed consent, and the risks and potential benefits of the research . . . consent [must] be documented and signed by subjects or their representatives."[13]

These three elements—review for appropriateness of research planned, a comparative assessment of risks and potential benefits, and informed consent— have remained the three fundamental components of U.S. policy for federally sponsored research involving human subjects. In 1991 applicability of the policy was extended from the Department of Health and Human Services (established in 1981 and incorporating the NIH) to cover all federally funded research.

The National Research Act of 1974 led not only to the codification of ethical principles governing medical research with human subjects funded by the federal government, but it reinforced the emergence of yet a new form of expertise—expertise in examining various cultural, legal, sociological, philosophical, and medical aspects of ethical issues posed by medical research and practice. This expertise found one of its first institutional homes in the United States at the Hastings Institute for Society, Ethics and the Life Sciences (now simply The Hastings Center), established in 1969 at Hastings-on-the Hudson in New York. While early gatherings at the Hastings center were truly interdisciplinary—that is, no one was a certified expert in the ethical aspects of biomedicine—by the time Congress in 1974 created the National Commission for the Protection of Human Subjects of Biomedical and Behavioral Research, NIH was able to appoint as one of the commissioners an academically trained "bioethicist."[14]

The extent to which this new form of expertise might replace the moral instincts of ordinary citizens and their physicians was reflected in the contrast between the commission's 1979 report, named "The Belmont Report" (for the historic mansion in rural Maryland where the commission had met) and the Helsinki Declaration's principles.[15] Whereas the Helsinki statement of basic principles is crisp and unambiguously referenced to actual clinical decisions, Belmont's discussion of both principles and applications reads like a distillation of several years' academic discussion among the eleven commission members—seven of whom were affiliated with academic institutions, two of whom were physician-administrators of a hospital and a medical school, a practicing attorney, and one individual doing double duty as distinguished citizen and member of a minority, the president of the National Council of Negro Women, Inc. (NCNW).[16]

The three fundamental principles in the Belmont Report—respect for persons, beneficence, and justice—are abstract and their meaning contingent on interpretation, while the Helsinki Declaration's twelve basic principles are concrete and unambiguous in their location of moral responsibility with "scientifically qualified persons" and the supervision of a "clinically competent medical person." Perhaps the best illustration of the difference in approach of the two landmark documents on the moral aspects of biomedical research is the way each addresses the question of the possible confusion of the responsibilities of physician and researcher. "In the field of biomedical research," instructs the Helsinki Declaration,

> a fundamental distinction must be recognized between medical research in which the aim is essentially diagnostic or therapeutic for a patient, and medical research, the essential object of which is purely scientific and without implying direct diagnostic or therapeutic value to

the person subject to the research. . . . In the purely scientific application of medical research carried out on a human being, it is the duty of the physician to remain the protector of the life and health of that person on whom biomedical research is being carried out. . . . In research on man, *the interest of science and society should never take precedence over considerations related to the wellbeing of the subject* [italics added].[17]

Here is the Belmont Report's discussion of the same issue:

It is important to distinguish between biomedical and behavioral research, on the one hand, and the practice of accepted therapy on the other, in order to know what activities ought to undergo review for the protection of human subjects of research. The distinction between research and practice is blurred, partly because both often occur together (as in research designed to evaluate a therapy), and partly because notable departures from standard practice are often called "experimental," when the terms "experimental" and "research" are not carefully defined.

For the most part, the term "practice" refers to interventions that are designed solely to enhance the well-being of an individual patient or client and that have a reasonable expectation of success. The purpose of medical or behavioral practice is to provide diagnosis, preventive treatment or therapy to particular individuals. By contrast, the term "research" designates an activity designed to test an hypothesis, permit conclusions to be drawn, and thereby to develop or contribute to generalizable knowledge (expressed, for example, in theories, principles, and statements of relationships). Research is usually described in a formal protocol that sets forth an objective and a set of procedures designed to reach that objective.

When a clinician departs in a significant way from standard or accepted practice, the innovation does not, in and of itself, constitute research. The fact that a procedure is "experimental" in the sense of new, untested or different, does not automatically place it in the category of research. Radically new procedures of this description should, however, be made the object of formal research at an early stage, in order to determine whether they are safe and effective. Thus, it is the responsibility of medical practice committees, for example, to insist that a major innovation be incorporated into a formal research project.

Research and practice may be carried on together, when research is designed to evaluate the safety and efficacy of a therapy. This need not cause any confusion regarding whether or not the activity requires review; the general rule is, that if there is any element of research in an activity, that activity should undergo review for the protection of human subjects.[18]

The Belmont discussion wends and winds its way through an opaque commentary on what words might or might not mean, with actual responsibility for ethical decision-making delegated to a committee, an impersonal entity that would—presumably with a greater understanding of the words and the circumstances at hand—ensure the correct moral guidance.[19] The commission's bioethicist approvingly confirmed as much when he reflected, nearly two decades later during a meeting of the Clinton administration's National Bioethics Advisory Commission,

> I believe that a redaction [of the Belmont Report] should encourage the sense that once principles are stated and their applications noted, the discussion only has begun. Ethics of research is a dynamic, casuistic activity. . . . Belmont is an essentially sound proclamation [of ethical considerations in biomedical research]. . . . [However] the written proclamation, what form it takes on paper, must be delivered to a body of responsible interpreters [i.e., an "Ethics Advisory Board"] who can make its words come alive in the particular circumstances of particular protocols, public policy, and the changing research enterprise.[20]

For the remaining decades of the century three ideas enunciated in the Belmont Report would provide the principal content for efforts by various presidential commissions to codify for policy the norms of ethical conduct for federally funded biomedical research involving humans and human biological materials. The first was the principle of "respect for persons," an elastic and socially contingent substitute for the notion of inviolability embodied in the categorical "do no harm" of contemporary international codes of medical ethics.[21] Belmont's "respect for persons" requires that individual autonomy only be "respected": "To show lack of respect for an autonomous agent is to . . . deny an individual the freedom to act on [his or her] . . . *considered* judgments, or to withhold information necessary to make a considered judgment, *when there are no compelling reasons to do so* [italics added]."

Second, Belmont advises that the treatment of individuals be *balanced* against the benefits of research to the "larger society" from "the improvement of knowledge, and from the development of novel medical, psychotherapeutic, and social procedures." The needs and interests of science, in short, are of an equal moral standing as the well-being of individuals. The ethical principle that allows this recalibration of the relative moral standing of individuals and scientific research enunciated in the Nuremberg Code is the principle of "beneficence," which "requires that we protect against risk of harm to subjects, *and also that we be concerned about the loss of the substantial benefits that might be gained from research*" [italics added].[22]

Third, the Belmont Report affirmed the principle that proposed research involving human subjects should be reviewed by an expert committee before proceeding as ethically acceptable, a concept incorporated in the NIH's 1966 internal guidelines and codified in 1974 for all federally sponsored research with human subjects. Such expert committees became formalized as institutional review boards (IRBs).[23] The Nuremberg Code had placed moral responsibility for the proposed experiment squarely on the principal individual or "scientist in charge" of the study. The three decades between 1947 and 1979 saw research-based and technology-intensive medicine, with its expanding institutional interests, evolve into yet another form of specialized expertise shielding itself from the harsh glare of public scrutiny. Of no less political importance was the notion that responsibility for moral or ethical choices can be shifted to committees—a significant departure in a society whose legal system holds only persons, including corporate persons, responsible for violations of civil or criminal law.

Questions about the effectiveness of IRBs and the adequacy of any institution's application of the principle of informed consent are raised continuously. The most fundamental question is the classic "who will guard the guardians?" As creatures of institutions conducting sponsored (externally funded) research, IRBs are likely to embody a bias in favor of the proposed research when attempting to balance its benefits against the possible risks to human research subjects. Moreover, the adequacy of procedures for obtaining voluntary informed consent of the sort envisioned by the notion of rational and informed choice is highly debatable. Those performing the research will (except in rare cases when physicians, researchers, or their peers themselves are the human research subjects) normally have a substantially greater understanding of the possible risks involved in any experimental biomedical procedure than could any candidate research subject. The likelihood that researchers will adequately communicate the same understanding, acquired through professional training and experience, to prospective research subjects while seeking their consent, is small if not negligible. Meanwhile, desperately ill persons, or the parents of desperately ill children, are more likely to agree to any experimental procedures as their last hope.

At the beginning of the twentieth century, as we saw in chapter 5, in federal law and policy making the authority of scientific and technical expertise has remained subject to the prior constitutional obligations of the judiciary and legislature. In the large and complex world of modern medicine, however, that authority has reached beyond the descriptive realm of research questions into the normative realm of moral choices. And, like science itself, these moral choices have become subject to the exquisite calibrations and social process of committee mediation and agreement. Ongoing political controversy over the volume of medical liability litigation and the size of awards to successful plaintiffs is not only a controversy about allegedly rapacious attorneys. It is also a debate about the locus of accountability for unnecessary injuries that occur when individuals turn to medicine for help in overcoming disease and other afflictions.

Feeding the Leviathan

Genuine progress in our understanding of disease and the therapies to ameliorate and cure them, and the fact that U.S. per capita health spending is the highest among member nations of the Organisation for Economic Cooperation and Development (13.9 percent of GDP), has led to a misperception that, as President Bush announced to a presidential debate audience on October 13, 2004, "our health care system is the envy of the world."[24] The fact that our system is also widely characterized as a free market based system, in contrast to single-payer systems common among the other industrialized nations, has distorted debates over reform of health care in the United States, in which individual patient choice is touted as its principal virtue. The reality is quite different.

While the United States may excel in the creation and marketing of new biomedical technologies, it ranks far from the best in a comparison of principal health indicators among other member countries of the Organisation of Economic Cooperation and Development (see Table 8.1). The United States ranks first in total health care expenditure as a percentage of GDP and first in total per capita expenditure on health. It is among the top third in the per capita availability of costly Magnetic Resonance Imaging (MRI) units. But the United States ranks near the bottom in the availability of practicing physicians per one thousand population, and at the bottom in the level of public support for health care, or to put it another way, places the largest burden for health care on private resources. This may help to explain why the U.S. ranks near the top—or sixth—in infant mortality, and the life expectancy of both men and women in the United States is among the shortest in all thirty OECD nations. Nor can our poor showing in on average health outcomes be blamed on the growing proportion of elderly in our population; compared to other countries the United States has among the largest proportion of people under sixty-five years of age.[25]

Not only does the U.S. health care system produce on average mediocre outcomes in comparison with other nations' systems, it is not an open system to which everyone has access—which is the essential characteristic of a system based on the principle of a free and open market. What's more, its size, complexity, and interlocking interests—e.g., hospitals, over 1,300 health insurance companies, the pharmaceutical industry, and physicians—ensure its opacity to ordinarily well-informed citizens. Thus reform efforts are typically aimed more at cost reduction than at the system's systemic weaknesses.

Medical care in the United States—which consumes roughly 14 percent of the nation's Gross Domestic Product—is among the most technology-driven economic sectors.[26] Its reliance on ever more advanced diagnostic and life-sustaining equipment (e.g., computerized tomography scanners, magnetic resonance imaging) is complemented by its use of drugs and surgical procedures (e.g., large joint replacements, cardiac by-pass surgery). Technology, according to the OECD, is "among the main drivers of rising health expenditures" in all OECD countries as well as the United States.[27] A 2002 survey by the international accounting firm of PricewaterhouseCoopers estimated that drugs, medical devices, and medical advances are the largest (22 percent) of seven factors driving the steady increase in health care premiums. In descending percentage of costs, the remaining six factors are: rising provider expenses (18 percent; largely negotiated hospital payments), government mandates and regulation at both state and federal levels (15 percent), increased consumer demand attributable to population aging and media-driven demand for expensive treatments (15 percent), litigation and risk management (7 percent), and other factors such as fraud and mismanagement (5 percent).[28] Historian of medicine David J. Rothman has shown how the U.S. public's infatuation with technology as the source of medical miracles (e.g., the iron lung machine for polio victims, kidney dialysis machines, and the respirator) inspired various efforts to make sophisticated new equipment available to more and more patients, in spite of their costs and doubts about their marginal usefulness.[29]

Modern medicine has become "the proverbial Leviathan," writes Roy Porter, "comparable to the military machine or the civil service, and is in many cases no less business- and money-oriented than the great oligopolistic corporations."[30] The growth of this modern leviathan was initially fueled by the medical needs of military personnel who had served in World War II (12 million on active duty in 1945), the Korean War (3.5 million in 1953) and the Vietnam war (3.5 million in 1968). The federal government attempted to meet those needs in an ever expanding network of veterans' hospitals, which in 2002 served nearly 565,000 patients. While it is customary when measuring the human cost of war to count battle deaths, in the annals of medicine it is the number and percentages of non-fatal casualties that tell the tale. Since few would argue that the weaponry of the wars in which the U.S. was engaged between 1917 and 1970 has become less lethal, the remarkable decline in battle deaths and non-mortal wounds after World War II, as well as the dramatic decline in deaths from infection and disease, illustrate the life-saving advances in trauma care, surgery, antibiotics, and management of infectious disease that have occurred since the 1930s (see Table 8.2). Treating hundreds of thousands of sick and wounded servicemen and women, both in the critical first hours and then as they returned from theaters of war between 1945 and 1973, necessarily required a transition from medical practice dominated by the individual physician at the bedside to hospital care dominated by specialist physicians,

TABLE 8.1

Thirty OECD* Countries: Rankings in Selected Health Care Indicators for 2003
Rank 1 = highest value in category

Country	Total Health Exp. as % of GNP	Total Health Exp. per capita	Public Exp. on Health as % Total Health Exp.	Total Exp. on Pharmaceuticals as % Total Health Exp.	MRI Units per Million Population	Practicing Physicians per 1,000 Population	Population aged 65 and Over	Infant Mortality: Deaths per 1,000 Live Births	Life Expectancy at Birth, Females	Life Expectancy at Birth, Males	Country
Australia	13	13	23	18	16	22	20	12	6	4	Australia
Austria	21	15	22	12	4	7	9	15	12	14	Austria
Belgium	10	10	n/a	n/a	12	3	6	17	15	18	Belgium
Canada	7	6	19	13	15	26	19	8	9	6	Canada
Czech Republic	22	24	1	7	21	6	17	23	26	72	Czech Republic
Denmark	14	11	8	25	8	17	14	16	24	6	Denmark
Finland	23	19	13	14	5	19	12	26	11	18	Finland
France	6	9	14	10	19	7	8	24	4	13	France
Germany	3	7	11	16	13	8	4	20	14	15	Germany
Greece	8	20	26	15	22	1	3	13	18	16	Greece

	Hungary	Iceland	Ireland	Italy	Japan	Korea	Luxemburg	Mexico	Netherlands	New Zealand
Hungary	15	25	16	4	20	12	14	4	29	29
Iceland	4	5	6	17	2	4	25	29	7	1
Ireland	24	14	12	21	n/a	19	26	10	22	17
Italy	16	16	15	6	6	2	2	18	4	9
Japan	18	18	9	11	1	27	1	28	1	2
Korea	30	26	27	3	9	28	27	n/a	21	24
Luxemburg	26	4	2	23	7	18	15	11	13	20
Mexico	28	29	28	9	25	29	29	2	28	25
Netherlands	9	8	24	22	n/a	14	18	14	17	11
New Zealand	17	21	10	n/a	17	24	23	7	15	10
Norway	5	2	5	26	n/a	15	16	25	10	8
Poland	27	28	19	2	23	22	21	5	25	27
Portugal	11	23	21	n/a	15	10	10	21	20	74
Slovak Republic	29	27	3	1	23	16	24	3	27	28
Spain	19	22	17	8	11	13	7	22	2	7
Sweden	12	12	4	20	n/a	11	5	27	8	3
Switzerland	2	3	25	24	3	5	11	19	3	5
Turkey	25	30	18	5	18	30	28	1	30	30
United Kingdom	20	17	7	n/a	14	25	13	9	19	12
United States	1	1	29	19	10	23	22	6	23	22

* Organisation for Economic Cooperation and Development, OECD Health Data 2005 (October 205).

surgeons, and paramedical personnel administering drugs and operating increasingly sophisticated medical equipment. A comparable transition occurred in civilian medicine, partly stimulated by depression-era construction of hospitals by the Public Works Administration (PWA) and the appearance of hospital-based private insurance programs to cushion the cost of medical care for patients—among which the most enduring has been Blue Cross/Blue Shield. Gradually, the hospital was transformed from a place of isolation for the poor and sick, to the preferred place of treatment for the middle classes, whose medical costs were typically met by employer-provided private health insurance.[31]

In the immediate post–World War II decades in the U.S., when "communism" and "socialism" were terms of political damnation, physicians, hospitals, and the pharmaceutical industry successfully fought efforts to correct the failure of the market to yield basic, affordable health care for all Americans by arguing that relying on the federal government as the single, third party provider of health care would lead to "socialized medicine." Only the reformist zeal of President Lyndon B. Johnson's Great Society, and assurances that what became Medicaid and Medicare would only affect the poor, the disabled, and the elderly—mostly retirees no longer able to afford private health insurance—enabled the passage of legislation in 1965 ensuring health care to all but the middle class. The middle class, along with its employers, would demonstrate the rewards of individual effort by insuring its own health care needs.[32]

Today, with the exception of hospitals operated or contracted by the military services and the Department of Veterans Affairs, actual medical care is provided in private physicians' practices, clinics, and hospitals, many of which are operated for profit. Physicians' offices and prescription medicines combined account for as much in health care spending as hospital in-patient and

TABLE 8.2

U.S. Military Casualties: World War I, World War II, Korean War, Vietnam War*

Conflict	No. Serving	Percent Battle Deaths	Percent Other Deaths	Percent Wounds not Mortal
World War II	16,112,566	1.8%	.7%	4.2%
Korean War	5,720,000	.6%	.05%	1.8%
Vietnam War	8,744,000	.5%	.12%	1.8%

* Source: U.S. Department of Defense, *Defense Almanac* (defenselink.mil, 2002)

TABLE 8.3

U.S. Health Care Spending by Type of Service, 2000*

Hospital Inpatient	Office-Based Medical Provider	Prescription Medicines	Dental Services	Hospital Outpatient	Home Health	Emergency Room	Other Medical
36.7%	20.1%	16.4%	8.9%	8.7%	4.1%	3.1%	2.1%

* Source: Agency for Healthcare Research and Quality, U.S. Department of Health and Human Services. *Medical Expenditure Panel Survey for 2000.*

emergency room care (see Table 8.3). Medical care paid for by employer-provided health insurance remains the principal source of funding for necessary medical care in the United States. In addition, the federal government provides federally funded health insurance to selected populations through the joint federal-state Medicaid (principally for the poor and disabled) program, and Medicare (for persons over sixty-five). Children of families whose income exceeds the limit for Medicaid eligibility are covered by the joint state-federal State Children's Health Insurance Program (SHIP), created in 1997.

In 2002 the cost of providing health care to the U.S. civilian non-institutionalized population reached $1.4 trillion, or nearly three times the after-tax profits of U.S. corporations for that year.[33] Somewhat less than 40 percent of this cost was paid for by workers and their employers through private medical insurers; 16 percent represented "out of pocket" expenses, and less than half, or 44 percent, was paid for by the federal government through Medicare, Medicaid, and other federal health care assistance programs (see Table 8.4). The cost of health care became one of the principal domestic policy issues during the 2004 presidential election year partly because the percentage of workers who received health insurance through their employers dropped from 63 percent in 1993 to less than half (45 percent) in 2003. At the same time, the percentage of workers with health insurance who were required to make $10 or higher co-payments for prescription drugs increased from 19 percent in 1993 to 79 percent in 2000.[34]

Contrary to the arguments of opponents of national single-payer (that is, publicly funded) health systems, which often characterize such systems as "socialized medicine," reliance on the private sector to provide—and on the market place to distribute—necessary medical care in the United States does not reflect the superiority of a "free market" in health services. The U.S. system of health care is neither open nor transparent—both essential characteristics of a

TABLE 8.4

U.S. Distribution of Health Care Payments by Source, 2002*

Out of Pocket	Private Insurance	Medicare	Medicaid and SCHIP	Other
16%	36%	19%	18%	7%

* Source: "A Data Book: Healthcare Spending and the Medicare Program," Medical Payment Advisory Commission (Washington, D.C.: 2004). "Other" includes programs such as workers' compensation, public health programs, active and retired military programs, Indian Health Service, and state and local government hospital and school health subsidies.

genuine free market economy. A standard of openness in any social system is universality of access. In the United States, access to health care is not universal. The highest percentages of the 45 million individuals lacking health care in 2003 (equivalent to the combined population of the states of California, Oregon, and Washington) could be found disproportionately among young people between the ages of eighteen to thirty-four years (57 percent), those who worked part-time or not at all during that year (50 percent), unrelated individuals not covered under family members' policies (48 percent), those earning less than $25,000 a year (24 percent; the median per capita income for 2002 was $22,974), persons of Hispanic origin (33 percent), blacks (20 percent), and persons living in central cities (20 percent).[35]

In short, access to necessary medical services in the United States is effectively rationed on the basis of individuals' ethnic, socioeconomic, or occupational circumstances. The relative certainty of health insurance as a benefit of full-time employment has disappeared, and health care benefits are rarely available to workers who cobble together full-time work with part-time or temporary jobs—which is typical of almost twice as many female as male workers.[36] Socioeconomic rationing of health care violates not only the open system principle fundamental to the U.S. political system, it also diminishes the human capital essential to growing the economy. Would-be independent entrepreneurs and small business owners—central players in the American success story—face discouraging health insurance costs. A 2004 study by the Institute of Medicine (IOM) of the National Academies, which urged that the United States adopt a system of universal health care by 2010, estimated that 18,000 people die prematurely each year because of lack of health coverage, while the country loses the equivalent of $65 billion to $130 billion annually because of the poor health or premature deaths of uninsured adults.[37]

Open systems thrive not only because of their inclusiveness and pluralism; they thrive because of transparency, or the availability to everyone of information enabling informed choices, whether those choices involve competing ideas, competing goods in the market place, or competing public policies and political leadership. But the U.S. health care system is opaque. Its inner workings are obscured by the special vocabularies and understandings of medical practice and commerce—administrators as well as practicing physicians, medical assistants, researchers, and commercial suppliers of drugs, medical devices, and hospital care. In another study released in 2004, the IOM reported that nearly half of adult Americans lack the level of literacy and numeracy necessary to understand consent forms, patient information sheets, advertising, and to use "medical tools such as a thermometer." Such difficulties are especially limiting among the elderly, the poor, those with less than a high-school education, or those with limited English proficiency.

But the problem of health literacy is not entirely attributable to the deficiencies of patients and consumers. Drawing on an analysis of more than 300 studies of health-related informational materials such as medication package inserts, the IOM found that "even highly skilled individuals may find the system too complicated to understand, especially when these individuals are made more vulnerable by poor health:

> Directions, signs, and official documents, including informed consent forms, social services forms, public health information, medical instructions, and health education materials often use jargon and technical language that make them unnecessarily difficult to use. In addition, cultural differences may affect perceptions of health, illness, prevention, and health care. Lack of mutual understanding of health, illness and treatments, and risks and benefits has implications for behavior for both providers and consumers, and legal implications for providers and health systems.[38]

The need for IOM's study was foretold by the death of eighteen-year-old Jesse Gelsinger on September 17, 1999 four days after he received an experimental genetic treatment offered by the University of Pennsylvania's Institute for Human Gene Therapy. He thus became the first known person to have been killed as a result of gene therapy. His father, Paul Gelsinger, a tradesman from Tucson, Arizona, said afterwards that he "would have liked to have had somebody there who was not affiliated with Penn that could have assisted in describing the whole process of gene therapy. . . . I didn't research it. But I shouldn't have to research it. I believed these guys, everything they were telling me."[39]

This grievous lapse in communication was not entirely the result of the senior Mr. Gelsinger's medical illiteracy. One reliable, recent estimate is that over 90 percent of the human research subject complaints brought to NIH's

Office of Human Research entail problems with the informed consent proce-dure.[40] Informed consent forms provided to patients by the medical industry and its affiliated research institutions are usually not deceptive in the sense that they state falsehoods; they are deceptive to the extent that their language is evasive. Guidelines, the alternative preferred by the research community to explicit prohibitions, can be riddled with equivocation and vagueness. Both researchers and practitioners may be asked to condition their decisions on such indefinite or subjective judgments such as "potentially harmful," "antici-pated benefits," "to the extent possible," "when appropriate," "respect for" in-dividuals, and whether a particular action involves more than "minimal risk." Of these elusive qualifiers only the last has a regulatory definition: "minimal risk means that the probability and magnitude of harm or discomfort antici-pated in the research are not greater in and of themselves than those ordinarily encountered in daily life or during the performance of routine physical or psy-chological examinations or tests."[41] But this, too, relies on the vaporous "prob-ability" (absent widely understood objective measures of harm), and overlooks great disparities among individuals at risk of harm "in daily life" or "routine" medical tests, which are hardly a universal experience in a society in which over 45 million people lack health insurance.

The IOM recommended that the federal Department of Health and Human Services and "other government and private funders" sponsor the necessary research to develop "culturally appropriate . . . National Health Education Standards," while professional schools and continuing education programs were "to establish the most effective approaches to reducing the negative ef-fects of limited health literacy."[42] Until the IOM's or comparable recommenda-tions are fulfilled, however, the vast majority of Americans will be at the mercy of the technocracy of modern medicine, a health system dependent upon "ex-perts who . . . justify themselves by appeal to scientific forms of knowledge. And beyond the authority of science, there is no appeal. . . . Within such a soci-ety, the citizen, confronted by bewildering bigness and complexity, finds it necessary to defer on all matters to those who know better."[43]

Health Care Reform

During the administration of President Bill Clinton growing complaints about the rising cost of health care, combined with the decline in employer-provided health benefits, led the White House to believe that the U.S. public would ac-cept some form of government-supported health care for everyone. The com-plex system it proposed, relying heavily on government-endorsed regional networks to manage competition among private health plans, and budget caps to control costs—was easy prey for its opponents' charges that the Clinton health care plan would lead to rationing in health care and loss of choice among physicians and treatments.[44] The charges, widely disseminated in a profusion of television and magazine advertisements, were effective and the plan was defeated in Congress.

Because of the continuing shift in age of the voting population from younger to older, whoever won the presidential election of 2000 could be expected to champion some form of relief from the single most outstanding medical expense faced by retirees and other persons over sixty-five—prescription drugs.[45] The pharmaceutical industry readied itself for what might have been a successful government effort to control the price of drugs for Medicare patients, fielding more than 600 lobbyists in 2001, while it spent $8.5 million in lobbying Congress during 2003.[46]

With the cooperation of President Clinton's successor, George W. Bush, congressional Republicans crafted Medicare reform legislation so convoluted that it defies ready comprehension. Part of the complexity of this mammoth piece of legislation (over 400 pages) was due to the phasing in and out (determined by specific years and cost thresholds) of various provisions—provisions designed principally to shift to Medicare most of the costs of prescription drugs, while promoting the movement of Medicare beneficiaries to private sector (profit-seeking) health plans.[47] Nonetheless, by December of 2003 the Congress passed, and President Bush signed, the Medicare Prescription Drug, Improvement, and Modernization Act of 2003.

The Act explicitly rejected the then current federal model for managing the cost of taxpayer funded prescriptions: doctors and pharmacists in the Department of Veterans Affairs research drugs for comparative effectiveness, and obtain discounted prescription drugs for the 4.5 million veterans receiving medical benefits through negotiated bulk purchases and competitive bidding. (The Department of Defense and private health maintenance organizations like Kaiser Permanente also routinely negotiate bulk purchases and compare drugs for effectiveness. In a similar vein, the House voted in July of 2003 in favor of several measures to authorize the Public Health Service and NIH's Agency for Healthcare Research and Quality to research the comparative effectiveness of prescription drugs.)

The drug industry, however, was able to defeat drug quality and price transparency initiatives in the Congress by pushing familiar political hot buttons: "Cost-effective analysis in the private sector can provide useful information. When employed by centralized decision makers, however, it often becomes just another term for *health care rationing.* . . . Cost-effectiveness studies show which drug works best, on average, for large numbers of patients, but the studies often overlook the value of specific medicines for individuals or *subgroups, like racial minorities* (italics added).[48]

By the time the Congress passed the 2003 Medicare reform bill, the pharmaceutical industry's lobbyists had done their work. Putative market place competition, not federal health officials, would determine the cost of prescription drugs—except that this principle was waived to allow the Secretary of Health to prevent the importation of cheaper drugs from Canada (where provincial governments negotiate drug prices), either by pharmacies or individuals. Furthermore, private sector prescription drug plan (PDP) providers—but not the U.S. department of Health and Human Services (HHS)—could negotiate prices with manufacturers of pharmaceuticals and biological products, while negotiated prices "shall *not* . . . be taken into account" by HHS in evaluating prescription drug prices charged by PDP providers.[49]

To further protect the pharmaceutical manufacturers' profits from any downward pull of bulk negotiated prices, the Act directs HHS to study "sales of drugs and biologicals to large volume purchasers, such as pharmacy benefit managers and health maintenance organizations," to determine whether prices charged to them (which are likely to be lower) should be "excluded from the computation of a manufacturer's average sales price[s]." And, in a provision of dubious constitutionality, the Act stipulates that no part of the administration of prices for the new Medicaid drug benefit will be subject to "administrative or judicial review"—a provision that effectively protects the drug industry's arrangements with PDPs, and the PDP providers' arrangements with the HHS, from appeal to U.S. courts.[50]

Finally, the Act directs HHS's Agency for Healthcare Research and Quality to "conduct and support research" to gather "scientific evidence and information" that might allow the HHS to compare the clinical benefits of various therapies. The utility of such studies, however, will be limited. Not only are the manufacturers of drugs exempt from furnishing "trade secrets, processes, operations . . . confidential statistical data, amount or source of any income, profits, losses, or expenditures" for the health care research agency's studies, but the results of such studies are explicitly prohibited from serving as a basis for mandating "a national standard" or requiring "a specific approach to quality measurement and reporting," while HHS is prohibited from using data collected in the studies "to withhold coverage of a prescription drug."[51]

The Medicare Prescription Drug, Improvement, and Modernization Act of 2003 failed to increase access to, or the transparency of, health care in the United States. The greatest beneficiary of the second Bush administration's venture in health care reform is most likely to be the pharmaceutical industry, which now enjoys a statutorily secured market at prices it is free to raise for drugs that have been oversold to physicians and the public, but may be of dubious benefit and possibly harmful—as drug manufacturer Merck's withdrawal in 2004 of its anti-inflammatory drug, Vioxx, attests.[52] Also sure to benefit are the hundreds of for-profit health plans, whose markets are likely to increase in response to enrollment incentives built into the legislation.[53] (The search among U.S. for-profit managed care plans for more enrollees has entered the global market place, where such businesses hope to operate in Latin America, the Far East, and European countries other than Great Britain, which has already incorporated the practice of managed care within its government funded national health system.)[54] As for the more than 35 million (14 percent) of the population eligible for Medicare, much of their prescription drug benefits are likely to be erased by steadily increasing prices for drugs most often prescribed to the elderly—prices ranging from two to nine times the rate of inflation during 2001–2003.[55]

In early 2003—as the Bush administration's Medicare reform plan was being debated in the Congress, Rep. John Conyers, Jr. (D-Mich.) introduced H.R. 676, a bill "To provide for comprehensive health insurance coverage for all United States residents, and for other purposes." The bill was co-sponsored by thirty-eight Democrats. Its supporters included the 10,000 members of Physicians for a National Health Program, as well as the United Steelworkers of America, United Mine Workers of America, Communication Workers of America, Coalition of Labor Union Women, the United Methodist Church, the Unitarian Universalist Church, National Organization for Women, Public Citizen, and the National Urban League.[56] If enacted in its original form, H.R 676 would create a publicly financed, privately delivered, single-payer health care system based on an expansion of the current Medicare program to all residents of the United States and its territories. The proposed U.S. National Health Insurance program (USNHI) would set reimbursement rates for physicians and other health care providers, and negotiate prescription drug prices. The cost, according to its proponents, would be roughly $1.8 trillion a year, which would be financed by (a) preserving federal, state, and local revenues that currently pay for Medicare; (b) an additional 3.3 percent payroll tax on all public and private employers (which would replace current premiums for private health plans); (c) an additional 5 percent income tax on those earning $140,000 to $250,000 annually, and an additional 10 percent income tax on those earning more than $250,000 annually; and (d) stock and bond transactions would be taxed at 0.25 percent of the purchase price. The USNHI would most resemble the national health insurance systems of Canada, Denmark, Norway, and Sweden.[57]

Opponents of federally funded universal health care argue, as they did in the case of the Clinton administration proposals and as President Bush argued during the 2004 presidential campaign debates, that a national health care system will "lead to rationing . . . less choice . . . more controls . . . [and] poor quality health care."[58] The experience of other nations that ensure universal health care suggests that these warnings are not justified. Life-expectancy, infant mortality, and the number of practicing physicians per 1,000 population are reasonable benchmarks for comparing the quality of health care provided by the current U.S. system with those of Canada, Denmark, France, Germany,

Japan, Norway, Spain, and Sweden. Only seven countries rank lower than the United States in the number of physicians per 1,000 population, only seven countries rank lower than the U.S. in female life expectancy, and five rank lower than the United States in infant deaths per live births and male life expectancy (see Table 8.1).

No proposal before the Congress, and neither of the presidential candidates in 2004, called for the creation of a national health system by which the federal government not only funds health care, but provides it through government-salaried medical personnel and publicly owned facilities. Most politically promising proposals to increase access to health care build, instead, on the current plurality of insurers and providers. The two major parties differ principally over (1) whether universal access to health care should be a federally guaranteed and funded entitlement (as in H.R. 676), (2) whether access should be increased principally by enabling broader participation in Medicaid and Medicare, or (3) whether broader coverage should be promoted by offering tax-advantaged "health savings accounts" to enable individuals to better meet the expense of health insurance themselves. The cost of medicines would theoretically be reduced by removing regulatory means that enable the makers of brand-name drugs to avoid competition with generic drugs, and allowing the importation of drugs from Canada.

Not surprisingly the trade groups with the largest stake in changes in the U.S. system of health care favor proposals that do not impede—and may benefit—the interests of their members. The Association of Health Insurance Plans (AHIP), for example, favors measures likely to increase their enrollments, such as "advanceable, refundable tax credits to assist uninsured, lower-income Americans in purchasing health coverage," "tax credits for small businesses . . . to offset a portion of the employer's premium contribution," and "flexible tax-advantaged health spending accounts." The AHIP erroneously (according to the PriceWaterhouseCoopers study which the AHIP funded) attributes increasing health care costs to—and opposes—"government regulation, runaway litigation, and increased bargaining power resulting from provider consolidations" (that is, negotiated prices for medicines and services).[59]

Meanwhile the American Medical Association, which has been among the top ten political donors in the United States since 1989, has opposed nationalized health care proposals.[60] The organization also favors tax credits—especially tax credits large enough to "enable individuals to choose coverage that reflects their health insurance and health care preferences and values," rather than being limited to employer-based and employer-chosen health insurance plans. Recognizing that in individual, rather than employer-based, health insurance markets high-risk individuals will be subject to what the AMA calls "adverse selection" (being channeled into high-risk, high-cost insurance "pools"), the organization proposes additional federal subsidies for such individuals or their insurers, funded from "general tax revenues to minimize the effect on the average risk members of the population."[61]

Access to increasingly costly health care and the comparatively high cost of prescription drugs in the United States were two of the leading issues during the 2004 national presidential campaigns. Both of the major parties staked out policy territory that was in step with the ideological tendencies of their members, with Democrats promoting federally guaranteed access to health care and federal reinsurance of the costs of catastrophic medical needs, and the Republicans favoring tax-based measures to reduce the financial burden of health care costs, as well as capping medical liability awards. Neither party proposed tampering significantly with the inherited system of providing health care in the United States, which is a pluralistic, employer-based system resting on private sector providers. Thus we find ourselves stalemated in a paradox: the principle of open systems prevails in the pluralistic means by which health care is provided, but the principle has not been extended to the other two measures of openness: universality of access and transparency.

The Promise and Perils of Biomedicine

Scientific research takes place within a political universe, trying to insulate itself from the uncertainties and messiness of politics while asserting that its aims are morally neutral (other than to improve the human condition through increased understanding of nature). Science aspires to find factual and material truths, below which moral choices—determined in the public arena by politics—can be subordinated. If scientific research must be regulated at all, it is normally regulated on the grounds of public safety, as with regulations for disposing environmentally hazardous laboratory materials. But when advances in science are likely to result in injuries to individuals, or are otherwise contrary to their moral beliefs, two profoundly different cultures must negotiate principles which, for each, are so fundamental as to be non-negotiable.

The prominence of religious organizations in the political controversies arising from the development and use of biologic abortifacients, embryonic stem cell research, fetal tissue research, and the possibility of human cloning has tended to obscure the extent to which deference to the inviolability of human life expresses a *secular* moral principle. Enshrined in U.S. constitutional law, the concept of the inviolability of the human person as a sentient and moral being is as much a product of that most secular revolution in ideas, the Enlightenment of the eighteenth century, as it is of the Judeo-Christian heritage of 97 percent of Americans claiming a religious affiliation.[62] Indeed, that concept was central to the legacy of European humanism which engendered the scientific revolution of the seventeenth century.

Thus policy issues associated with the use of biological materials for medical therapies, research, and other possible social objectives are especially challenging because they are inevitably intertwined with religious and moral issues. At the same time, most citizens lack the background to question the arguments of both supporter and opponents of such questions as federal funding for stem cell research. To an unparalleled degree such issues expose the vast chasm between the ability of scientific and technical experts, and the ability of their challengers, to contest policy issues on a level playing field of information and understanding.

By the early 1970s molecular biologists had developed techniques enabling them to unravel the two halves of an organism's DNA strand and splice a short sequence of "foreign" nucleotides into it, thereby introducing a deliberate gene mutation into the original organism. In 1971 the possibility that scientists might experimentally introduce the DNA of a monkey virus into the cell of an *E. coli* bacterium (*E. coli* normally live in the intestinal tract of animals and man, aiding in the process of digestion), led a number of scientists to call for a moratorium on recombinant DNA (rDNA) experiments. Responding to mounting concerns, in 1975 the NIH created an advisory committee—the Recombinant DNA Advisory Committee (RAC)—to address the future of rDNA research, and convened a conference on the issue at Asilomar, California.

The conference resulted in the creation of guidelines to ensure the safe handling and containment of biological materials used in rDNA research conducted with federal (predominantly NIH) funds. Both the U.S. Food and Drug Administration and the Environmental Protection Agency would require compliance with the guidelines as a condition of regulatory approval for the use of products resulting from rDNA research. The conference and its outcome marked the beginning, and created the structure for, the federal oversight of research with potentially harmful consequences for society as well as individuals.

The array of possible issues arising from genetic research and its possible applications is enormous. Genetic engineering offers the hope of not just relieving, but curing, genetic disease; but will the possibility of genetic "improvements" create a climate of social intolerance for imperfect children? And whose notions of human perfection will prevail? How will deliberate genetic reduction of birth-defects—whether of mind or body—differ from eugenics? Might new artificially created organisms become pathogenic themselves? Genetic "typing" might offer a precise identification of individuals that would help in the pursuit and conviction of criminals, but will insurance companies and employers use the same information to discriminate against individuals with genetic propensities toward particular diseases?[63] Individuals among us may have answers to these questions, but will all accept the same answers?

In November of 1978, during the Carter administration, the Congress authorized the president to establish a special commission for the study of ethical problems in medicine and biomedical and behavioral research.[64] The measure provided broad statutory authority for the executive branch to create and maintain (through the budget of the Department of Health and Human Services, if not by special appropriation) presidential commissions to examine the unfolding suite of ethical and public policy issues accompanying the rapid growth of biotechnology, a suite that would be dramatically increased in 1997 by the first publicly announced creation of an artificially cloned animal, Dolly the sheep.[65] At the end of 1979 President Carter issued Executive Order 12184 creating the President's Commission for the Study of Ethical Problems in Medicine and Biomedical and Behavioral Research. During the next four years (the commission's charter expired in 1983) the commission produced eleven reports on a variety of issues confronted by patients and physicians, researchers, and citizens.[66]

Following a hiatus during the Reagan and George H. W. Bush administrations, President Clinton revived the White House bioethics commission in 1995 with the creation of the National Bioethics Advisory Commission (NBAC). The Clinton NBAC reflected the growth of the field of biomedical ethics in the composition of its membership: Its sixteen members other than the chair, Harold T. Shapiro (president of Princeton University), comprised seven bioethicists, five medical or public health faculty, one provost emerita from a major state research university, the chief business officer of a Boston area pharmaceutical firm, one jurist, and one spokesperson for the mentally ill.[67]

In the autumn of 1998 a small biotechnology company in California announced a discovery that would generate one of the more contentious issues of the 2004 presidential campaign. The Geron Corporation had successfully isolated and derived human embryonic stem (eHS) cells. This breakthrough was of historic importance, because such cells are not only pluripotent (they may grow into many different types of tissues) and immortal (able to continue dividing indefinitely while retaining their genetic character), they are also malleable, in that they can be manipulated without losing their cell function. The possible medical applications of this discovery seem limitless, given the prospect of being able to repair damaged organs. The prospect of removing cells from fetal tissue has alarmed some observers, however, who see this violation of the fetus as diminishing their prospects for reversing the Supreme Court's ruling in Roe v. Wade (1973) affirming a woman's constitutional right to an abortion.

The executives at Geron appreciated the controversial nature of what they had undertaken, and enlisted the Graduate Theological Union in Berkeley for help in exploring the ethical (and thus inevitably policy) issues likely to arise from their work. The Clinton administration's NBAC also examined the ethical ramifications of stem cell research, and concluded that federally funded "research that both derives and uses stem cells from cadaveric fetal disuse and from embryos remaining from fertility treatment," might continue, "if certain guidelines and safeguards are in place and if there is an appropriate and open system of national oversight and review."[68]

Among the safeguards proposed by the NBAC were a proposed prohibition on the buying or selling of "embryos and cadaveric fetal tissue," and the application of federal protections for human subjects research to research conducted with coded or identified samples of human biological materials. However, the commission recommended against federal funding for the creation of embryos for research by either in vitro fertilization (IVF) or somatic cell nuclear transfer cloning. In part because of the evolving science and ongoing societal conversation about ethical issues, NBAC did not suppose that it could offer the final word on the ethics of human stem cell research. Especially daunting was agreement on *the best possible balance* of ethical and medical considerations, or how to resolve the tension between proper respect for cadaveric fetal disuse and embryos remaining after IVF, *on the one hand*, and promoting research that could relieve much human suffering, *on the other hand* [italics added].[69]

The NBAC's principal reports, issued from 1999 to 2001, addressed research involving human participants, research involving persons with mental disorders, ethical aspects of human stem cell research, research with human biological materials, and ethical and policy issues associated with clinical trials in developing countries.[70] Like all presidential bioethics commissions through the Clinton administration, the NBAC affirmed the principles of the Belmont Report, elaborating on the application of the Belmont principles in more complex circumstances than its predecessor commissioners may have experienced. Oversight by Institutional Review Boards and informed consent remained the twin cornerstones of the commissions' principal recommendations. The Clinton bioethics commission also reaffirmed Belmont's argument that moral concerns arising from biomedical research must be *balanced* against the promise that scientific research holds out for curing disease and improved quality of life.

The NBAC also counseled against allowing human cloning, thus affirming a step the Clinton administration had already taken when, in 1994, it prohibited the NIH from funding the creation of human embryos for research purposes. The commission did, however, acknowledge the new philosophical and legal issues raised by the "commodification" of human biological materials. For example, to what extent, if any, is an individual "entitled to some or all of the profits that are realized from a product in the development of which his or her biological specimen played a role?"[71]

On one matter the NBAC could be unequivocal: ethical guidelines that apply only to federally funded or federally performed research pertain to a diminishing portion of all research conducted in the United States involving human subjects or human biological materials. "As a result, people have been subjected to experimentation without their knowledge or informed consent in fields as diverse as plastic surgery, psychology, and infertility treatment. This is wrong. Participants should be protected from avoidable harm, whether the research is publicly or privately financed. We have repeated this assertion throughout our deliberations, and recommendations in this regard appear in four previous reports."[72]

The White House was returned to a Republican president with the election of George W. Bush in 2000, and the charter of the National Bioethics Advisory Commission was allowed to expire on October 3, 2001. The Bush administration enjoyed more political support from groups characterized by the media as "the Christian right" and others who oppose the use of tissue from aborted fetuses for moral or religious reasons. To little surprise, Bush issued a highly controversial order limiting the use of human stem cells in federally funded research to existing stem cell lines, the relative usefulness of which was disputed by many scientists. Bush also announced the creation of a new council on bioethics headed by a University of Chicago bioethicist (Leon R. Kass) known to place relatively greater emphasis on moral restraint in the conduct of biomedical research.[73]

Embryonic stem cell research was thrust into the maelstrom of the 2004 presidential election campaign partly as a result of the death that summer of former Republican president Ronald Reagan, who had suffered from Alzheimer's disease. Believing that eHS cell research would lead to a cure for Alzheimer's, Mrs. Reagan became an advocate for federal support for the research. It seemed in time that all of Hollywood had become champions of eHS cell research: one actor (Michael Fox) was afflicted with Parkinson's disease; another actor, Christopher Reeve (completely paralyzed as a result of a fall from a horse), who died from heart failure a few weeks before election day, advocated embryonic stem cell research as a possible source of renewed neural-spinal chord tissue; and nearly everyone seemed to know someone with diabetes, for which eHS research also promised a cure. California's Republican governor, Arnold Schwarzenegger, also favored public support for the new research.[74]

Nonetheless, incumbent candidate George W. Bush and most of his Republican supporters held firm in their opposition to further federal funding for eHS cell research, while the Democratic candidate, John Kerry (and the Democratic party in general) favored it. Kerry's loss of the election did not, however, mean a loss for eHS cell research advocates. Voters in California passed the state's Proposition 71, which would commit the state to funding around $3 billion in eHS research. Needless to say, Democratic California's many research institutions, led by the University of California, would be the principal beneficiaries, while those with ailments that might be cured as a result of the research would continue to wait.

Cloning now having been shown possible, a related issue has been whether scientists should be allowed to clone human embryos in order to obtain embryonic stem cells. Cloning was not lawful for federally funded researchers under the Clinton administration's order of 1994, nor is a Bush administration likely to lift the ban. Meanwhile, since 2001 at least four bills have been introduced into the Congress to prohibit human cloning and numerous hearings have been held in both the House and the Senate to examine these legislative proposals. Some have contained exemptions for "somatic cell nuclear transfer technology to clone molecules, DNA, cells, or tissues" for "important areas of medical research, including stem cell research."[75]

While organized advocacy for human cloning is not noticeable, political battle lines have been drawn over the continuation of "research cloning," or nuclear transplantation research.[76] Organized supporters of research cloning include the Biotechnology Industry Organization, the Association of American Medical Colleges, and the Association of American Universities. Opponents include the Family Research Council, the National Right to Life Committee, and the U.S. Conference of Catholic Bishops. The middle ground in the debate, which accepts a moratorium on research cloning, is represented, among others, by the International Center for Technology Assessment, Boston Women's Health Book Collective, and the Center for Genetics and Society.

Even if cloning research were eventually to produce medically practical substances, under the Public Health Services Act of 1944 and the Food, Drug, and Cosmetic Act of 1938 (as later amended), the FDA's regulatory authority includes requiring clinical trials to establish the "safety, purity, and potency" of "somatic cell products"—which are considered "biological products . . . intended to prevent, treat, or diagnose diseases or injuries," which are subject to licensure.[77] The FDA has asserted the same authority over clinical research using somatic cell nuclear transfer to create a human being, which requires the prior submission and FDA approval of an "investigative new drug application."[78] Because of the rapid growth of the biotechnology industry, the FDA will be among our most important policy tools for managing the future of bioengineering for human applications in the United States. In the absence of changes in the FDA's priorities and funding, how effective is it likely to be?

A $100 Billion Pharmacopoeia and Too Little Flu Vaccine

The policy regime into which tomorrow's biotechnology innovations will flow is one in which the public supports biomedical research through the NIH, while it looks to the FDA to ensure that only effective and safe medicines and devices appear in the market place. It is a compromised regime; but it is the regime that will determine who, beyond the biotechnology industry and its affiliated research institutions, will benefit most from funding—whether public or private—for research in stem cells and other advanced biotechnologies which their advocates promise will bring new remedies and cures to the practice of modern medicine.

The health care issues that received special media attention during the 2004 presidential election campaigns included not only whether the federal government should fund stem cell research (Republicans said no; Democrats said yes), and the high cost of prescription drugs, but the failure of the U.S. drug industry to produce sufficient flu vaccine for the 2004 winter flu season. The production of seasonal flu vaccines is costly—vaccines must be produced anew each season, tailored to the specific flu threat, and what is not used must be discarded. There remains, as with all pharmaceuticals, the risk of adverse effects for which manufacturers may be held legally liable. What's more, flu vaccines are relatively unprofitable because they do not lend themselves to costly marketing, while their market is dominated by group purchasers likely to want to negotiate prices. Meanwhile, in an average year 51,000 die from flu and flu-related illnesses. Vaccines for childhood diseases such as tetanus, whooping cough, and chicken pox attract few makers, and thus have been in periodic short supply.

Thus when the British government found safety violations in the factory of one of the two makers of flu vaccines used in the U.S. and shut it down, the resulting shortages led to vaccine rationing, limited to children six to twenty-three months, persons sixty-five years and older, and any others with chronic illnesses. Even these priority groups could not be guaranteed vaccines; further de facto rationing occurred as those desiring vaccination, regardless of which category they occupied, experienced long waiting lines, price gouging, and reminders from the Centers for Disease Control that people who have the flu should cover their mouths and noses when coughing and sneezing, wash their hands, and stay home from work. The economic impact of widespread public health failures is not trivial. One estimate of the possible productivity lost from the increased incidences of flu and flu complications in the winter of 2004–2005 was as high as $20 billion.[79]

The stark contrast in the availability of heavily marketed and costly proprietary medicines in the United States and vaccines for viral diseases with pandemic potential uniquely illuminates the most fundamental policy challenge of modern health care in the United States: we have come to equate modern health care with novel bio- and medical technologies, while the ordinary and often widespread medical needs of large numbers of people—such as eye glasses and vaccines for common childhood diseases—go begging. Commercially produced prescription drugs are today's wonder-working technology, promising to alleviate not only disease, but to cure an increasing list of what, in an earlier day, might have been considered the quotidian afflictions of daily living and advancing years.[80] Every discomfort is a candidate to become a newly named syndrome, its pharmacological remedies touted by pharmaceutical companies in the mass media. Advertisements urge consumers to consult with their physicians who alone can write the necessary prescriptions, and thus bear much of the responsibility for what one doctor has described as "over-dosed America."[81]

During the last decade the number of prescriptions written for drugs increased by nearly 50 percent to 3.1 billion in 2001. This increase pushed sales during the same period from $61 billion to $155 billion. Television advertising of prescription drugs (allowed by the FDA after 1997), with its rapid-fire warnings to consult with one's doctor about possible side effects, surely accounts for a good portion of that increase. Drug companies market aggressively to doctors, as well. A former editor of *The New England Journal of Medicine* and professor emeritus at Harvard Medical School, Arthur S. Relman, lamented:

> Most medical practitioners nowadays learn which drugs to use, and how to use them, mainly from teachers and educational programs paid for by the pharmaceutical industry. . . . Medical schools, professional associations and hospitals that offer continuing education programs accept grants from the pharmaceutical industry and frequently allow the industry to suggest topics and speakers and help with preparation of the programs. . . . As for the doctors attending these industry-sponsored educational programs, they like the slick presentations, which often use industry-supplied educational materials . . . they are confident that their own independence is wholly unaffected by all of this—although surveys reveal that they are less sanguine about other doctors' ability to resist industry's blandishments.[82]

The pharmaceutical industry defends regular increases in the prices of prescription drugs—an increase that outpaces increases in health care costs generally—as necessary to fund the heavy investment the industry makes in research. The industry claims that it spends, on average, $802 million on R&D per each new drug on the market—including the cost of unsuccessful investigations into other new drug candidates. This figure also includes the "opportunity cost" of the money spent, or what drug companies might have earned if their R&D dollars had been invested in the stock and bond market instead. But the $802 million does not include the dollar value of federal corporate tax avoided as a result of being able to deduct R&D expenses from a company's tax base. Nor is the pharmaceutical industry clinging to the edge of profitability: since the 1980s the median net returns to the top ten pharmaceutical companies have been more than triple the median net return of other industries on the Fortune 500 list, whether measured as a percent of revenues, assets, or shareholder equity.[83]

Much of the innovation in pharmaceuticals touted by the industry actually derives from publicly funded biomedical and drug research sponsored by the NIH and conducted either in NIH or university and hospital laboratories. Government-funded scientists also normally conduct pre-clinical tests of a new drug on animals for toxicity and metabolic properties, as well as initial clinical tests in small numbers of volunteers with the condition the drug is intended to ameliorate or cure—first for safety (phase I) and then for safety and effectiveness (phase II). Phase III clinical trials with thousands of people are then conducted, largely by pharmaceutical or commercial drug trial firms, as part of the process of obtaining FDA approval to market the product.

Federal policy to promote the commercialization of federally funded research has promoted the pairing of public and private resources in new drug research and development. The Stevenson-Wydler Technology Innovation Act of 1980 and the Bayh-Dole Act of 1980, respectively, directed federal R&D agencies to promote the commercialization of results of their research, and to allow small businesses, universities, and other non-profit organizations to seek patents on inventions produced under federal contracts and grants.[84] In 1986 Stevenson-Wydler was amended to authorize federal agencies to enter into cooperative research and development agreements with private companies.[85]

Three important new drugs that have come on the market in recent years illustrate the way in which the pharmaceutical industry has benefited from publicly funded R&D and federal technology transfer policy. Erythropoietin (marketed by Amgen as Epogen), which replaces a protein hormone necessary to proper kidney function that is inadequately produced by a damaged kidney, resulted from NIH-funded research at Columbia University in the 1960s. Today Amgen receives more than $2 billion from annual sales of Epogen, much of it coming from publicly funded Medicare, which pays for treatment for kidney failure.[86] Zidovudine (or "AZT"), a molecule originally synthesized at the Michigan Cancer Foundation as part of a search for cancer treatments, was found in the early 1970s to be effective in suppressing viral infections in mice. In the early 1980s government-funded scientists at the NIH and Duke University discovered that the molecule also suppressed the AIDS virus. Zidovudine has been marketed in the United States by Glaxo-SmithKline as an effective AIDS treatment since 1987 under the brand name Retrovir.[87] Bristol-Meyers Squibb (BMS) has made over $9 billion in sales of the cancer-fighting drug Taxol (paclitaxel), based on NIH-funded research, during 1993–2002.[88]

In 2003 the U.S. General Accounting Office (GAO) issued a report on the technology transfer process that had enabled the commercialization of NIH's work on paclitaxel—an extract from the bark of the Pacific yew tree—which the National Cancer Institute (NCI) had been investigating, along with over 30,000 other plant species, since 1958. However, the ability of the General Accounting Office (renamed Government Accountability Office in 2004), or anyone else, to fully assess the contribution of publicly funded biomedical research is handicapped by the fact that federal laws generally "prohibit [federal] agencies from disclosing information that concerns or relates to trade secrets, processes, operations, statistical information, and related information."[89] Broadly written as this provision is, it ensures that the real extent to which the pharmaceutical industry's disproportionate profits are attributable to taxpayer funded research will be protected from public scrutiny.

The Food and Drug Administration has played an important role in enabling physicians to prescribe, pharmacists to dispense, and consumers to buy drugs with some assurance that those drugs are effective and will not harm, rather than help, those who use them. The mistaken belief of many that FDA approval guarantees safety has been exploited by the industry in its campaign against measures to increase market competition for its products—for example, allowing the importation from Canada of prescription drugs by individuals and state governments. Imported medications, the industry rightly asserts, will not have undergone FDA scrutiny for efficacy and safety. The pairing of the terms "efficacy" and "safety," however, perpetuates the illusion that a drug can be proven safe in the same way that it can be demonstrated to be efficacious.

Clinical trials have become universally accepted as the scientific (based on systematic empirical investigation) and thus sole authoritative means of establishing drug efficacy and safety. The FDA's insistence on such trials is its most powerful tool for ensuring the public credibility of its approval process. However, clinical trials cannot demonstrate safety in the same way they can demonstrate efficacy. In the case of demonstrating efficacy, the test drug and a placebo are given to a controlled population, and if those who receive the test drug experience an improved condition or a cure, the test drug is reasonably considered efficacious. But testing for safety is another matter. In engineering, for example, the load bearing tolerance of new structures can be tested under uniform conditions until one or more of the structures fail. The point at which a structure fails becomes its "limit." We cannot determine the safety of particular medicines in the same way by administering a test drug to a controlled population until a noticeable number of its members suffer and die. In 2002 a major NIH clinical trial of risks and benefits of estrogen plus progestin hormone therapy in healthy menopausal women was stopped when early results indicated that the therapy increased these women's risk of invasive breast cancer, coronary heart disease, stroke, and pulmonary embolism.[90]

Today, the safety of a new drug cannot be absolutely known for the simple reason that not until a drug has undergone long-term post-market clinical use among varieties of large numbers of people can its capacity to produce side effects in some persons be detected. Some adverse effects may take years to appear. Occasionally errors in prescribing drugs and administering them can cause serious side effects. The seriousness of an individual's illness, or the number of medications being taken, can also cause adverse drug reactions. These variables are difficult to isolate in large populations. A recent study by the GAO confirmed that of the prescriptions taken by nearly half of all Americans, we have less useful data on the frequency of adverse side effects than we do on specific drug-related side effects themselves. Even that information is sparse, since it tends to come from studies of patients in controlled settings such as hospitals and nursing homes, while out-patient drug consumption is more common.[91]

While reporting adverse drug events is mandatory for drug and biologic manufacturers and packers, it is provided only *voluntarily* by physicians, consumers, and veterinarians. What's more, the FDA "considers safety not in absolute terms but as a *balance* of risks and benefits. For example, a new drug may have serious adverse effects on some patients but still win FDA's approval because of its overall effectiveness in treating certain conditions relative to alternative therapies [italics added]."[92] The FDA currently receives about 275,000 adverse drug events reports annually, and estimates that around 100,000 deaths annually can be attributed to drug side effects.[93]

The enactment in 1992 of the Prescription Drug Users Fee Act (PDUFA) program has not improved the perception that the FDA is captive to the industry it was created to regulate. For years the drug industry had complained that FDA slowness in approving drugs was an intolerable burden on the industry's efforts to provide the public the benefits of new life-improving and life-saving drugs. At the end of the 1970s the industry complained that some drugs had required three years for FDA approval. A GAO investigation found that much of the delay was due not to "excessive caution and over-regulation," but understaffing at the FDA; meanwhile, administration of the approval process (rather than review of drug tests) often consumed as much as 40 percent of the staff time at the FDA.[94] PDUFA was the Congress's response.

PDUFA (which requires re-authorization every five years) allowed the FDA to collect fees from drug and biologic firms which it could then use to increase the personnel necessary for more rapid drug evaluation and approval. During 1992–1997 it collected $87.5 million annually, which it used primarily to increase its staff by 56 percent. During 1997–2002 the fees it collected rose to $161.8 million, which was again used primarily for staff increases.[95] Thus, by an act of Congress, a federal regulatory agency's staffing (and thus its employees' job security) has become dependent upon the industry it was mandated to regulate. The industry obtained some of the result it sought. Between 1992 and 2002 the median time required for the FDA to approve new drug applications (NDAs) and biological license applications (BLAs) initially fell from twenty months to six months for priority applications and ten months for standard applications. More recently, approval times for these substances have increased slightly to 14.8 months and 9.8 months, respectively. The FDA attributes this increase to an increase in the number of NDAs and BLAs it received after PDUFA was enacted.[96]

Meanwhile, seven of the drugs that the FDA has had to withdraw from the market since 1993 (Latronex, Redux, Raxar, Posicor, Duract, Rezulin, and Propulsid) generated $5 billion in sales before they were withdrawn.[97] In 2004 Merck withdrew from the market its widely prescribed anti-inflammatory pain killer Vioxx (typically prescribed for arthritis pain), after Merck's own studies showed an increase in adverse cardiovascular events (e.g., heart attack, stroke) among individuals taking Vioxx compared to those taking a placebo. Sales of Vioxx, on which Merck spent no less than $100 million a year for direct consumer advertising, reached $2.5 billion before it was withdrawn.[98] Critics of the drug industry cite these instances, the aggressive marketing of new and "me too" drugs (multiple drugs with only minor molecular differences as remedies for the same malady), and high prescription drug prices as proof of an industry consumed by greed. But drug companies do not prescribe drugs; physicians do.

A HALF CENTURY AGO, when historians could hazard cultural generalizations about American society that would not survive today's appreciation of the varieties of traditions that constitute our social landscape, the historian David Potter reflected that material abundance was the single most formative influence on the American character. While we might dispute the causes of this abundance, he wrote, "the importance of it has never been doubted, and a long procession of travelers, other observers, and social analysts have pointed to it as a basic and conspicuous feature of American life."[99]

For Potter's "people of plenty" upward mobility for enterprising individuals was the certain fruit of the garden. Otherwise unexplainable, poverty must therefore be attributed to individual rather than broad social or economic failure, and success applauded as an individual rather than community achievement. Of all the virtues in the American pantheon, individualism thus rose above the rest. It was and is what gives urgency to the values of liberty, justice, and government accountable to the governed. It is why open systems—politically manifest in accountability, pluralism, and transparency—matter. It is why the historical secularization of divine understanding through the natural sciences—which required the assertion of individual understanding over received priestly wisdom—matters. The widespread assumption of unending abundance and its ideological correlate, the enterprising individual, explain our tenacious regard for the "middle class"—not yet rich, but never again to be poor—whenever large questions of American public policy are in play.

David J. Rothman attributes to this middle class the failure of the United States to develop a system of universal guaranteed health care. Publicly provided services are for the poor and elderly (e.g., Medicaid, Medicare); for others they are a mark of failure. Whether or not government-provided health care actually or necessarily will lead to official rationing and loss of choice—each representing a denial of individualism—the ideology of the enterprising middle class requires the expectation that it will. But in an unsystematic and unofficial way health care is rationed according to socioeconomic variables, while choice is denied to 45 million people in the United States, most of whom have not yet and may never ascend securely into the middle class. Too bad for them; it is not our loss, the opponents of a national health system seem to say.

This chapter argues otherwise.

In keeping with the tendencies of experts in most professions, the basis of decision making in the biotechnology-intensive medical industry is opaque to outsiders. Ordinary citizens must defer to an exceptional degree to the authority of experts in obtaining medical information or making health care decisions—for example, they depend upon a physician to obtain prescription drugs or receive laboratory analyses of their own blood. With today's Internet one can get some information about medications and procedures, but most physicians do not, in turn, defer to information obtained by patients. In the case of biomedical research opacity obscures a danger that extends far beyond the scope of health care policy into the very core of our politics.

The commerce in medical technologies and therapies that thrives on the inadequately informed choices of individual consumers is less transparent than commerce in any other economic sector in the United States. With the expansion of this economic sector, and the biomedical research institutions with which it has a symbiotic relationship, the exercise of moral restraint by expert medical or scientific authority has itself become a field dominated by experts. Reliance on committees to weigh choices in the putative balance between the desires of science and the needs of patients or human research subjects only further serves to defeat transparency, for committees do not make decisions; they only aggregate the decisions of individuals.

If necessary medical care in the United States was provided and allocated as an open system—embodying principles of transparency, universal access, pluralism, and accountability—what might it look like? First, access would be universal, and the standard of care would be consistent across all recipients, according to their individual medical needs. It would be funded by a universal tax based on income, while payments could be administered at the state or federal level. Medical care itself would be provided by the same combination of private sector health plans and individual practitioners that provide health care today. Individuals who prefer optional medical services such as cosmetic surgery would be free to seek it individually from the private sector providers—but they would not be allowed to opt out of the public sector program to ensure that support for essential health care remains a shared responsibility of all citizens.

Similar programs already exist: they are the federal health insurance programs that provide health insurance benefits for members of Congress, the federal judiciary, employees of the executive branch, active duty and retired military personnel, and comparable programs for public sector employees throughout the individual states. Indeed, universal health care in the United States could be achieved relatively simply by allowing every U.S. citizen to participate in a federal (or state) program serving public sector employees.

But a truly open system would go further: it would ensure transparency by directing *and funding* the Department of Health and Human Services to make readily available, in plain language, on the Internet and through public libraries comparative assessments of all diagnostic procedures and pharmaceutical, surgical, and other therapies, *by provider name and including clinical success rates, regardless of their proprietary status.* Widespread availability of this information—thus providing the system with the transparency necessary for patients and consumers to make meaningful medical decisions—would introduce genuine competition into the market for medical goods and services. Genuine competition in a fully transparent health care market is more likely than the much maligned government price controls to introduce cost discipline into the system. It will also lead consumers to the most effective, rather than the most heavily marketed therapies.

9

Fossil Fuels and
Clean Air

While energy and environmental policy are made and carried out in the United States by different legislative and administrative structures at both state and federal levels, their intimate connection is reflected in the fact that no energy policy initiative today can be made without addressing its probable impact on the environment. This connection was acknowledged in the May 2001 report of the Bush administration's controversial National Energy Policy Development Group (see chapter 5), which supported incentives to increase the use of fuel-efficient motor vehicles, and devoted a chapter to the necessity of "protecting America's environment" with reduced emissions from electric power plants of sulfur dioxide, nitrogen oxides, and mercury. The report echoed the consensus of most industrialized nations which, as members of the International Energy Agency, consider "the maintenance of the basic ecosystem and a healthy environment" integral to energy policy.[1]

This linkage is especially critical in the United States because of our reliance on fossil fuels. The extraction of coal, whether from deep mines or strip mines, takes a toll on the landscape through polluted streams and the wasting of ecosystems and areas of natural beauty. The movement of oil and liquefied natural gas through pipelines and by ocean tankers conveys with it risks to natural habitats on land and sea. Oil refineries and natural gas terminals mar the appearance of the communities that contain them. And then there are the by-products of the combustion of fossil fuels—carbon, nitrogen, sulfur oxides (CO, CO_2, SO_2), and mercury—which produce smog, acid rain, toxic emissions and, in the case of CO_2 especially, impede radiation of the Earth's heat, thus altering the Earth's temperature.

In this chapter we will examine the foundations of contemporary U.S.

energy and environmental policy, focusing on the use of fossil fuels and its environmental consequences. We will see how the natural abundance of fossil fuels in the American landscape not only reinforced an inherited conception of nature as an open system, but also prevented Americans from recognizing, much less accepting, the certainty that our most difficult energy policy challenge might be to manage scarcity. Thus our effort to do so during the energy crisis of the 1970s fell short of the measures necessary to adequately recondition American consumers to the importance of reducing oil consumption, for example, an increased federal tax on gasoline to fund alternative fuels research.

The legacy of the open systems ideal of Adam Smith's laissez faire market place set boundaries for government interactions with the oil industry during the Great Depression and World War II, resulting in an ad hoc and sometimes paradoxical sequence of government efforts to ensure sufficient petroleum supplies while regulating demand. Then, after two decades of post-war growth and abundant cheap gas, the energy crisis of the 1970s compelled policy makers finally to attempt to forge a comprehensive national energy policy. The mixed results of that effort were characteristic of the pluralism embedded in public policy making and administration at the federal level.

We then turn to the environmental hazards of our use of fossil fuels for virtually all of our energy needs, and examine briefly three touchstone issues at the beginning of the twenty-first century: oil drilling in federally protected natural areas, fuel economy standards for motor vehicles, and the reduction of greenhouse gases to address the phenomenon of "global warming." Launched in the late 1960s, federal policies to reduce damaging emissions and effusions from the country's fossil-fueled industries and transportation sector were not only predicated on accumulating scientific data and everyday experiences of polluted air. They also embodied a tacit view of nature that was the opposite of that which informed evolving national energy policy. That view was pastoral, venerating nature as a pristine world of delicately balanced ecologies—a world whose survival was threatened by the incursions of technology and industry. Thus the initial model chosen for environmental policy was a frontal or regulatory control model: the government sets standards, and industry complies, or is penalized.

The 1980s, when the administration of Ronald Reagan began to undo much of the edifice of federal regulation that had been built during the 1960s and 1970s, witnessed the reassertion of laissez faire as the guiding model for the federal government's relationship with industry, a relationship that was central to both energy and environmental policies. The concept of open systems also underlay a rethinking of the best approach toward environmental regulation. This reconsideration resulted in the easing of the tension between industrial interests and the future of the natural environment, giving way to the overarching policy theme of "sustainable development."

Sustainable development requires a balancing of economic and environmental interests. The only way such a balancing can take place in the course of policy making is through negotiated (rather than directed) targets for reduced emissions and other environmental goals. Regulatory standards that result from negotiations are the administrative equivalent of laws that result from participatory government, both of which together represent the most fundamental manifestation of the open systems concept in political life.

Nature as an Open System and the Foundations of U.S. Energy Policy

Until the successful adoption in the United States during the 1820s and 1830s of the coal-fired steam engine for rail and water transportation, the use of energy—principally water, wood, and animal and human muscle—was largely limited to the locations in which it could be found. Coal, in tandem with the steam engine originally developed in England during the eighteenth century by Thomas Newcomen (1663–1729) and James Watt (1736–1819), added energy *portability* to the mix of essential elements that enabled this country's industrialization in the nineteenth century. The importance of this new source of energy that could move people and commodities as well as run machinery was proudly announced during the first centennial of the Declaration of Independence in Philadelphia, when the thirty-foot high, 700 ton, and 1400 horsepower steam engine built by George Corliss was the featured exhibit at the 1876 world's fair.

The marriage of coal to motion had already laid the legal groundwork for the federal government's authority to regulate the extraction, distribution, and use of the growing nation's energy resources. That authority would become essential to the government's power over interstate commerce, which required a broad interpretation of the Constitution's commerce clause. Most readers of American history have some familiarity with the landmark Supreme Court case, *Gibbons v. Ogden* (1824), in which the Federalist Chief Justice John Marshall wrote the majority opinion invalidating a monopoly the state of New York had given to Aaron Ogden for steamboat operations on the Hudson river, including between New York and New Jersey. Ogden's former business partner, Thomas Gibbons, had obtained a coasting license from the federal government—whose authority over coastal navigation was undisputed—to operate steamboats between New York and New Jersey. Ogden petitioned the supreme court of New York to order Gibbons to stop his steamboat operations on the Hudson, and Chancellor James Kent, chief justice of the New York Court, ruled in Ogden's favor. Gibbons appealed to the U.S. Supreme Court, with the now well-known outcome.

Suppose, however, that Chancellor Kent's opinion had not been appealed or had been upheld. Kent was no less a Federalist than Marshall, and was certainly his peer as a jurist.[2] Where Kent and Marshall differed was not only on whether state and national jurisdiction in navigation between states might be concurrent (and thus, in the absence of a contradictory federal statute, state jurisdiction might prevail), they also differed on whether New York's monopoly grant to Ogden to operate boats powered by *steam* (as distinct from navigation in general, to which Ogden's monopoly did not apply) was well within the state's authority to regulate internal trade and navigation. Marshall dismissed New York's distinction between steam and other forms of navigation. Congressional authorization for the federal licensing of boats operating in the rivers, bays, or coastal waters of the United States, argued Marshall, does not specify the means by which such boats are propelled. Thus energy technologies for transportation were swept up with commodities in Marshall's broad interpretation of the "commerce clause" of the U.S. Constitution.

Had Chancellor Kent's judgment prevailed, state after state might have continued to grant exclusive franchises (as Connecticut, Ohio, Georgia, Massachusetts, Pennsylvania, Tennessee, New Hampshire, and Vermont had done) for the operation of various modes of transportation solely on the basis of the technologies by which they were propelled. The commercial and industrial development of the United States, for which a broad and open national market was a key factor—as Hamilton, Marshall and most of their Federalist colleagues understood (see chapter 2)—would have labored under state-by-state barriers limiting the spread of new transportation technologies and their associated industries.

Integral to the Hamiltonian vision of an expansive and open national market, unencumbered by local or regional barriers, was the vast and open American landscape, endowed with virgin soils, running streams, dense forests, and vast plains. By funding various exploratory expeditions the federal government ensured that the continent's vastness would not remain forbidding for long.[3] What these explorations seemed to confirm was the centuries' old conception of nature as autogenic and divinely blessed plentitude, a universe of endless diversity.[4] Beneath the surface of this richly blessed continent were vast deposits of coal, oil, and natural gas.

The significance of the abundance of the new republic's natural resources was not only economic. Nature's plenty spread across the American landscape of the nineteenth century providing the material substance for tens of thousands of newcomers to realize the ideals of individual freedom and social mobility. While democratic politics would help those Americans to realize those ideals, the same political system was a poor vehicle for forming policies predicated on scarcity. Few will accept any form of rationing as long as continued abundance seems "natural."[5] Furthermore, Americans' faith in the open system of the free market as the best guarantor of prosperity through the efforts of private enterprise would not be shaken on a national scale until the Great Depression. The corresponding faith in the capacity of nature to provide indefinitely the fuels that powered the modern economy would not be shaken until the 1970s.

Coal

Not the least of the natural resources to be uncovered in the early nineteenth century was the United States' rich deposits of coal. Coal could be found not only in western Pennsylvania, West Virginia, and Kentucky, but Ohio, Michigan, Indiana, and Illinois, and the foothills of the Rockies. The inauguration of the first successful rail line in 1830 (the Baltimore and Ohio Railroad) ushered in not only the settling of the west by rail, thereby ensuring the creation of a national market, but the beginning in the United States of the age of coal, as mining companies began to appear in Ohio and Illinois and the Appalachian mountain states.

By the end of the nineteenth century coal provided heat and fired the steam engines used to power the nation's rail networks, machinery, and pumps to enable deeper coal mining to obtain the best coal to transform iron ore into steel. The emergence of the electrical industry in the 1890s added a new market for coal, while the post–World War I strategic decisions of General Electric and Westinghouse to develop consumer appliances to expand the market for electricity initiated the ever expanding use of electricity—a user as well as supplier of energy.

Coal continued to dominate energy production in the United States until World War I, when it was overtaken by petroleum, which yields more power per volume, and natural gas. Between 1920 and 1960 coal production declined by around 65 percent while petroleum production increased eightfold and natural gas production increased by a factor of sixteen. Coal today represents a third of all energy produced in the United States. It supplies almost all of the energy used by the electric power sector, a sector that has seen a tenfold increase in the consumption of electricity since 1950 (compared to the threefold increase in overall energy consumption in the United States during the same period). Declining prices of electricity have undoubtedly been a factor: except for a brief rise in the early 1980s, the price in constant 1996 dollars of retail residential electricity has declined steadily since 1960.[6]

Oil

At the middle of the nineteenth century American machinery was moved by water and coal, homes were heated by wood and coal, and lamps lit with animal grease, the highly flammable turpentine derivative "camphene," or for those who could afford it, sperm whale oil. Gas sufficient to light street lamps and the homes of the well-to-do was distilled from coal. Some oil was harvested from ground seepage around coal deposits and salt mounds (natural geologic companions of oil fields), as it had been harvested in the Middle East since antiquity, and was used both in Europe and the United States for illumination and as a medicinal unguent and tonic. Only limited quantities of oil could be obtained from ground and stream seepage.

During the mid-1800s eastern Europeans used refining techniques borrowed from the Middle East to produce kerosene oil, while a Polish pharmacist developed a lamp design—using the now familiar pear-shaped glass chimney—that would virtually eliminate the smell and smoke from burning kerosene. The new kerosene lamp soon found its way to the United States, expanding the market for a fuel to replace sperm whale oil, which had grown increasingly expensive as the whaling industry depleted whale schools in the North Atlantic.

It is easy to take artificial illumination for granted, but cheap lighting that reached beyond the simple lamp on the parlor table or street corner was as essential for industrialization as the steam engine. No less essential were lubricants for machines composed of ever more rapidly moving and finely tooled components. Both needs would be met in abundance after Edwin L. Drake's successful effort in August 1859 to drill for oil in the oil fields (so named for the oil that seeped from the coal beds) around Titusville, Pennsylvania.[7]

In a little over a year after Drake's well first produced oil, seventy-five productive wells—and not a few dry holes—were scattered around Titusville. In April 1861 the first flowing well spurted upward and began to produce 3,000 barrels a day; a year later Pennsylvania wells were producing 3 million barrels a day. Along with the wells came refineries to turn crude oil into kerosene. Still the market could not always keep up, and "boom and bust" became as common as oil wells as the price of oil fell from $10 a barrel in the beginning of 1861 to 10 cents a barrel at the end of the year. The Civil War would play its role here, as it did in other areas of American life, stimulating yet more speculative activity. Southern supplies of turpentine (source of lamp camphene) dwindled to a trickle, war material and railways needed ever more lubricants, and northern revenues from exported cotton could be (and were) replaced by revenues from oil exports to Europe. The demobilization of thousands of troops boosted the supply of hungry speculators, who, in turn, snapped up land left and right in hopes of drilling the well that would make them rich.

Boom and bust was the order of the day for refineries too, which mushroomed to 300 firms in 1863 and then shrank to 150 by 1870. Competition was vicious in this new capital-intensive industry, one in which refining capacity could reach twice the receipts from current sales—a recipe for financial disaster. Into this fray in 1862 stepped John D. Rockefeller (1839–1937), who arrived in Cleveland with his family at the age of fourteen and by the age of twenty had set up his own successful commission business in hay, grain, meats, and other basic goods. At twenty-four Rockefeller built his first refinery, and began steadily buying up other refineries or forming partnerships with their owners until in 1869 he was able to form the Standard Oil Company of Ohio.

Through a combination of efficient management and a strategic approach toward business—carried out with the acquisition of pipelines, terminal facilities, competitors, and favorable rates (and rebates) from his shippers—Rockefeller owned or leased by 1878 nearly 90 percent of the refining capacity of the United States. Within four more years, by consolidating the stock of Standard Oil and its affiliates in other states under the control of a single board of trustees headed by himself, Rockefeller created the Standard Oil "trust," which now enjoyed a virtual monopoly of the oil business in the United States. His edifice would become the classic model of "horizontal integration," a business strategy for achieving stability and assured profits in an industry selling an essentially undifferentiated product.

The Congress's response to the Rockefeller monopoly (the Standard Oil Trust) signaled the first effort since the Federalist era to reinforce an openly competitive economic system—not against the power of government (which for most of the nineteenth century had little power at all) but of market predators. The response came in the form of the Sherman Antitrust Act of 1890, which made it a federal crime to enter into any "contract, combination in the form of trust or otherwise, or conspiracy, in restraint of trade or commerce among the several States, or with foreign nations." Violators could be prosecuted by the Department of Justice, competitors who could prove injury could sue for damages, and firms could be ordered dissolved or dismembered—as indeed the American Telephone & Telegraph Company (for example) was forced to do in 1984.

But what exactly did it mean to "restrain trade?" Presumably the statute outlawed cartels and price-fixing agreements, but the language of the act did not actually say so. Was it realistic to expect a capital-intensive industry like petroleum drilling and refining to depend on a host of small refiners adjusting their production and prices to vacillating market demand? When did a firm become so large that the production and distribution advantages—and resulting market share—it enjoyed by virtue of its size make it a *de facto* monopoly? Could a firm be prosecuted because of a monopoly won through successful business practices, for example, but in the absence of any "conspiracy" or other agreement to "restrain trade?" Federal regulators and the Department of Justice have wrestled with these questions since 1890 (see chapter 6).

Meanwhile, Congress in 1903 created a Bureau of Corporations within the then Department of Commerce and Labor to investigate possible antitrust violations. The bureau would become the Federal Trade Commission created in 1914 by the Clayton Act. The Clayton Act also specified the kinds of practices covered by the Sherman Act (that is, price discrimination among buyers, exclusive selling and buying contracts, acquisition of competitors' stock, and interlocking directorates among competing firms, *if* their effect was to lessen competition).

But how could anyone establish whether and how competition was affected? The idea of market competition, for all the power it has held over U.S. policy making, was turning out to be a greased flag pole, hinting at a complex sport with many possible players, and no opening bell or end-of-game buzzer to guide the referee. When and where did the oil market "begin?" At what point was consumer impact assessed—only when consumers complained? And were consumers *necessarily* harmed by monopoly prices? The presumptive answer was "yes." For over a decade antitrust cases brought before the Supreme Court faltered on these and other interpretive weaknesses in antitrust law.[8] By the 1930s, when the tide of bigness in business could not be reversed, only Standard Oil and the American Tobacco Company (another horizontally integrated business giant) were dissolved, only to be replaced by today's oil and tobacco oligopolies.

Natural Gas

Natural gas is a relative newcomer among the fossil fuels that heat, illuminate, and provide power to U.S. homes and industry. This highly flammable type of petroleum—a gaseous hydrocarbon composed largely of methane and ethane—is difficult to capture, store, and transport, and thus its use (which is traceable back to ancient Persia and China) was limited to the locations in which it was found until the nineteenth century.[9] With twice the heating value of its synthetic predecessor (gas from carbonate coal was used in the United States beginning in the late eighteenth century to light houses and streets), widespread use of natural gas depended on advances in piping and cryogenic technologies.

The availability of steel piping with gas-tight couplings at the end of the nineteenth century enabled the growth of long-distance gas transmission in the 1920s to more than ten major systems in the United States by 1931. However, gas can be stored and transported most easily—though not necessarily more cheaply—in double-walled cryogenic tanks, which keep the gas cooled below its boiling point (−259° F). As with pipelines, the ability to transport natural gas in this more compact form (at 1/600th of the gas volume) had to await the necessary advances in technology. Cryogenics—cooling between −238° and −460° F (absolute zero, when molecular motion theoretically ceases)—had to await the efforts of Louis Caillete and Raoul Pictet of France, and the Scotsman James Dewar, to liquefy oxygen, which they succeeded in doing in 1877–1878.

Relatively clean burning, the combustion of gas produces virtually none of the soot, carbon monoxide, nitrogen oxides, or sulfur dioxide that are released by the combustion of coal and oil. Natural gas is also relatively plentiful—and often found with oil fields—with the largest fields in Russia, the Arabian-

Iranian basin, northern Europe, North America (in Alberta [Canada], Kansas, and the Oklahoma-Texas panhandles, and the Permian Basin of western Texas and southeastern New Mexico), North Africa, and Indonesia. Nonetheless, it was not until after World War II that pipelines over four feet in diameter and 3,400 miles long were laid, and these were first laid in Russia, designed to carry natural gas across the Urals and into Europe.

The real spur to extensive use of natural gas as a national energy source in the United States came with the crude oil shortages and price increases of the 1970s, which made the production and use of natural gas more economically competitive. Nonetheless, like oil, natural gas is a diminishing resource, and federal policy (the Fuel Use Act of 1978) prohibited new power generators from using gas and set a timetable for shutting down existing gas-fired plants. Gas prices were later deregulated, resulting in increased production, and the Fuel Use Act was repealed.

EXCEPT FOR THE OIL crisis of the 1970s, wastefulness in our use of energy in comparison with European countries (where automobiles are comparatively small and lights that shut off automatically are ubiquitous) has commanded less policy attention than the costs—whether in dollars or in actual or probable environmental damage—of increasing energy supply. While the availability of fossil fuel energy at any cost is determined not only by the supply of fuel in the ground, but by the cost of extracting, refining, and distributing it, our deep-seated belief in nature's plenty has limited our ability to devise an energy policy based on scarcity. And, indeed, *on a global scale* fossil fuels *are* abundant.

In addition to coal, which can be found throughout the world, improvements in exploration and production technologies have led to an increase in the proven reserves of oil of roughly 50 percent since the 1970s. The largest oil deposits are found in the Middle East, eastern Europe (largely Russia), and North America, while smaller oil fields lie under the North Sea and areas of Africa, Asia, and South America. Various estimates of undiscovered oil resources vary, but experts agree that they are substantial. "The availability of oil in terms of reserves and geology isn't an issue," observed the chief economist of the International Energy Agency; the issue is whether the profitability of exploration and production will be enough for the industry to meet growing demand.[10]

Estimates of projected coal, oil, and natural gas reserves in the United States vary widely, dependent as they must be on current geological surveys and a host of assumptions about market conditions, projected demand, and productive capacity.[11] Proven U.S. reserves of crude oil (about 22 million barrels in 2003) have been declining, as have been reserves of natural gas liquids (7.5 billion barrels in 2003). Reserves of dry natural gas, however, have increased (to 190 trillion cubic feet). Unlike reserves of oil and natural gas, coal reserves are abundant throughout the world, with the United States having an esti-mated 213 billion tons (or 28 percent) of the world's recoverable coal deposits.

Fossil fuels provide over 85 percent of all energy used in the United States.[12] Hence our ability to satisfy the demands of the four principal energy use sectors—commercial, industrial, residential, and transportation—depends on our ability to satisfy demand either with the quantities of coal, gas, and oil we are able to produce, or by importing them. Many of us may wish to satisfy some of our energy needs through alternative sources such as solar, wind, and hydro power; but these sources represent less than 10 percent of our energy use today and, absent significant changes in the relative cost of building and main-taining the regional infrastructure necessary to distribute solar, wind, and hy-droelectric power, that percentage is not likely to increase.[13]

While nuclear generation of electricity increased substantially during 1970–2003 (as late as 1975 236 plants were being planned, built, or operating) less than 20 percent of our electricity today is generated by the 104 nuclear reactors operating in thirty-one states. (Illinois, Pennsylvania, South Carolina, and New York have the largest number of working nuclear reactors.) Fewer nuclear power plants are being brought on line because of the high cost of licensing and building them, and public opposition to nuclear power after the 1979 cooling valve malfunction in the Three Mile Island nuclear power plant in Harrisburg, Pa. and the 1986 explosion and partial meltdown of the nuclear power plant at Chernobyl in the Ukraine. The remaining sources of energy for generating electricity are coal, natural gas, hydroelectric, oil, and geothermal—at about 52, 17, 7, 3 and 2 percent, respectively.[14]

As we can see, those venturing onto the terrain of national energy policy run the risk of becoming awash in numbers and a variety of measures that can render straightforward observations difficult to make. Table 9.1 attempts to simplify the flood of energy statistics available from the federal government by comparing the increases (or decreases) in our consumption of fossil fuels since 1960 by residential, commercial, and industrial consumers, and for transporta-tion and electricity. What we learn is that our consumption of natural gas has increased steadily for all categories of users. Meanwhile our consumption of coal has declined for all categories *except* electric utilities, where coal con-sumption has increased fivefold and the electric power sector consumes over 90 percent of the coal consumed in the U.S.[15] (Electricity also wastes energy during the process of generation, transmission, and distribution—and signifi-cantly so, losing twice the amount of energy generated by natural gas for the electrical utilities' residential and commercial consumers.)[16]

While the residential and commercial use of petroleum has decreased since 1960, our overall consumption of petroleum has doubled, thanks to sig-nificant increases in consumption by the petrochemical industry and the elec-trical utilities and a two-and-a-half times increase in the transportation sector,

TABLE 9.1

U.S. Energy Consumption by Type and End-Use Sector, 1960–2002*
Shown as constant, or in decreases or increases as multiples of n1
Number in ( ) = quadrillion BTUs consumed in 2002
Total U.S. consumption of petroleum, natural gas, and coal in 2002
= 83.6 quadrillion BTUs.

Consumption	Petroleum	Natural Gas	Coal
Residential	Declined × 1.5 (1.49)	Increased × 1.6 (5.02)	Declined; relative amounts negligible
Commercial	Declined × 1.7 (0.73)	Increased × 3 (3.20)	Declined; relative amounts negligible
Industrial	Increased × 1.6 (9.15)	Increased × 1.4 (8.47)	Declined × 2.2 (2.092)
Transportation	Increased × 2.6 (26.12)	Increased × 1.8 (0.66)	Declined; relative amounts negligible
Electric Utilities	Increased × 1.6 (0.91)	Increased × 5.4 (5.66)	Increased × 4.8 (19.99)
Total 2002:	38.40 quadrillion BTU	23.06 quadrillion BTU	22.19 quadrillion BTU
Total 1960:	19.92 quadrillion BTU	12.39 BTU	9.34 quadrillion BTU
Total Increases:	× 1.9	× 1.9	× 2.38

* Source: U.S. Department of Energy, *Monthly Energy Review.*

which relies almost entirely on petroleum. Today our consumption of petroleum equals roughly 85 percent of our consumption of natural gas and coal combined. From the economic perspective, however, one might note that during the same period that the total U.S. consumption of petroleum, natural gas, and coal doubled, the U.S. gross domestic product in constant 1996 dollars quadrupled.[17] Thus the nation's GDP has grown twice as fast as its energy consumption.

That growth in the U.S. GDP, however, has not included a growth in domestic energy production sufficient to meet demand. Annual energy production lags behind consumption by about 25 quadrillion BTUs a year.[18] If we refer again to Table 9.1, we can see that 25 quadrillion BTUs is only slightly less than the entire increase in petroleum use for transportation in the U.S. between 1960 and 2002. As with other fossil fuels, the production of oil depends not only on industry capacity and business decisions, but on underground reserves, and our reserves of oil (unlike coal reserves) are declining.

Crude oil now represents less than 20 percent of U.S. annual energy production. (Coal constitutes 32 percent and dry natural gas slightly under 30 percent; the remaining annual production is of liquid natural gases, nuclear electric power, and renewable energy.)[19] U.S. domestic petroleum production, which peaked in 1970 at 9.6 million barrels per day, has declined steadily to 5.8 billion barrels per day; an increasing share of production has come from Alaska, which now contributes 24 percent to the total annual domestic production of oil.[20] Alaska—as well as Canada—is also the likely source for our increased consumption of natural gas. All of which results in the single most critical fact for U.S. energy policy: the nation's dependence on imported foreign oil to meet demand. For the United States—absent substantial changes in energy consumption—*the continued ability to import oil, as well as natural gas, absolutely requires that the global market place for these fossil fuels remains an open system.*

Energy Supply and Demand: Policies and Paradoxes

Because fossil fuels—and since World War I most notably oil—have become the single most critical natural resource for the U.S. economy and national security, federal regulation of the extraction, production, and distribution of petroleum, coal, and natural gas products can serve as the acid test of the ideological and political boundaries of U.S. policy making. Ideologically, the perception of nature as a plentiful and self-generating system has conspired with the abundance of fossil fuels to impede government efforts to manage demand as well as supply. A no less tenacious belief in the open system of a laissez faire economy has further complicated the making of energy policy in the United States, with the result that more often than not it appears composed of ad hoc responses to near-term crises, resulting in national policies more remarkable for their paradoxes than for their coherence.

Because the petroleum, coal, and natural gas industries arguably require the largest outlay of initial capital investment (in exploration, transportation, processing, and distribution) of any modern industry, market stability—and thus predictability—is critical to its survival. Thus the oil industry in the past has welcomed government intervention in the energy market place as readily as the most ardent environmentalist—but it has done so in the interests of a stable business environment.

The close involvement of energy industry executives in the development of federal energy policy—an involvement that led to a public interest group lawsuit against the Bush administration's National Energy Policy Development Group for alleged violations of the Federal Advisory Committee Act (see chapter 5)—dates back to World War I. Tanks, trucks, and motor vehicles were faster and more versatile than the railways that had been commandeered for earlier wars, while aircraft saw their first use not only for surveillance, but for combat. All ran on the internal combustion engine, as did the increasing number of private automobiles, which was both cause and effect of suburbanization. Both Britain and the United States had already (in 1911) decided to convert their navies from coal to oil. In Britain the Royal Dutch/Shell Company, in which the government had acquired shares in order to ensure oil supplies, "became integral to the Allies' war effort . . . acquiring and organizing supplies around the world for the British forces and . . . ensuring the delivery of the required products from Borneo, Sumatra, and the United States to the railheads and airfields in France."[21]

When, in 1918, the United States was drawn into the war partly because of the German U-boats' ravages of trans-Atlantic shipping, France and Italy joined the United States and Britain in forming an inter-allied Petroleum Conference to acquire, coordinate, and distribute oil to allied forces. Much of the actual work of the conference was carried out by personnel from Royal Dutch/Shell and Standard Oil of New Jersey. Once the Russian revolution ended access to that country's plentiful oil fields, the Allies came to depend almost entirely on U.S. oil. In order to ensure sufficient supplies not only for war needs but the rapidly expanding domestic market, Woodrow Wilson created a Fuel Administration, whose Oil Division worked hand in glove with the industry's National Petroleum War Service Committee to coordinate supplies. The large U.S. oil companies worked as one, the prohibitions of the Sherman and Clayton antitrust legislation having been overlooked in the interest of national necessity.[22]

During the Great Depression of the 1930s a glut of oil in Texas resulted in a collapse in prices, which state-imposed production quotas ("prorationing") had not alleviated, thanks to the producers of bootleg ("hot") oil secretly siphoned from pipelines. The Independent Petroleum Association of America and its members' representatives in Congress appealed to the White House for help. Help came in the form of President Franklin D. Roosevelt's secretary of the interior, Harold L. Ickes, a progressive minded lawyer out of Chicago. While Ickes had little love for the oil industry, he recognized that "without oil, American civilization as we know it could not exist."[23]

Under the provisions of the Interstate Commerce Act, Congress authorized the president to ban bootleg oil from interstate commerce. Then, under the National Industrial Recovery Act (1933, which authorized a new National Recovery Administration, or NRA, to develop codes with industry to set prices, wages, hours, and limit output) Ickes worked with the oil industry to devise an Oil Code that set state-by-state production quotas. Only the fear that oil might eventually be treated like another public utility prevented the industry from supporting government-set prices, but it welcomed federal prorationing (limiting production) where state prorationing had failed. Thus it was that the U.S. government, in peacetime, intervened in the nation's (and by then the world's) energy market place in one of the most substantial ways conceivable, short of nationalization. And it did so at the behest of the industry, whose interests at that time happened to coincide with the government's.

The New Deal's intervention in the workings of an open market for oil may have been accepted by the industry as the lesser of two evils, but the Supreme Court found that the Congress's delegation of authority to the president to set mandatory industry codes through the NRA represented not only an unconstitutional violation of the separation of powers, but an abuse of the administrative powers normally granted the executive to implement congressional legislation. Both principles were (and remain) fundamental to the open system of government crafted by the founding generation of United States (see chapters 1 and 4). In January 1935 the court ruled that the oil provisions of the NRA forbidding the interstate transportation of bootleg oil left too much to the administration's discretion. Six months later, when the NRA was about to expire, in a case involving not hot oil but sick chickens (which had been sold locally after being transported across state lines) the court ruled that the NRA's effort to control business practices exceeded the scope of the Congress's power over interstate commerce.[24]

Fearing another oil glut and collapse of prices, the oil producers were willing to do "voluntarily" what the government could not require them to do: they would adhere to "suggested" production quotas based on the Bureau of Mines' (not coincidentally a bureau of Ickes' Department of Interior) estimates of current demand. A tariff on imported crude, fuel oil, and gasoline passed by Congress (and signed into law) in 1932 also began to have its effect, so that by 1934 price *stability*, which the industry wanted, was achieved without price *controls*, which it dreaded. Between 1934 and 1940 the average price of a barrel of oil in the United States ranged only between $1.00 and $1.18.

By World War II, when the adequate supplies of oil drove major strategic decisions on all sides, to ensure adequate oil for their war effort the British reorganized their oil industry (consisting in 1940 of Shell, Anglo-Iranian, and Standard Oil of New Jersey's British subsidiary) as a government agency, the Petroleum Department. Rationing soon followed. The United States, by contrast, had a substantial surplus—a result of the glut of the 1930s—and by 1940 was producing about two-thirds of the world's oil. President Roosevelt appointed his now well-seasoned oil czar to the new position of petroleum coordinator for national defense (shortly thereafter renamed head of the Petroleum Administration for War, or PAW), where Ickes wrestled with the latest paradoxes in the nation's ad hoc energy "policy." Before, he had to control overproduction; now he had to ensure an abundant supply of oil and prevent scarcity. Before, the government was happy to organize the industry into a cartel to manage production; now the Department of Justice, having rediscovered the Sherman and Clayton antitrust acts, was busily prosecuting the American Petroleum Institute (comprised of twenty-two major oil companies and over 300 smaller independents) for innumerable violations of the antitrust law.

Ickes convinced the Department of Justice to exempt the oil industry from its antitrust prosecution, so that the companies could coordinate operations and pool supplies. When Ickes was criticized for the high proportion (over two-thirds) of PAW managerial and technical personnel who came from the industry, he justified this melding of public and private interest as necessary to provide the PAW the necessary competence to do its job. This argument would be repeated again and again in later years as federal agencies sought to justify their reliance on "experts" from their constituencies as members of their advisory committees (see chapter 5).

As Germany became more successful at sinking oil tankers traveling up and down the eastern seaboard, two pipelines were constructed between 1942 and 1944 to carry oil from Texas and the southwest to the east coast. The pipelines—"Big Inch" and "Little Inch"—totaled over 2,700 miles in length, and gave a boost to pipeline technology that would prove valuable in the years ahead. Ickes was able to persuade the Congress to allow oil exploration companies to deduct drilling costs from their taxes. Gasoline rationing to reduce oil consumption was justified to the public not on the grounds of scarcity, but of patriotism. Citizens should conserve rubber for tires for the military, now that the Japanese had cut off 90 percent of the world's natural rubber supply from Indonesia and Malaya. Most importantly, the PAW succeeded not by commandeering the oil industry, but with its cooperation. Bountiful nature, which had first formed and filled the country's vast oil fields in the distant past, also helped. Six of the seven million barrels of oil consumed by the United States and its allies during World War II came from the United States.

THE ENERGY POLICY model forged during the decades embracing World War I, the Great Depression, and World War II illustrated the power of the concept of open systems—reflected in the ideological hold of laissez faire economics and sanctity of private property—to frame the boundaries of government intervention in private enterprise. First, unlike most other industrialized nations, private individuals in the United States could, and did, own mineral rights as well as the federal government. Thus the ability of the government to commandeer oil and coal fields during national emergencies was limited. Nationalization not being a realistic option during a period when communism became the bete noir of national politics, the government needed the good will of the oil, coal, and natural gas industries as much as those industries needed a friendly government.

Second, when national emergencies forced the federal government to act, its actions were ad hoc and relied heavily on measures negotiated with the industry. In the mechanized and mobile warfare of the twentieth century, government had to do what it could to ensure oil supplies, while during the Great Depression it was called on to restrain over-production. At the same time that the federal government had been intent on prosecuting antitrust violations in the oil industry, it was assisting in the forming of cartels in order to enlist the help of the industry in managing oil supplies and distribution.

The end of war in 1945 did not mean the end of the need for oil supplies. As the domestic economy of the 1950s churned out tens of millions of automobiles to satisfy an entire society that had taken to the road, heading toward suburbs, fast-food eateries, or the motels—traveling over more than 100,000 miles of interstate highways built largely with federal dollars—the U.S. lost its self-sufficiency in oil.

Cheaper oil beckoned from the Middle East, especially Saudi Arabia, and to help the U.S industry retrieve it the federal government assisted Standard Oil of California, Texaco, Standard Oil of New Jersey (later Exxon), and Socony (later Mobil) with the legal and financial hurdles necessary to form the Arabian-American Oil Company, or ARAMCO. The large American oil companies, now multinational, were soon joined by thousands of smaller companies seeking federal help in obtaining oil from cash-strapped countries in the developing world. Meanwhile, the Petroleum Industry War Council of World War II was replaced in 1946 by the National Petroleum Council (NPC) which consisted entirely of industry executives who picked their own members and set their own agenda, and made recommendations for the appointment of federal officials. The NPC remained in business until the creation of the Department of Energy in 1977 incorporated its functions within a new Energy Research Advisory Board, re-chartered in 1990 as the Secretary of Energy Advisory Board, or SEAB and subject to the Federal Advisory Committee Act of 1972 (see chapter 5).

Third, ad hoc policy remedies for inadequate or excess production in an industry that has different sized players, both exporters and importers, can and did foster plural constituencies, and that pluralism would be mirrored in the Congress and the federal bureaucracy. Large multinational oil companies like ExxonMobil do not necessarily share the same interests as smaller independent oil producers and refiners such as Hunt Oil Company, a Dallas-based company that led the development of candidate George W. Bush's energy plan. For example, from 1959 to 1973 oil importers were subject to import quotas that were a form of protectionism for the domestic oil industry. The Mandatory Oil Import Program was justified as promoting U.S. energy independence from political disruptions in the Middle East (the Suez Canal was bombed in 1956 during fighting between Israel and Egypt). The large importing companies ("big oil") had argued that continued imports would help to postpone the depletion of domestic oil reserves, but the interests of domestic producers prevailed. Pluralism can also lead to a mélange of oblique regulatory measures, and subsidies buried deep within the tax code (such as oil and gas drilling tax incentives and the oil depletion allowance), instituted at the behest of one or more industry interests.

The extent to which anything remotely resembling national energy policy was fragmented among multiple economic and political interests is readily apparent from the number of organizations that were swept up under the bureaucratic umbrella of the Department of Energy, created in 1977 as a response to the "energy crisis." In October 1973 Egypt and Syria launched a surprise attack on Israeli forces on the Golan Heights and at the Suez Canal. Though reluctant to intervene, the United States began to supply Israel after the Soviet Union began to reinforce Egyptian and Syrian forces. Not surprisingly, the Arab-led Organization of Petroleum Exporting Countries (organized in 1960 by Saudi Arabia, Iran, Iraq, Kuwait, and Venezuela) retaliated with an oil embargo on the United States.[25] As first-come, first-serve rationing occurred at gasoline stations around the country, President Richard M. Nixon (1969–1974) urged the public to drive less, lower thermostats, and dim the lights. He also obtained congressional approval to allocate all petroleum-based fuels and set temporary prices. An unusually cold winter in 1976–1977 added to the crisis—though opinion polls repeatedly found that most Americans blamed the government, the oil companies, and OPEC, rather than their own fuel consumption, for the shortages.

As the country saw with the "war on terror" following the September 11, 2001 terrorist air attack on New York's twin World Trade Center towers, policy crises in Washington typically result in paroxysms of reorganization designed to effect centralized, coordinated, and comprehensive bureaucratic remedies for systemic contradictions in policy making and the administrative confusion that results. The energy crisis of the 1970s was no exception. In 1974 the Congress combined the energy research functions of the Atomic Energy Commission (est. 1946) and the interior department's Office of Coal Research and Bureau of Mines' research centers into a new Energy Research and Development Administration (ERDA), and created the Nuclear Regulatory Commission to license and regulate private nuclear facilities. It also authorized Nixon to create the Federal Energy Administration (FEA) to manage petroleum allocation and pricing, the national strategic oil reserve, and federal programs to promote energy efficiency and conservation. (These functions were transferred in 1977 to the Economic Regulatory Administration.) Three years later, with dependence on foreign oil increasing and influence in the oil fields of the Middle East waning, the Congress relented on its "big government" scruples and approved the creation of a new Department of Energy (1977).

The DOE was born as a mammoth conglomeration of ERDA and FEA, along with energy management functions from the Interstate Commerce Commission; the interior department's energy information and analysis function, as well as its Alaska, Bonneville, southeastern and southwestern electric power marketing administrations; the Department of Defense's Navy oil and oil shale reserves; the Atomic Energy Commission's national laboratories; the National Science Foundation's solar and geothermal research programs, and the Environmental Protection Agency's (see the following paragraphs) automotive systems research program.

Each and every one of these bureaucratic units had its patrons in the Congress, who could be counted upon to protect jobs at home and power in Washington. But no genius of public administration could overcome the basic paradox of U.S. energy policy during the 1970s. The ideological trope of a free market was becoming an illusion, buffeted by the political cross winds of the oil crisis of 1973–1977. Public opinion and members of Congress blamed OPEC and "big oil" for the apparent shortages and price increases, which deflected political energies from serious efforts to reduce demand. Responses thus focused on supply and prices, beginning with an economy-wide price freeze in the autumn of 1971. Two years later Congress allowed price increases, which could reflect cost increases, but imposed profit limitations. An oil "entitlements" program introduced in 1974 restricted oil refiners to a certain number of barrels of processed oil in proportion to their refining capacities. By 1975, reacting to legitimate complaints that gasoline and heating oil prices disproportionately harmed the poor, Congress (in the Emergency Petroleum Allocation Act) directed the White House to create a Federal Energy Agency to allocate all petroleum products equitably and efficiently around the country.

The regulatory campaign to replace the supply/demand function of the market with federally controlled oil allocations and prices turned out to be an administrative nightmare and, as complaints mounted, bureaucratic intervention in oil allocations and pricing made the free action of the market place seem beautiful indeed. It did not help the pro-regulators that a surplus in world oil supplies seemed to remove the need for such extensive intervention in the market. In 1975 President Gerald Ford persuaded the Congress to approve a gradual elimination of administrative controls of allocations and prices, establish a strategic petroleum reserve which would store 750 million barrels of oil in salt caverns in the Gulf Coast states to protect against major disruptions in U.S. oil supplies, and delay compliance deadlines under the Clean Air Act of 1970.[26] It was already evident that further complicating the challenge of creating a coherent and comprehensive national energy policy was the growing realization that burning fossil fuels clouds the air with substances harmful to humans and other living things.

Environmental Policy and the True Costs of Fossil Fuels

The origins of the U.S. Environmental Protection Agency illustrate the incremental nature of government regulation in a society deeply imbued with the ideology of laissez faire. The EPA was cobbled together in 1970 from pre-existing programs in other agencies. These included air quality, solid waste, and drinking water regulation in the Department of Health, Education, and Welfare (HEW); water quality and pesticides research in the Department of Interior; and pesticides regulation in the Department of Agriculture. Similarly, the Congress built the EPA's portfolio of regulatory authorities gradually, beginning with air pollution (Clean Air Act of 1970), surface water pollution (1972), drinking water quality (1974), the generation and disposal of hazardous wastes (1976), the manufacture and marketing of toxic chemicals (1976), and cleanup of abandoned waste sites (1980). The Congress has been willing to give the EPA authority to set and enforce regulations only gradually, with regulations specific to an industry or type of emissions often requiring separate statutory authority. Moreover, arguably more than in any other instance of science and technology policy examined in these pages, the making of environmental policy since 1970 has had to engage a highly polarized constellation of constituencies and opponents. This challenge is intensified by these constituencies' and opponents' divergent but largely tacit views of the nature of the natural environment.

To many supporters of the energy and manufacturing industries most affected by environmental restrictions on emissions, nature is robust and resilient—an open and plastic system that continuously absorbs change as an ordinary dynamic of its vitality over time—much in the same way that Darwinian evolution functions through continual adaptation to a changing environment by random variations in an open-ended universe of varied species. This perception of nature is consistent with the economic ideology of the open market (see chapter 2): the market, if allowed to operate naturally (according to the law of supply and demand), will, more effectively than any government policy, stimulate and sustain a continuing flow of new and better things. In this ideal world industry and entrepreneurship, drawing on nature's bounty, can ensure the continuation of American prosperity.

To many proponents of strict environmental regulation, on the other hand, the natural world—save for the depredations of humans and their technologies—is a pristine garden at the edge of a wilderness, thriving on the delicate ecological balances that enable the continued existence of great quantities and varieties of animal life and vegetation. The "natural" place humans have in such a landscape is one of simple and contemplative living, producing and consuming only what is necessary to life and the innocent pursuit of non-material pleasures. Both views of nature are embedded in environmental policy debates, in which they are rarely, if ever, joined. Leo Marx concluded his classic study of the leading figures of nineteenth-century American letters with the observation: "There is nothing in the visible landscape—no tradition, no standard, no institution—capable of standing up to the force of which the railroad is the symbol. . . . The contrast between the machine and the pastoral ideal dramatizes the great issue of our culture. It is the germ, as [Henry] James puts it, of the most final of all generalizations about America."[27]

Environmental regulation in the United States begins with the environmental movement, which was both cause and effect of the popularization of the view that the natural world is composed of ecosystems which shape individual life forms' relationships with their environment—an idea familiar to natural scientists since the late nineteenth century. A gifted writer and aquatic biologist with the U.S. Fish and Wildlife service, Rachel Carson published essays and two books about the richness of natural life under the sea—one of which received the National Book Award and has been translated into thirty languages—before warning about the danger of pesticides in *Silent Spring* (1962).[28] *Silent Spring* explained for wide audiences just how toxic chemicals— in this case dichloro-diphenyl-trichloroethane (DDT), a by-product of petroleum—accumulate within the tissues of animals and humans. Already (beginning in 1957) the U.S. Department of Agriculture (USDA) and Department of Interior had begun to phase out use of chlorinated hydrocarbons (which include DDT) on federal lands, and on the eve of the EPA's creation in 1970, the USDA began to cancel DDT registrations for use on shade trees, aquatic areas, house and garden, tobacco fields, food crops, cattle, goats, sheep, swine, lumber, finished products and buildings, and commercial, institutional, and industrial establishments. DDT was the first pesticide denied federal registration by the EPA, though its action was hastened by a 1971 court order following a suit by the Environmental Defense Fund, organized in 1967. The Federal Environmental Pesticides Control Act of 1972 expanded EPA's regulatory authority, allowing it to issue temporary DDT registrations for use under certain emergency conditions, when no alternative was available.

While reliance on non-polluting pesticides, herbicides, and fertilizers continues to be a challenge, public concerns over the environmental costs of meeting the nation's energy demands have crystallized around issues stemming from our use of fossil fuels, in particular gasoline and diesel fuels for motor vehicles, and the use of coal to power electrical generation plants (see Table 9.1). Three touchstone issues at the opening of the twenty-first century have been (1) whether to permit oil drilling in the Arctic National Wildlife Refuge (ANWR); (2) increasing fuel economy standards for automobiles and light trucks; and (3) U.S. policies to reduce greenhouse gases, including participation in the international Kyoto Protocols setting international targets for emissions reductions.

Oil Drilling in the Alaska National Wildlife Refuge (ANWR)

Drilling for oil in the Arctic National Wildlife Refuge in Alaska has been proposed repeatedly since the mid-1980s. Championed by Alaska's congressional delegation as a way to boost the state's economy, drilling in the ANWR has been estimated by the U.S. interior department as likely to produce lease revenues for state and federal governments of about $2.4 billion in 2007, to be divided between state and federal governments. However, a major oil spill from the tanker Exxon Valdez in 1989 off the coast of Alaska galvanized opponents, who feared damage to the region's aquatic ecosystem from further accidents. A proposal to lease ANWR for drilling was vetoed by the Democratic president Bill Clinton in 1995, while his successor, George W. Bush, included oil drilling in the ANWR as part of his administration's national energy plan. The Bush administration, the Alaska congressional delegation, and other supporters of ANWR drilling argue that the amount of recoverable oil under the ANWR coastal plain (estimates vary widely, from less than 10 percent to more than 25 percent of the nation's domestic oil reserves) is not only essential to achieving energy independence, but that modern technology enables drilling to proceed without harming the area's wildlife.

Despite appearances, ANWR is not simply another instance of big oil and its Republican allies struggling to satisfy the country's energy greed at the expense of pristine natural areas. In fact, the major multinational oil companies—British Petroleum (BP), Chevron Texaco, ConocoPhillips, and ExxonMobil—which are importers as well as exporters of crude oil, no longer lobby for ANWR drilling, aware from exploratory drilling in the 1980s that the yield is not likely to be worth the high capital and transportation costs. But "big oil" is not the only player in an industry that contains numerous other smaller and independent companies, such as Hunt Oil Company of Dallas, Texas, whose top executive Hunter Hunt was among those who advised the Bush administration in the development of its national energy policy. The ANWR issue remains alive also because its proponents see it as an opener for the release of restrictions on oil drilling in more promising waters off the coast of California and Florida.[29]

How the issue is resolved was likely to depend on the outcome of fierce congressional battles over the Bush administration's federal budget for the 2006 fiscal year, a budget that became a flash point for other controversies, such as the war on terror and the continuing presence of U.S. troops in Iraq; Medicare reform, especially its prescription drug benefit, which in early 2005 was estimated to cost nearly twice what the administration said it would when the bill was passed in 2003 (see chapter 8); and the administration's proposal to reform Social Security by allowing younger enrollees to divert some of their Social Security taxes to private retirement investment accounts. The resolution of the ANWR controversy—and similar controversies to come—will be negotiable, which is the way of all tenable solutions in open political systems.

Increasing Mandated Fuel Economy Standards for Automobiles and Light Trucks

Petroleum use in the United States has declined since 1960 in the residential and commercial sectors, which have turned increasingly to natural gas for their energy needs (see Table 9.1). But any general reduction in petroleum consumption represented by these declines has been erased by the moderately increased use of petroleum in industry and to generate electricity, and the nearly *threefold* increase in petroleum use for transportation. While increased petroleum use by industry and electric utilities is almost proportional to the 1.6 times increase in the U.S. population since 1960 (from around 180 million to 282 million), it is much less than the fourfold increase in the nation's GDP during the same period. In short, were the production of petroleum to keep pace with demand without increasing social, economic, or political costs, oil use might not be the national energy policy issue that it has become.

But petroleum supply has not kept pace with the threefold increase in petroleum use in the nation's transportation sector, an increase that represents more than two-thirds of the total increase in the nation's consumption of petroleum since 1960. What's more, motor vehicles consume over two-thirds of all transportation fuels; the rest is consumed by aircraft and diesel-powered vehicles in marine shipping and commercial trucking.[30] Substantially reducing gasoline and diesel fuel use by motor vehicles would have a dual benefit by lessening both U.S. dependence on foreign oil imports and environmentally harmful emissions.

Contrary to what one might conclude from the amount of traffic congestion on the nation's inter-urban highways, the number of motor vehicles owned by Americans has increased, but not dramatically, since the 1950s. In 1960 there were two registered motor vehicles for each three persons over the age of eighteen residing in the United States. By 2000 there was one registered motor vehicle for each person over eighteen years. Since the post–World War II economic expansion, when annual new passenger car sales jumped from 3.7 million in 1940 to 6.6 million in 1950, sales of new passenger cars have *declined* steadily to just under 5 million in 2001. *However*, annual sales of new trucks and buses (which also almost doubled between 1940 and 1950), have increased steadily from 1.3 million in 1950 to 7 million in 2000.[31]

Reacting to the OPEC oil embargo, the Congress in 1975 passed the Energy Policy and Conservation Act, which (among other things) mandated Corporate Average Fuel Economy (CAFE) standards for automobiles and "light trucks" (gross vehicle weight rating, or GVWR of 8,500 lbs. or less). The standards would be administered by the National Highway Traffic Safety Administration (NHTSA) which, like other regulatory federal agencies must maintain a cooperative working relationship with the industry it attempts to regulate in order to succeed. Thus the fuel economy standards the NHTSA devised apply to an automakers' average fleet mileage, rather than the mileage of particular vehicle categories, gave the industry long lead times to meet their CAFE standards, and allowed manufacturers to accumulate CAFE credits. Four years later the NHTSA also set mileage standards for light trucks, minivans, and sport utility vehicles (SUVs) which were more lenient than those for passenger vehicles. The industry responded by producing new vehicles with an average fuel economy by 1980 of 23.5 miles per gallon—which exceeded the CAFE for that year by over 3 mpg.

The market pressure of consumers coping with high gas prices during the 1970s alone may have been enough to stimulate improved fuel economy, and as gasoline prices began to fall again in the 1980s, the CAFE standards served as a non-market constraint on the industry. However, effective marketing and consumer taste conspired with the NHTSA's own definition of a "truck" to enable manufacturers to launch new product lines that could side-step the CAFE standards: minivans and SUVs. Since a truck, according to the NHTSA, is any vehicle in which all seats (besides the driver's) can be folded or removed to create a cargo area, minivans and SUVs could pass muster as light "trucks," and thus need only meet the CAFE standard of 21 mpg, rather than 27.5 mpg for passenger cars. Heavy duty trucks—those over 8,500 GVWR—are generally exempt from fuel economy standards, which helps to account for the appearance of such large "sport utility vehicles" as General Motors' "Hummer H1" and Ford Motor Company's "Excursion."

In 2003 the NHTSA began a review of the CAFE standards, with a view toward redefining "truck" to close loopholes that allow large gas-thirsty SUVs to masquerade as trucks, and raising the weight limit for "light trucks" to 10,000. Whether such revisions, if they take place, will be circumvented by the same ingenuity in automotive design and marketing, and a willing consuming public, remains to be seen. The larger question, however, is whether the approach represented by CAFE standards can be as effective as the simple pressure of higher prices on consumer demand, and the industry's market-driven response. Given that consumer demand for gasoline (and smaller cars with better gas mileage) was shown to be price-sensitive during the 1970s, the long-term downward trend in gasoline prices, *adjusted for inflation*, has been modulating the market mechanism that might otherwise lead to more fuel efficient motor vehicle technologies.[32] Indeed, the CAFE standards may be functioning as fuel inefficiency *allowances*, rather than fuel efficiency ceilings.

Meanwhile, the automobile industry's own innovations suggest a recognition that the regulatory process for fuel economy (whether federal or state) is unpredictable and costly, just as unpredictable as geopolitically sensitive oil supplies. At the same time, environmental regulations restricting allowable emissions of nitrous oxides, carbon monoxide, and unburned hydrocarbons led the industry to explore ways to design and build "green" automobiles. An early industry response to environmental restrictions on motor vehicle emissions was the catalytic converter, developed in the mid-1970s by General Motors. Positive crankcase (PCV) valves, which reduce unburned hydrocarbon exhausts, and unleaded fuels, which reduce damage to catalytic converters from lead, soon followed. Thanks to these and other innovations today's cars produce 60 to 80 percent less pollution than cars available in 1960.[33]

By the 1990s various alternative-fueled vehicle designs were in production. As a result the total number of alternative-fueled vehicles (which do not include gasoline/diesel/electric hybrid vehicles) available in the U.S. nearly quadrupled between 1992 and 2003, when they numbered slightly over 880,000.[34] Led by Japanese automakers Honda and Toyota, manufacturers have added over 50,000 hybrid gasoline/diesel/electric vehicles to this number; both Ford and General Motors now include them in their fleets. Ford, Chrysler, and General Motors have also introduced cylinder deactivation systems, which electronically allow some cylinders in larger V-8 engines to idle when they are not actually needed for acceleration.

The Environmental Protection Agency has also sponsored research and development on cleaner diesel engines which are being adopted by Ford for some models, as well as hydraulic transmission systems which can increase fuel efficiency by as much as 60 percent. These alternatives to mechanical transmissions are ideal for the stop-and-go driving done by delivery trucks, and are now being tested by United Parcel Service on its own trucks. Ford Motor Company is also working with the EPA on a hydraulic transmission system for its SUVs, aiming for production in 2010.[35]

From Clean Air to Clear Skies: Open Systems for Sustainable Development

The federal government's first efforts to reduce air pollution occurred with the passage in 1963 of the first Clean Air Act, which created a voluntary program of environmental research funding and encouraged the creation of state pollution control agencies. The first real regulatory step was taken two years later, when the Act was amended to require the Department of Health, Education, and Welfare to create and enforce automobile emission standards. Further amendments in 1970—the same year that EPA opened its doors for business—directed EPA to develop National Ambient Air Quality Standards to protect humans as well as the natural environment; establish "new source" standards for setting pollution allowances for industry; and required the development of state pollution reduction plans to be approved by the EPA.

Post-1990 advances in cleaner or alternative fuel burning automobile engines are generally invisible until they are actually introduced onto the market and advertised. This is not the case, however, with SUVs, which have surely attracted as many critics as buyers, with critics scorning SUVs as dangerous, gas-guzzling road hogs, symbolic of the worst excesses of American material-ism. Still, in 2003, SUVs, pickups, and vans comprised 48 percent of all sales, more than twice their market share in 1983.[36] Only increases in gasoline prices beginning the summer of 2005, resulting partly from interruptions in supply thanks to Hurricane Katrina, could reverse the increase in SUV sales.

The 1970 Clean Air Act was noteworthy for its limited effectiveness. Indus-tries questioned specific requirements and penalties and the EPA was soon embroiled in costly litigation over numerous actions that delayed enforce-ment. Companies would file suits arguing that the mandated levels were too high; environmental groups would counter that the levels were too low. Dozens of cities failed to meet national standards for the reduction of ozone (to reduce smog), carbon monoxide, and particulates—thus demonstrating the weakness of "command and control" approaches in an open system of government that is designed for negotiated resolutions of policy disputes.

In response to the mixed results of the 1970 amendments, the 1990 Clean Air Act amendments introduced a "market-based" open-systems innovation into environmental regulation. The EPA credited the system's design to the Environmental Defense Fund, whose chief economist, Daniel Dudek, has been an effective advocate for allowing firms to accumulate and trade emissions "al-lowances." The EPA would provide "incentives, or 'credits' for companies which act quickly to reduce toxic emissions" or exceed them and allow "tradable emission credits for producers of certain kinds of reformulated fuels, for manufacturers of clean-fuel vehicles, and for vehicle fleets subject to clean fuel requirements."[37] In introducing the 1990 changes, EPA administrator William K. Reilly promised that "the market-based approach to environmental pro-tection . . . will serve as a model for other [George H. W. Bush] Administration proposals in the future . . . the key is to devise . . . programs that put the mar-ketplace to work on behalf of the environment."[38] And use the model it did; the George W. Bush administration's proposed Clear Skies Act of 2004 was still be-ing hotly debated in 2005, with critics arguing that it was a gift to industry in that it reduced some of the mandated pollutant levels from the 1990 Clean Air Act amendments and lacked mandated reductions of carbon dioxide emissions.

Supporters of the proposed Clear Skies legislation pointed to its "cap and trade" provisions, which expand the market-based feature of the 1990 law. In-dustries as a whole would be given targets, and companies could pursue their own strategies for reducing emissions, including trading permits among them-selves. According to the EPA, the nearly one-third reduction of sulfur dioxide emissions from coal-fired power plants since 1990 can be attributed to a pilot "cap and trade" program for the electric power industry that was authorized in the 1990 law. Opponents and supporters alike pointed—for opposite reasons—to the likelihood that the Clear Skies act, if passed, would serve as the model for future efforts to reduce carbon dioxide emissions.[39] Meanwhile, states—such as those in the northeastern United States—can act, and have acted, to reduce carbon dioxide emissions, and in doing so have adopted "cap and trade" approaches.

Awareness of the phenomenon of atmospheric heating due to trapped "greenhouse" gases dates back at least to 1896, when the *Philosophical Magazine* published a paper by the Swedish chemist Svante Arrhenius entitled "On the Influence of Carbonic Acid [i.e., carbon dioxide] in the Air Upon the Temperature of the Ground." By the mid-1950s scientists had begun to collect data systematically on atmospheric concentrations of CO_2 with many becoming convinced in the 1980s that the Earth was undergoing a man-made process of "global warming." Skeptics questioned whether changes in global temperature during even several centuries could be generalized into millennial trends, while others speculated that the widespread extinctions of ocean and land species 250 million years ago were most probably due to the greenhouse effect caused by CO_2 released into the atmosphere from volcanic fissures. How well could scientists isolate man-made CO_2 emissions as singularly responsible for global warming, compared to other sources? And how much was "too" much, with what certain consequences for whom, and when? While the federal government has funded, and continues to fund, the collection of data about the Earth's climate from the ground, oceans, air, and space, how to interpret much of the data, and what anyone should do as a result of conclusions drawn (even if they were unanimously agreed to) continue to bedevil the issue.

Nonetheless, enough of the 178 national governments represented at the 1992 United Nations Conference on Environment and Development in Rio de Janeiro had been convinced that global warming constituted a serious threat that they agreed to set voluntary targets for reducing emissions of carbon dioxide, methane, and other greenhouse gases. Three years later a scientific panel organized by the United Nations reported that available evidence indicates man-made emissions were affecting climate. When the parties to the Rio de Janeiro agreement met again in Kyoto, Japan, in 1997, they agreed to mandated emission targets by industrial nations—all, that is, except the United States.[40] The U.S. Senate was unwilling to ratify an agreement that would bind the country to environmental regulations not of its own making and thus probably harm U.S. industry. Rejection by the U.S. senate, however, was not the only sign of a lack of international unanimity on the issue. A diplomatic fissure appeared between the relatively wealthy industrialized nations of Europe and North America, and the poorer developing countries of Africa, Latin America, the Middle East, and parts of Asia—who saw environmental restrictions as an unfair burden on their own struggling economies. This burden might be alleviated however, they argued, by increases in financial aid.

When the Kyoto Protocol of 1997 went into effect in February of 2005, it lacked some critical players. While Russia ratified the Protocol in 2004, the United States, China, and India are not parties to it. Nor is the United States likely to subscribe to the Kyoto Protocol as long as China and India do not. The absence of these three large countries will handicap the Kyoto Protocol's effectiveness, because the United States produces about a quarter of the world's greenhouse gases, while 60 percent of the increases in the burning of coal during the next 25 years have been estimated to come from the United States, China, and India.

Another signal feature of the Kyoto Protocol is that it adopts the same market-based approach introduced into U.S. environmental policy in 1990. The thirty-five industrialized countries that agreed to the protocol may exchange credits for exceeding emissions targets with those who have failed to meet their own. Countries can also offset against their own targets successful emissions-reduction projects located in other countries.

These provisions reflect a significant conceptual affirmation in current environmental policy of an approach toward government regulation that reaches beyond basic notions of international law or cooperation, and mirrors the change in U.S. environmental policy in 1990. This is the concept of "sustainable development," a term that was articulated in the 1987 report of the United Nations Commission on Environment and Development (appointed in 1983). The concept acknowledges the conflict between economic development and environmental protection, or the needs of today's and future generations. The mediating principle between these dual objectives is *balance*—a principle we encountered in chapter 8, where bioethicists offered it as a way to mediate the needs of biomedical research and the moral issues surrounding research with human subjects.

However, balancing opposing social, moral, and political interests is not like balancing physical objects, for there is no quantifiable way to calibrate their relative weights. Moreover, in the arena of human conflict, balance, once achieved, cannot remain static for long because of the constancy of changing circumstances. Hence efforts to correct an imbalance of interests is a necessarily dynamic process which may occur destructively (through coercion), or peacefully—through bargaining, "log rolling," or, more elegantly, negotiations. This was the great insight James Madison brought to politics in his Tenth Federalist paper (1787), comparable in its historic importance to the insight Adam Smith brought to economics slightly over ten years earlier. Whether in the 1780s or today, approaches to policy differences that rely on successful negotiations require open systems to ensure the greatest number of options from which opponents can choose, be those options regulatory limits, emission allowances, or competing goods in the market place. This is the compelling pragmatic argument for participatory decision making: the more who come to the table, the more open the negotiations, the greater the likelihood that a mutually agreeable and thus enduring set of options can be found.

THESE TOUCHSTONE ISSUES—oil drilling in federally protected wildlife areas, reduced use of petroleum in the transportation sector for environmental as well as economic and foreign policy reasons, and the nature and extent of U.S. regulation to achieve clean air and retard global warming—are especially contentious for several reasons. First, they involve the economic interests of the oil, coal, gas, and electric power generation industry; the automobile industry; and all the states in which these industries are important to the state's economy—for example, Alaska and Alabama (see discussion of the Shelby amendment in chapter 5). Second, the current and anticipated environmental costs of failing to reduce greenhouse gases, or the national security consequences of U.S. dependence on foreign oil—such as U.S. involvement in historic power struggles in the Middle East—may not be burdensome to those not immediately affected by them. But they will be burdensome to future generations, a prospect that arguably has not significantly influenced U.S. domestic or foreign policy making to date, and is not likely to influence it in the near future.

Third, these issues necessarily engage tacit (and sometimes explicit) attitudes toward technology, and any litany of virtually unquestioned truths in U.S. policy making will include the notion that technological advance drives the economy by enabling industry to produce new or more advanced products in more efficient ways (see chapter 2). Industrialization and modern technology have always had their detractors, detractors who often ally themselves with "back to nature" movements and "green" politics. This phenomenon was especially pronounced in the United States during the 1960s and played no small role in the emergence of sufficient public support for environmental legislation during the 1970s.

Fourth, pro-industry and agriculture opponents of ambitious environmental restrictions on industry emissions or agricultural effluents do not lack an appreciative view of the natural world. Where they differ is in their understanding of nature's plasticity, or ability to adapt to exogenously induced changes—in other words, whether nature is truly an open system. The devastating earthquake and tsunami that ravaged Indonesia and the southeast coast of India in December 2004, taking over 130,000 lives, was but a reminder of the catastrophes our world's nature inflicts on the Earth as she constantly remakes herself.[41] By contrast, nature as a universe composed of fragile and delicately balanced ecosystems is a closed system, one presumptively imperiled by changes introduced by exogenous factors—principally technology and human behavior. Those who fear global warming do not fear global warming per se; they fear changes they expect to result from global warming. Thus the controversy—one in which the parties identify themselves by the choice of the phrase "global warming" or "climate change"—is not a matter of inadequate or conflicting scientific data, though both sides to the debate will contest the other side's data. It is a debate over what the data *means* in the broadest possible sense, and thus what should be done about it.

Fifth, environmental policy issues are not socially or economically neutral. Political choices made in the interest of ensuring a clean and unspoiled natural environment impact not only industries and the people they employ, but entail "quality of life" choices. In a vast nation with a largely capitalist economy, numerous cultural differences and socioeconomic strata, "quality of life" differences quickly translate into economic disparities and consumer preferences. As every automotive marketing executive knows, the make and model of any given private automobile signifies social status and "taste," by which status is most commonly advertised. Thus the rise (and now fall) in popularity of relatively heavy, fuel-inefficient, and not necessarily safe SUVs are as much a function of socioeconomic aspirations of those who drive them as they are a function of the attitudes of those who won't. The growing preference for smaller SUVs does not portend a widespread rejection of big vehicles; many consumers who might have bought a large SUV are now switching to pick-up trucks, which have grown heavier and *less* fuel efficient during the last twenty years.[42] But the automobiles we drive are not always a matter of personal consumer preference. In many rural areas, where the pastoral ideal is often a mockery of soul grinding poverty, a functioning car or truck decides whether one works or not. Thus it is that opinions surrounding policies to protect the environment from Earth-fouling emissions and effusions tend to be divided along partisan lines not only for ideological reasons, but for sociological, cultural, and economic reasons as well.

IN 1980 RONALD REAGAN, successfully campaigning against "big government," was elected president, and thus began two terms of an administration that was determined to bring deregulation to Washington in the name of restoring free-market competition to U.S. industry. A regulatory regime that had extended to airlines, telecommunications, securities, trucking, railroads, buses, cable television, oil, natural gas, financial institutions, and public utilities was gradually—but not completely—dismantled. For example, the restoration of laissez faire did not extend to the oil and gas industry subsidies buried in the federal tax code's oil drilling and oil depletion allowances. (Out of fairness to Reagan, he had wanted to see all such provisions eliminated in the tax reform act of 1984, but lost out to the oil industry's lobbyists and their protector in Washington, Texas attorney and White House Chief of Staff, James Baker. Baker would become secretary of the treasury the following year.)[43]

But the regulatory impulse that was challenged by Reagan's election was not entirely defeated. Congress was able to pass the Comprehensive Environmental Response, Compensation, and Liability Act (CERCLA) of 1980, or "Superfund" act, which requires parties responsible for non-federal polluted sites to clean up the contamination or pay the EPA to do it; the Hazardous and Solid Waste amendments of 1984; and the Safe Drinking Water Act amendments of 1986. The Federal Energy Regulation Commission (FERC, created in 1977) continues to regulate the transmission of electricity, oil, and natural gas; wholesale electrical rates; and the location and abandonment of pipelines and other oil and natural gas facilities. At the same time, again to promote greater competition, the FERC in the early 1990s oversaw the "unbundling" of the vertically integrated electric utilities, thus allowing the emergence of a host of separate generation, transmission, and supply companies; it required a similar unbundling in the natural gas pipeline, transmission, sales, and services industry.

In retrospect, the federal response to the fuel crisis of the 1970s represented an aberration in a half-century of public policy. The Congress was willing to help the industry stabilize its business environment by managing supplies through pro-rationing and import quotas. It had interfered in the market to this extent before—but only during wartime. However, price controls and supply allocations—tried for several years as a political reaction to OPEC's own oil geopolitics, and then abandoned—went too far. At the same time, supporters of the eastern coal industry promoted a revival of a government program to produce synthetic fuels (or "synfuels") first begun in 1960 with the creation of the Office of Coal Research (OCR) in the Department of Interior. (Eastern coal, which caked readily, was less well suited than western coal to the most commercially viable technologies for gasification.)[44]

The Energy Research and Development Agency (absorbed into the Department of Energy in 1977) began a program for synthetic fuels research in 1976, when the program—which required joint funding with industry—was given a political boost by the prospect of continuing interruptions in imported oil supplies. Thus the Congress was able to create the Synthetic Fuels Corporation (SFC) in 1980 to commercialize synthetic fuels by investing in plant development, to be financed by "windfall" profits taxes on other forms of energy. Plagued by cost-overruns and production short-falls, the SFC was cancelled in 1985; as oil prices fell dramatically in 1980–1985, so also did support for the program. The synfuels program represented another instance of energy policy at cross-purposes, because a new and by no means cheap fuel was less likely to prove "commercially viable" in an energy market in which prices for oil were artificially low.

The federal government's approach to the fuel crisis of the 1970s also represented an important missed opportunity. The energy crisis might have been used to accustom the public to adjusting to shrinking supplies at higher prices by consuming less. But too many members of Congress found it easier to resuscitate the populist figure of the rapacious businessman conspiring to profit at the expense of ordinary people. Ordinary people, of course, outnumber oil industry executives—whether rapacious or not—at the polls. What's more, after having enjoyed abundant and cheap oil (not to mention wood, water power, and coal) for so much of their history, the American people had come to believe that full enjoyment of abundant energy was their natural birthright.[45] It was left to the environmental movement, which was premised on a view of nature that was less open to change, to lead the country down the path of more efficient and cleaner fuels.

Significant increases in the cost of petroleum and gas at the pump during the late summer of 2005, at the same time that the U.S. was embroiled in a bloody conflict in Iraq, emboldened some observers in the political mainstream to call for an increased federal gasoline tax to dampen demand and thus reduce the country's costly dependency on foreign oil.[46] Low-income individuals, who also tend to be dependent on older, less fuel-efficient vehicles, might be assisted through tax credits. Given the likelihood that petroleum-exporting countries would reduce the cost of oil proportionately as a countermove to sustain sales, the net effect of an increased tax on gasoline and heating oil would be small. Meanwhile, a greater proportion of the dollars spent at the pump could fund the development of alternative fuels and fuel plants for the U.S transportation sector and industry. However, the probability of such proposals surviving political campaigns during the 2006 and 2008 election years was small. Rare is the candidate for political office who can muster winning support for an increase in federal taxes of any stripe.

10

═══════════════════════════════════════

Epilogue

From Woodrow Wilson to George W. Bush, U.S. foreign policy has aspired to illuminate the world with a combination of liberal political and capitalist values more commonly understood as an inseparable pairing of democracy and prosperity. This simple formulation overlooks the importance of a nation of laws constrained only by a written constitution in ensuring both political freedom and an orderly and predictable environment for economic activity. It also overlooks the central role of science and technology in generating the prosperity that is presumed to result from "democracy." While the iconography of American freedom in the United States is dominated by references to the growth of the American republic, abroad the iconography of American freedom consists as much of the cellular telephone, television sets, and the computer. Ironically, these wonders of modern technology are less and less likely to have been manufactured in the United States.

Thus accounting for the relationship between political systems, culture, and environment on the one hand, and national prosperity on the other, remains a central question of science and technology policy. Conventional wisdom over the years has offered various explanations for American prosperity. They include Max Weber's "protestant ethic" (*The Protestant Ethic and the "Spirit" of Capitalism* [1905]); the vast expanse of land which gave substance to the material and social ambitions of millions of internal and foreign migrants; and the governing institutions we have crafted for ourselves—institutions which, as Winston Churchill reminded the House of Commons in 1947, may on occasion seem to have created the worst form of government, except for all the others. But not one, or even all three, of these explanations is sufficient. What *can* adequately account for American prosperity is the common ancestor of most such explanations: a proclivity toward openness in the way we go about our business, whether that business is creating and maintaining our institutions or designing technological systems. As we saw in chapter I, that proclivity—by a fortuitous accident of historical timing—was the natural inheritance of the individuals and communities that initially populated and created the United States at the end of the eighteenth century.

In chapter 2 we observed the emergence of a philosophy that saw human happiness (and material prosperity) as the result not of restraints on human sinfulness by metaphysical authority, but of allowing natural human tendencies full play. The genius in Adam Smith's moral and economic philosophy lay in the *dynamism* of the underlying principle of Smith's challenge: an open economic policy—laissez faire—stimulates creative activity. The more who invent, and the more who bring their inventions into the market place, the more prosperous their society. Smith's essential principle was echoed in the political philosophy that shaped the new federal Constitution of the United States, most notably in the belief that political freedom, with its capstone right of "freedom of speech," entails not merely restraint on government, but an *active* process of corrective political discourse. The more who participate, and the more who bring their knowledge into the "market place of ideas," the richer the civic culture of the governed.

Among the powerful ideas that seeded themselves in the new landscape of America's frontier was the ideology of science. With its own seemingly limitless capacity to enrich and inform human experience, science—especially physics in the twentieth century—acquired an authority of its own. This authority, grounded in the exceptionalism of its practitioners rather than the experiences of ordinary people, would in time challenge the openness of the institutions those ordinary people used to govern themselves. Those institutions included a complex array of administrative mechanisms to carry out not only the policies of the Congress and White House, but the principles embedded in the Constitution. We observed this challenge, and the responses, in some detail in chapters 4 and 5.

In the remaining chapters we saw the extent to which a proclivity toward open systems in national law, policy making, and public administration has helped to shape the emergence of four of the nation's most expansive technological systems. In the case of the development of the Internet, an open systems approach toward the "architecture" of the new digital communications networks, as well as the federal government's open (or pluralistic) procurement strategy for building networks to support military and scientific research, enabled the rapid global expansion and popularization of the Internet. At the same time, efforts to regulate the *content* of these communications have been repeatedly frustrated by the federal judiciary, intent upon preserving the same open market place of ideas that has generally prevailed for print and telecommunications.

The early success of the United States' space program was also a function of the principle of open systems, in this case manifest in Eisenhower's strategic insight that "open skies" for everyone provided the best guarantee of the United States' ability to exploit the military as well as scientific possibilities of being able to operate spacecraft in Earth's orbit. The survival of the space program, long after its Cold War purpose had been met, is also due to NASA administrator James Webb's shrewd strategy for carrying out the research and development necessary for the program, namely, to procure R&D, as well as hardware and operations, from the private sector, with contracts liberally scattered throughout the country.

Whether this approach obtained the "best" R&D is impossible to know, absent an alternative example. But each and every NASA technical failure (most recently the failure of a fuel gauge on the Space Shuttle *Discovery*, preventing an early July 2005 Space Shuttle return mission to the International Space Station after the *Columbia* disaster of February 2003) raises anew questions of contractor accountability and NASA oversight. More important for the long-term survival of NASA, Webb's procurement strategy ensured the creation of a vested economic interest in the program in virtually every congressional district, and especially in Texas, California, and Florida—key states for the "manned" space program, the most costly space enterprise, with contracts many times as remunerative as those typically awarded for space science projects.

Arguably the most complex of the nation's policy challenges is that vast leviathan, feeding on medical and pharmaceutical technologies, known as the U.S. health care system. Here the authority of scientific expertise—in this case, medical science—has virtually triumphed over the need of ordinary people to act as informed buyers in what might otherwise be an open market for medical products and services. Opposition to a single-payer form of national health care (government sponsored, to ensure the availability of adequate health care for everyone, regardless of income or employment status) has so far rested on the belief that free market "choice" among providers and insurers will guarantee the continued availability of the "best" health care.

In fact, however, the U.S. does not have the "best" health care by internationally accepted measures. Health care consumers in the United States know less about what they are buying when they purchase a life-sparing surgical procedure or prescription medicines than they do when they buy the family automobile. The health care market is not, in fact "open"; the choices that do exist can rarely be exercised in a meaningful way—that is, with a full command of the necessary information and ability to compare options. Chapter 8 examines why this is so, and offers an open systems remedy. Whether consumers buy health care individually, or aggregated into large (and possibly government-sponsored) organizations, informed choice is as basic to the long-term effectiveness of the nation's health care system, as of any other economic or political system. However, little significant improvement is likely to occur until a critical mass of American people, corporate employers, and the health care industry accept that both equity and economic sanity demand a national, all insured, health care system.

In chapter 9 we observed how the use of fossil fuels in the United States and the nation's effort to prevent the degradation of its natural environment are inextricably interlinked. Unfortunately the marriage is rarely a happy one, partly because the partners espouse opposite views of the natural world upon which they both depend. The country's profligate use of fossil fuels assumes nature is an open system, eternally adaptive and self-regenerating. And, in fact, the global abundance of oil, coal, and natural gas would give some support to this notion *as a broad generalization*.

But the economy of the United States is not a broad generalization. Its continuing dependence on oil—and more recently *imported* oil—and the political heft of the petroleum industry (itself divided among small U.S. producers and international "big oil" importers), threatens the independence of the United States to a degree not seen since the War of 1812. Government intervention has, since World War II and especially during the "energy crisis" of the 1970s, protected American consumers from the painful vagaries of the global energy market, and thus policy makers have wasted a historic opportunity to accustom Americans to bearing the true cost of their consumption of oil. That cost includes not only the petroleum products themselves, but their environmental costs and research and development into alternative fuels and more efficient fuel plants. To date it is not clear that the production and distribution of alternative fuels (other than nuclear) can be done at less cost than the cost of fossil fuels replaced by alternatives. Meanwhile, as new nuclear generated power systems come on line, France is leading an international consortium to develop the world's first nuclear fusion reactor as an alternative means of generating electrical power.

Congressional and White House efforts to develop a national energy policy have broken little new ground, emphasizing as they do a mix of incentives (e.g., investment tax credits, limited liability for nuclear plant operators) to increase exploration of oil and gas supplies, promote nuclear power generation, and foster the development and use of alternative fuels such as solar and wind power. The burden of energy conservation, especially in the use of petroleum, has fallen largely on the supporters of environmental protection policies, if only because cleaner energy use tends to translate into less petroleum use.

Meanwhile, the environmental movement is animated by a different view of nature, a view that sees nature as a relatively closed system—one sustained by delicate ecological balances threatened by the depredations of modern industrialization. Thus it finds federal mandates a congenial regulatory tool for disciplining industries that pollute our air and waters. But here, too, is an open systems–inspired policy opportunity in danger of being missed. Advocates of aggressive pollution reduction targets tend to focus on the targets, rather than the likelihood of their being met. And that likelihood has been reduced by the ability of targeted industries to delay compliance through administrative protests and litigation. As most judges would probably agree, plaintiffs who can be persuaded to devise their own settlement, rather than having one imposed upon them, are more likely to adhere to its terms. Hence the importance of "cap and trade" features of recent legislative proposals. "Cap and trade" transforms environmental targets into commodities, lawful objects for the natural human "propensity to truck, barter, and exchange." This is what business people understand, and do naturally.

LIKE THE INDUSTRIES in Joseph Schumpeter's explication of capitalism (see chapter 2), open systems contain within themselves the seed of their own destruction. For Schumpeter the destruction was creative, essential to the periodic renewal of firms and industries. Schumpeter understood the process of creative destruction to be part of the imperfect equilibrium of economic development, in which technological innovation plays a central role. As purchasers are offered new options, they make different choices and the nature of the market changes with those choices.

In analogous (though not necessarily identical) ways, although open systems welcome newcomers (be they new ideas, talented individuals, well-run companies, or well-tooled components), some of those newcomers will be more successful than others. "Nothing succeeds like success," thanks to the confidence and experience contenders acquire as their research produces valued findings or their businesses prosper. The sociologist Robert K. Merton described the phenomenon, when it occurs in science, as the "Matthew Effect" (see chapters 1 and 3). In the arena of politics the phenomenon is known as the "advantage of incumbency," while in business one reads of the "winner take all" effect. Thus open systems generate forces tending to enclose them, as successful individuals and institutions use their relative economic and political power to maintain the advantages of success against other newcomers.

In the case of technology-intensive industries or scientific research, experience usually translates into increased know-how, new research, and innovative engineering. Whether successful newcomers, as they become winners, use their advantage to solidify market power, or to invest additional capital (intellectual as well as material) in new ideas and products, is the critical policy question, as we saw in the case of Microsoft's antitrust troubles in chapter 6.

But the Microsoft case is by no means the only example. The phenomenon of "winner take all" has resulted in an unprecedented period of business consolidation, producing oligarchies in key technology and research-based industries.

The last decade of the twentieth and early years of the twenty-first century witnessed the largest industrial mergers and acquisitions in modern economic history, as over twenty-five firms acquired or merged with competitors for settlement amounts ranging from $40 million to at least $16 billion, what SBC Communications paid for AT&T, itself once the subject of a successful federal antitrust breakup. The industries tending toward oligopolistic structures were the electronics, communications, oil, and pharmaceutical industries. The acquired or enlarged companies included such familiar names as Compaq Computer, Chevron, Texaco, MCI, Oracle, Sprint, Nextel, Semantec, Pfizer, Inc., and Warner-Lambert. As an indication of the extent to which the new oligopolies were becoming global in reach, the U.S. oil firm Amoco, which began its corporate life in 1911 as Standard Oil of Indiana, was bought by British Petroleum, and IBM sold its personal computer business to China's largest maker of PCs, Lenovo.[1]

Recent consolidations have been encouraged by the presence in the White House of a Republican president who, like Ronald Reagan, vowed to reduce government regulation of industry. At first blush a policy of deregulation might seem wholly consistent with an open systems approach toward economic policy, namely, that the government should not interfere in the normal workings of the market place. But in the case of the telecommunications industry, federal regulation throughout much of the later twentieth century was designed to promote competition, prevent consolidation among more successful companies, and to make it possible for competitive newcomers to enter the telecommunications market. Hence deregulation during the Bush (George W.) presidency has resulted in greater consolidation, not less, as evidenced by the FCC's tacit approval of SBC Communications' acquisition of AT&T, Verizon Communications' acquisition of MCI, and Sprint's acquisition of Nextel. Similarly, during the Bush administration the FCC unsuccessfully sought to loosen restrictions on the number of television and radio stations, broadcast cable services, and newspapers a single company could own in the same city, or the number of TV stations (one) that a single company could own in one TV market area.

These measures, however, prompted opponents to appeal (successfully) to the federal courts in 2005, arguing that the FCC's approval of greater consolidation in the broadcast industry threatened diversity and competition in "the market place of ideas" (see chapter 6). Similarly, in July 2005 the U.S. Court of Appeals for the District of Columbia struck down FCC efforts to manipulate Internet communications technology in the interests of the motion picture and broadcast network industries. The FCC had issued a regulation requiring

computer makers to install software that would prevent users from downloading and copying digital programming distributed by radio and television
broadcast. The court ruled that the regulatory agency had exceeded its authority, limited by Congress to radio and wire transmission.

This episode in the checkered history of the FCC reveals not only the federal judiciary's interest in preserving a constitutional separation of powers in
the workings of the U.S. government, but also the interplay of interests among
its three branches which shape the rise (and occasional fall) of national science and technology policy initiatives. As we saw in chapter 6, the radio spectrum, portions of which are licensed by the federal government to purveyors of
wireless telecommunications (which include the Pentagon as well as the local
cellular phone company) has become an increasingly scarce public commodity. Both the Congress and the executive branch have been eager to promote
the adoption of digitally transmitted high definition television to reduce demand on analog (radio) airwaves, thus making more portions of the spectrum
available for lucrative federal licensing to commercial wireless communications companies. Hence the FCC's interest in promoting HDTV, a commercial
medium, which challenges the decades-old dominance of broadcast news
and entertainment—open access to which has been regulated as a public
service.

Meanwhile, the aerospace and defense industry had already undergone its
own consolidations in the 1990s, as Boeing acquired the defense and space
units of Rockwell International and McDonnell Douglas; the Lockheed Corporation merged with Martin Marietta and then acquired the space systems division of General Dynamics, GE Aerospace, and the Loral Corporation; and
Raytheon acquired the defense businesses of Texas Instruments and General
Motors' Hughes Electronics. Only the opposition of the Clinton administration's justice department prevented the further merger in 1998 of Lockheed
Martin with Northrop Grumman. Whether mergers and acquisitions in technology industries have led to improved products and greater innovation—
rather than greater market power and political influence—is debatable.
Industry analysts appear to agree that, with few exceptions, the economic or
technological gains of the new combinations have not been impressive. One
analyst observed, "there have been very few mergers that have succeeded in
maturing technology sectors. Large companies have very established cultures
and merging them in a fast-moving business like information technology is
very hard."[2]

The phenomenon of concentrations of power emerging from open systems
is evident as well in the academic research sector which, despite its characterization as "non-profit" for taxation and other policy purposes, behaves much
like any other industry, as we saw in chapter 5. Thanks to the U.S. Supreme

Court's 1980 ruling in *Diamond v. Chakrabarty* (1980) that genetically engineered life forms are patentable, the Stevenson-Wydler Technology Innovation Act of 1980, and the Bayh-Dole Act of 1980, research-performing academic institutions have accumulated a significant amount of publicly financed intellectual property with immeasurable potential commercial value. These institutions have also become active politically in order to ensure their interests among federal policy makers, with the twenty most successful of such institutions spending nearly $7 million on lobbyists and federal political campaign contributions during 2000 alone (see Table 10.1). Acting very much like any other industry, research-performing academic institutions are beginning to see a return on their political investments. "Earmarked" research—federally funded research designated for specific institutions—has grown, so that a recent study finds a $1.56 increase in earmarked research funding for every $1 spent on lobbying, while institutions in the districts of congressional appropriations committee members can expect to see a return of $4.50 for every $1 spent.[3]

The unequal distribution of intellectual property among the 190 academic institutions receiving patents since Bayh-Dole went into effect has produced a new oligopoly in the "knowledge-based global economy."[4] Half of the patents awarded to academic institutions during 1981–2001 went to only slightly over 10 percent of those 190 institutions. This new oligopoly may or may not be based entirely on merit, but it clearly wields economic power, as became apparent when President Bill Clinton and Prime Minister Tony Blair announced the completion of the joint U.S. and British Human Genome Project in 2000. Their statement that "the genome should be made freely available to scientists everywhere"—a declaration of an open systems policy if there ever was one—sent bio-technology stocks plummeting as investors concluded that the federal government would oppose patenting human genomic data. The White House subsequently had to reassure the securities markets that it would seek no change in patent law allowing the patenting of life forms.[5]

For the academic scientists whose work is increasingly funded by commercial firms, or who may have a profit-sharing arrangement with their commercial sponsors, avoiding "conflict of interest" in the representation of their research findings to help protect future profits is likely to be more and more difficult. A 1996 analysis of 789 articles in medical journals found that 34 percent of the authors had a financial interest (normally patent pending) in the subject of the article.[6] As the hazards of FDA-approved antidepressants and arthritis medications began to receive broad media attention during 2004, the extent to which academic scientists had signed non-disclosure agreements with pharmaceutical companies sponsoring their research received considerable exposure as well.

TABLE 10.1

Top Twenty Academic Patent Holders (1981–2001), Amount Spent on Lobbying (2000), Amount Spent on Political Contributions (2000), and Rank among Top 100 Academic Institutions in R&D Expenditures of Federally Originated Funds (2001)

* Total No. Patents Awarded, All Institutions, 1981–2001 = 32,710. Largest No. Institutions Awarded Patents in Any Year = 190. Top 20 Institutions as Percent of Total Institutions = 10.5%. Top 20 Institutions' Percent of Total Patents Awarded, 1981–2001 = 49.9%

Institution	Patents Received (1981–2001)	Lobbyist Spending (2000)	Federal Political Total Contributions (2000)***	Total Political Spending	Rank: Acad. Institutions' Exp. of Fed. Originated Funds****
1 University of California	3411	$388,599	$575,271	$963,870	1 (8)
2 Massachusetts Inst. of Tech.	1908	$214,622	$96,855	$311,477	15
3 University of Texas	1189	$180,000	$147,088	$327,088	34 (45)
4 Stanford University	1051	$180,000	$465,287	$645,287	8
5 California Inst. of Technology	906	$80,000	< $95,789	$80,000	51
6 University of Wisconsin	874	$120,000	< $95,789	$120,000	2
7 Cornell University	705	$220,000	< $95,789	$220,000	14
8 Johns Hopkins University	679	$260,000	< $95,789	$260,000	7 (8)
9 University of Florida	604	$120,000	< $95,789**	$120,000	26
10 University of Minnesota	575	$0	< $95,789	$0	11
11 Iowa State University	569	$85,228	< $95,789	$85,228	62
12 State University of New York	555	$96,200	< $95,789	$96,200	59 (64)
13 University of Michigan	505	$260,000	< $95,789	$260,000	3
14 University of Pennsylvania	496	$280,000	< $95,789	$280,000	10
15 Columbia University	457	$1,200,000	$146,023	$1,346,023	27
16 Michigan State University	389	$140,000	$136,274	$276,274	38
17 University of Utah	386	$240,000	$114,403	$354,403	55
18 Harvard University	378	$440,000	$530,314	$970,314	23
19 University of Washington	349	$140,000	$103,795	$243,795	4
20 Duke University	341	$0	N/A	$0	22
TOTALS:	16,327	$4,644,649		$6,959,959	

* Sources: National Science Foundation, *Science & Engineering Indicators, 2004*, Appendix Table 5–54; www.opensecrets.org, downloaded October 20, 2004. ** Listing for University of Florida Health Sciences Center *** "< $95,789" = These institutions' contributions do not rank within top twenty of contributors to federal candidates and parties for the year, all of whom gave $95,789 or more. The amount shown for Washington University, which ranked twentieth for the year 2000. **** First number is rank for lead campus. Number in parentheses () = average for institutions with several qualifying campuses.

IN SPITE OF THESE TRENDS toward consolidation and information enclosure in technology- and research-based industries, the principle of open systems appears still to be able to inspire correctives to the disequilibrium introduced by the "winner take all" phenomenon. The U.S. Court of Appeals for the Third Circuit of Philadelphia in June 2004 ordered the FCC to reverse its action relaxing restrictions on the number of media outlets any single company could own. The court agreed with a group of small broadcasters that the FCC had been arbitrary in relaxing the old rules, which promoted competition in the dissemination of news and entertainment over the airwaves and through the print media. The broadcasters who challenged the FCC included Latino, Asian-American, African-American, and Native American organizations. Of equal importance, the Bush administration withdrew its planned appeal of the decision to the U.S. Supreme Court. Some have argued that the administration's acquiescence was due merely to a fear of exposing the FCC's new rules against "indecency" in television and radio broadcasting to further Supreme Court scrutiny. If so, the administration would have also been conceding the power of the idea of openness to resist efforts to control the flow of ideas through American public life.[7]

While Microsoft may still enjoy the lion's share of PC software sales, the information technology industry continues to reflect a preference for open systems engineering as a way to reach the broadest possible market. For example, virtually all of today's computer peripherals, such as printers and "smart card" readers, can be connected to any maker's processing unit with a second generation universal serial bus (USB 2.0), which is roughly forty times faster than its predecessor "high-speed" USB. The USB 2.0 came on the market not as a proprietary product, but an industry *standard* developed jointly by Compaq, Hewlett Packard, Intel, Lucent, Microsoft, NEC, and Philips.

This adaptation of open systems engineering may have been reinforced by the circumstance that the information technology industry is arguably the most "globalized" of all technology-driven industries—a fact which may explain why Microsoft has had to yield to antitrust pressures originating from Europe. In March of 2004 the European Commission ruled Microsoft in violation of its antitrust laws, thus stifling competition. (Microsoft was fined 97 million euros and ordered to reveal more programming information to its competitors, as well as release versions of Windows without Media Player. Microsoft lost its subsequent appeal of the commission's ruling to the Court of First Instance in Luxembourg.) No doubt reading the handwriting on the wall, Microsoft has been negotiating settlements of antitrust suits with AOL-Time Warner (owner of Netscape), Novell, the Computer and Communications Industry Association, and Sun Microsystems—all of whom do business in the global market place.

Meanwhile, the open systems software engineering model has spread from Linux to Sun Microsystems, which will release its Solaris 10 operating system on an open-source basis. That same model has also spread to the world of biotechnology. Responding to the patenting of tools for genetically engineering crops by companies such as Monsanto, Syngenta, and Bayer CropScience—which opponents say impedes the spread of improved crops in developing countries—a group of researchers from Australia announced in 2005 an "open source" method for creating genetically modified crops. The method enables the transfer of desirable genes among plants, and the Australian group will make it available for any and all to use, provided the resulting improvements are also freely available. The group calls itself the Biological Innovation for Open Society (BIOS).

Another challenge—albeit a belated one—to the information enclosure trend in biotechnology occurred in early 2005 when the NIH acknowledged, as a result of a Freedom of Information Act request from the Associated Press, that it had failed to require its scientists to disclose royalty agreements with the companies for whom they were conducting research trials with human subjects. Although the NIH has had a policy requiring scientists to do so since 2000, the agency only began to enforce it recently. Around the same time, the *Los Angeles Times*'s revelations of consulting agreements between NIH scientists and pharmaceutical and biotech companies resulted in hearings before the U.S. House committee on commerce and energy. The hearings in turn forced the NIH to ban its scientists from entering into outside consulting activities. The NIH ban, however, may not prove as effective in reducing conflicts of interest among biomedical researchers as it might have been only a few years ago, thanks to a momentous decision on the part of the citizens of California on election day in 2004.

The voters of California endorsed Proposition 71, a heavily campaigned measure endorsed by Governor Arnold Schwarzenegger to spend $3 billion (and eventually more, to be sure) over ten years on human embryonic stem cell research. While the long-term consequences of Proposition 71 remain to be seen, it surely represents a historic reversal in the nation's dependence on the federal government as the principal public sponsor of biomedical research and development. The movement behind the measure was prompted by the Bush administration's restrictions on research using stem cells from human embryos, and may be one of the first assertions of "states' rights" in the area of science and technology policy. The new California Institute for Regenerative Medicine (CIRM), which will be the largest single backer of stem cell research in the United States, is likely to be the center of a new California industry comparable to the fabled Silicon Valley of the computer age.

And yet the CIRM may also become a center of troublesome compromises of public interest for private gain, and the absence of federal oversight of an undertaking with such broad ethical and policy ramifications has worried some critics (see chapter 8). What's more, the prospect of losing biotechnology research talent, status, and private sector partnerships to California has alarmed leaders of research institutions in other states like Massachusetts and New Jersey. Has California launched a new era of unprecedented competition in research? Will the knowledge generated by the resources of the people of California be available to everyone in the interest of the medical advances for which the research was touted? Or will it be consumed and hidden away by the proprietary interests of the new firms joining in what California's Lt. Gov. Cruz Bustamante has called "this century's gold rush"?[8]

If and when the cures promised by embryonic stem cell research materialize, health care consumers may become far better informed about not only available choices of therapies, but about their comparative merits and side effects. Since the passage of the 2003 Medicare reform bill (see chapter 8) several organizations have stepped forward to provide the public the critical information that a heavily lobbied Congress has had political difficulty requiring. Prompted by upward spiraling drug costs, Oregon and twelve other states (all states share in the funding of Medicaid, not to be confused with Medicare), established an initiative to compare the safety and effectiveness of hundreds of drugs as reported in the medical literature. The results of these comparisons are then disseminated, increasing the chance that patients can actively choose the most effective as well as least costly drug therapies.

While the pharmaceutical industry opposes such initiatives as "designed to disguise cost containment as credible scientific analysis," the Oregon project uncovered the increased risk of heart attack and stroke associated with Vioxx in 2002, two years before widespread media coverage of the same risks forced Merck & Co. to withdraw the medication from the market. The American Association of Retired Persons (AARP) and Consumers Union have both launched websites offering comparative safety, effectiveness, and cost information on hundreds of drugs.[9] The American Medical Association, a powerful player in the shaping of federal health policy, has advocated both the creation of a federal data bank of mandatory information about medical devices and drugs and an end to contract provisions that prevent researchers who conduct drug trials from disclosing the trials' results.

The combination of AARP and the AMA has finally proven a political match for the pharmaceutical industry. And yet, the Pharmaceutical Research and Manufacturer's Association's (PhRMA) plan to provide voluntarily supplied trial information on its own website is of dubious merit. Voluntarily supplied drug trial results would render the value of comparing drug data useless, since comparisons would be based on incomplete information. The industry also opposes disclosure of preliminary drug trials on the grounds that it would reveal proprietary information as well. Meanwhile, the National Institutes of

Health, whose parent federal Department of Health and Human Services (DHHS) also oversees Medicare and Medicaid, has launched websites providing side-by-side comparisons of hospitals according to various measures, using information supplied voluntarily by most of the hospitals in the United States.

AND SO, we come full circle. The principle of open systems is an idea which, like the unity of the universe, has a life of its own. First found in the philosophy of the ancient Greeks, it has been the necessary means by which the human spirit expresses itself within its social settings, using its incomparable tools of mind, heart, and hand. Belief in the liberating capacity of open systems forced sea changes in human affairs—witness the democratic revolutions around the Atlantic basin at the end of the eighteenth century, and later democratic movements in other parts of the world. But unlike the romantic, transcendental stirrings that have appeared from time to time, inspiring poetry, art, and authoritarian politics, the principle of open systems has provided the dynamic element in the design of liberating social arrangements, political systems, economic policies, public administration, and our constitutional law. It has also found its way into arguably the most successful approach toward designing technological systems capable of virtually infinite adaptability and expansion. But the life of the idea of open systems has not survived without challenges. That these challenges occur is a sign not of the weakness of open systems, but of their strength.

Will its centuries old proclivity toward open systems ensure the United States a competitive advantage in the global market place of the twenty-first century? Can a nation today succeed economically (as China appears to be doing) without an institutional and political infrastructure built on open systems principles? Both of these compelling questions are perhaps best addressed by observing that the "nation" as a coherent economic unit, is in danger of becoming a thing of the past, thus undermining the ability of a single national government to forge its people's economic destiny. Assuming that advances in science and technology can generate significant economic expansion anywhere on the globe, the extent to which the phenomenal economic growth of a country like China—hardly an open systems stalwart—is due to research and innovation is not yet clear.

The Chinese economic "juggernaut" is also fueled by that country's enormous population (and thus relatively cheap labor pool) and its large trade surplus, which is partially supported by what was, until July 2005, an officially deflated value of the yuan tied to the U.S. dollar (at 8.277 yuan = $1). Large Chinese investments in the dollar, which help finance U.S. budget deficits, have reinforced this relatively low exchange rate. China's relatively cheap currency attracts foreign investment and intensifies the competitive disadvantage of U.S.-produced goods in the global market place. In July 2005 the Chinese government announced a new currency policy, by which the value of the yuan would be determined daily, based on the value of a "basket of currencies." The yuan, at the same time, was revalued to 8.11 to $1, a change regarded by most observers as token, at best.[10]

The vulnerability of the U.S. economy to Chinese currency policy illustrates the second observation one might make about the capacity of the principle of open systems to ensure the United States' national wealth and well-being for the foreseeable future. In strict economic terms, our view is that the institutional boundaries of the U.S. market place have become so porous that the benefits of open systems described in the preceding chapters can no longer guarantee U.S. hegemony in science, technology, or economic enterprise. However, that this is so by no means spells the end of the power of open systems to shape evolving political systems or foster scientific discovery and industrial innovation. To the contrary, open systems have been one of the principal drivers of "globalization," or the emergence of a truly trans-world market place.

This phenomenon, well documented by Thomas L. Friedman, has been propelled—thanks to the expansion of virtually instantaneous global communications—by the ability of producers and consumers anywhere to obtain goods and services from anywhere else on the globe with much greater speed than was possible in the pre-Internet age.[11] Thus components of "American-made" automobiles and computers (for example) come from the Philippines, Costa Rica, Korea, Taiwan, Germany, etc. while Internet and telephone consumers of banking and business services are as likely to communicate their needs with "customer representatives" in Bangladesh as in Nebraska. Microsoft has formed cross-licensing agreements with non-U.S. giants like Toshiba and, partly in response to adverse antitrust rulings in the U.S. government and by the European Union, the software giant has begun to collaborate with Sun Microsystems to enable their respective software programs to communicate.

What all this means for "the wealth *of nations*" can be disputed, and is. Supporters of the European Union faced two important setbacks in 2005: France and the Netherlands declined to ratify a new EU Constitution, while EU members failed to agree on the union's next (2007–2013) budget. In the United States, social mobility—which since the country's beginning has sustained the belief that open systems are the best route to "the greatest good for the greatest number"—has been declining. The rich get richer, the poor get more plentiful, and the middle class—whose property interest in social stability has been critical to the success of the democratic nation state—has begun to shrink. Thus the current Bush administration struggles to shore up support for the Central American Free Trade Agreement (CAFTA).

While the economic stakes of CAFTA may be relatively small, the measure has failed to attract the support of centrist Democrats who supported the previous Clinton administration's successful North American Free Trade Agreement with Mexico and Canada.[12] The opposition of sugar and textile industries, as well as organized labor, can be assumed. But increasing numbers of policy makers are doubting the ability of the United States to mitigate the costs to their constituents of free trade with countries that are not required to adhere to U.S. labor and environmental standards. Economist Robert J. Samuelson rightly cautions us against assuming that globalization is destined to triumph; national psychology and national politics do matter.[13] As we have seen, the liberating—and hence creative—power of "open systems" is as much a product as a source of the historically special nature of the national "psychology" and politics of the United States and its industrial wealth. How effectively, and how widely, that power prevails in the coming decades, remains to be seen.

APPENDIX:
ESSAY ON SOURCES

Whether one explores issues in U.S. science and technology policy for academic or for practical reasons, the first step is to learn "the lay of the land." However, because an individual's interest in this subject is rarely neutral, the danger is great at the outset of becoming captive to special pleading (or one's own biases). Policy advocates as well as academic scholars will be more effective if they fully appreciate every salient position on the controversy at hand. Thanks to the developments discussed in chapter 6, Internet access to congressional deliberations and policy development in the executive branch is substantial. Readers should start with the following websites (most of which have site maps with links to more specific sources of information or opinion): <http://www.house.gov/science>, website for the U.S. House of Representatives Committee on Science; <http://www.senate.gov>, for links to the U.S. Senate committees on commerce, science, transportation, and energy and natural resources; and then <http://www.ostp.gov/index.html>, the Office of Science and Technology Policy (OSTP) for links to White House policy position papers (many of which can be in the form of speeches made by the president's science advisor or reports by the National Science and Technology Council and President's Council of Advisors on Science and Technology).

Links on White House, OSTP, and Office of Management and Budget (OMB) website home pages vary somewhat from administration to administration. The OMB's website at <www.whitehouse.gov/omb>, which tends to be the most consistent over time, is an essential tool for students of federal science and technology policy. Currently (during the second George W. Bush administration) OMB's home page has direct links to White House policy on energy, environment, transportation, health, science, and space policy issues. There are also direct links to current budget documents and executive branch testimony before and reports to Congress. The page's search feature allows readers to see administration statements and reports from federal agencies involved in science and technology, by entering in their acronyms, e.g., NASA, EPA (Environmental Protection Agency), and NSF (National Science Foundation). Meanwhile, each executive branch agency has a website that can be accessed by entering its name into a browser's search field, and virtually all executive branch agency websites have links to their internal policy development organizations and publications. Also an essential bookmark for Internet users is the federal government's web portal, "FirstGov" (<www.firstgov.gov>), which has an excellent reference section with links to federal data, statistics, libraries, laws, and regulations.

The largest challenge facing first-time researchers is the surfeit of policy-relevant information. A good screening rule is to limit one's initial searches to organizations that have gained some traction in the policy-making process. Some of the many organizations and interest groups that cluster around science and technology policy agendas carry more weight than others in the White House and the Congress, whether for substantive or financial reasons. A good way to identify those organizations and individuals is to scan the hearing calendar of the pertinent congressional committees to determine the organizational affiliations of scheduled witnesses; having done so, visit the organizations' websites for more information about them and their policy views.

For the executive branch, from the OSTP website use the site map and outreach/reports links to reach a list of reports on various subjects, most of which name the participating individuals and organizations. Among the biggest non-government players are the National Academy of Sciences/National Research Council, the National Academy of Engineering, and the American Association for the Advancement of Science, at <http://nationalacademies. org> and <http://www.aaas.org/pp>. For more critical views visit the websites of Public Citizen (<http://www.citizen.org>), the Center for Science in the Public Interest at <http://www.cspinet.org>, and the Center for Responsive Politics, which collects and publishes information on campaign contributions and lobbying at its website, <http://www.opensecrets.org>.

In addition, federal agencies responsible for promoting U.S. research and development publish annual or periodic surveys which, when consulted together, can provide a necessary anchor for the generalizations we rely on to frame valid policy questions or proposals. These are the National Science Board's annual *Scientific and Engineering Indicators*, published by the National Science Foundation in both print and online versions (Washington, D.C.: National Science Board, annually); the Office of Management and Budget's annual "Analytical Perspectives, Budget of the United States Government" most readily accessed online at <http://www.whitehouse.gov/omb/budget/>; and the U.S. Department of Commerce (DOC), Office of the Secretary annual "Summary Report on Federal Laboratory Technology Transfer [calendar year]: Report to the President and the Congress Under the Technology Transfer and Commercialization Act," available online at <http://www.technology.gov/Reports.htm>.

In using these reports, which rely almost entirely on quantitative data, researchers must be mindful not only of the limits of quantitative data in economic and social policy development, but of their comparative context. For examples, see Will Lepkowski, "Science and Engineering Indicators—Of What?" *Science and Policy Perspectives*, Vol. 2 (Center for Science, Policy and Outcomes, <www.cspo.org> [2001]); National Research Council, *Quantitative Assessments of the Physical and Mathematical Sciences, A Summary of Lessons Learned* (Washington, D.C.: National Academy Press, 1994); National Research Council, "Measuring the Science and Engineering Enterprise" (Washington, D.C.: National Academy Press, 2002); Organisation for Economic Co-operation and Development (OECD), "Fiscal Measures to Promote R&D and Innovation" (Paris, 1996) and "Research and Development in Industry: Expenditure and Researchers, Scientists and Engineers 1976–97" (Paris, 1999). The OECD offers numerous statistical reports pertinent to comparative energy, health, environmental, and information policy issues that can be accessed from links at <http://www.oecd.org/home>.

HAVING SKETCHED OUT a topographical map of the science and technology policy issues that interest one the most, researchers can begin to burrow down and make some geological observations of those issues' political, institutional, and historical terrain. Any list of sources must necessarily be somewhat idiosyncratic, for the simple reason that the quantity of publications, from good journalism to the most obscure doctoral dissertation, is so great that to include them all would consume the pages of another book. Therefore no reader is likely to find the following suggestions complete, but this author has found the following published resources especially valuable.

For general historical frameworks and survey treatments, see Don K. Price, *Government and Science* (New York: New York University Press, 1954); A. Hunter Dupree, *Science in the Federal Government: A History of Policies and Activities* (Cambridge, Mass.: Harvard University Press, 1957, 1986); Michael Polanyi, "The Republic of Science: Its Political and Economic Theory," *Minerva*, Vol. 1 (1962); Don K. Price, *The Scientific Estate* (Cambridge, Mass.: Harvard University Press, 1965); Daniel Greenberg, *The Politics of Pure Science* (New York, N.Y.: Penguin, 1967); Harvey Brooks, *The Government of Science* (Cambridge, Mass.: The MIT Press, 1968); W. Henry Lambright, *Governing Science and Technology* (New York, N.Y.: Oxford University Press, 1976); Derek de Solla Price, *Little Science, Big Science . . . and Beyond* (New York, N.Y.: Columbia University Press, 1986); Bruce L. R. Smith, *American Science Policy Since World War II* (Washington, D.C.: The Brookings Institution Press, 1990); Daniel Kleinman, *Politics on the Endless Frontier* (Durham, N.C.: Duke University Press, 1995); Bruce L. R. Smith and Claude Barfield, eds. *Technology, R&D and the Economy* (Washington, D.C.: The Brookings Institution Press, 1996); and Daniel Kleinman, *Science, Technology, and Democracy* (Albany, N.Y.: State University Press of New York, 2000). See also Michael Crow and Barry Bozeman, *Limited by Design: R&D Laboratories and the U.S. National Innovation System* (New York, N.Y.: Columbia University Press,

1998), and U.S. Library of Congress, Congressional Research Service, *Linkages Between Federal Research and Development Funding and Economic Growth*, Report No. 92-211 SPR (Washington, D.C.: U.S. Library of Congress, 1992).

To the extent that science policy decisions have been premised on the relationship between science and government posited in 1945 by Vannevar Bush's report to the president, *Science—the Endless Frontier* (New York, N.Y.: American Council of Learned Societies-ACLS History E-Book Project, 2001)—or the Bush paradigm—the principal issue in science policy has been providing sufficient funding for university-based (academic) research while allowing federally funded researchers maximum programmatic and administrative latitude. A good overview of issues in federally funded university research can be found in David H. Guston and Kenneth Keniston, eds. *The Fragile Contract: University Science and the Federal Government* (Cambridge, Mass.: The MIT Press, 1994). See also Claude E. Barfield, ed. *Science for the Twenty-First Century: The Bush Report Revisited* (Washington, D.C.: The American Enterprise Institute Press, 1997). For critical treatments of the linear model of the influence of scientific research on technological advance and subsequence economic growth see Daniel Sarewitz, *Frontiers of Illusion* (Philadelphia, Pa.: Temple University Press, 1996) and Donald Stokes, *Pasteur's Quadrant* (Washington, D.C.: The Brookings Institution Press, 1997).

The tension between scientific and political authority is as old as modern science itself. Those wishing to dig more deeply into the ideological foundations of contemporary "science v. politics" controversies should become familiar with the notion of "scientific revolution" as the leading force for change in modern life. Begin with Thomas S. Kuhn, *The Structure of Scientific Revolutions* (Chicago, Ill.: University of Chicago Press, 1962) and John Henry, Roy Porter, and John Breuilly, *The Scientific Revolution and the Origins of Modern Science* (New York, N.Y.: St. Martin's Press, 1997). Roy Porter, in *Enlightenment* (New York, N.Y.: Palgrave Macmillan, 2001), provides good European intellectual context for the ideas that would flourish and migrate to the new republic in the early nineteenth century. Daniel J. Kevles's *The Physicists: The History of a Scientific Community in Modern America* (Cambridge, Mass.: Harvard University Press, 1971) offers an unparalleled, intimate look at the way questions of authority and political influence operated among the professional aspirations and successes of the arguably most important community of scientists in the United States in the twentieth century.

A good sampling of commentary on the varied ways in which scientific and cultural authority have competed in twentieth-century America can be found in Ronald G. Walters, ed. *Scientific Authority in Twentieth-Century America* (Baltimore, Md.: The Johns Hopkins University Press, 1997). See also Kleinman's, *Politics on the Endless Frontier*; Sheila Jasanoff, *The Fifth Branch: Science Advisers as Policy Makers* (Cambridge, Mass.: The MIT Press, 1990); Bruce L. R. Smith, *The*

Advisers: Scientists in the Policy Process (Washington, D.C.: The Brookings Institution Press, 1992); and Gordon Adams, *The Iron Triangle: The Politics of Defense Contracting* (New York, N.Y.: Council on Economic Priorities, 1981). While Adams's study is now over twenty years old, its observations and conclusions are as valid today as they were in the early Reagan administration.

An old joke among lawyers about *pro se* litigants warns that "he who represents himself has a fool for a client." That may be so, but the survival of constitutional government requires that ordinary citizens have a general familiarity with constitutional principles so that they can recognize when those principles may be in jeopardy. Moreover, constitutional law sets both the boundaries and possibilities of creative and constructive policy making in the arena of science and technology policy no less than in any other policy arena—as we have attempted to illustrate in the preceding chapters. For an overview of the relations of science, technology, and law, see Sheila Jasanoff, *Science at the Bar: Law, Science and Technology in America* (Cambridge, Mass.: Harvard University Press, 1997). Three sources in particular are invaluable in offering general discussions of constitutional issues and the significance of critical decisions by the federal judiciary. For historical Supreme Court cases into the 1950s the best summaries can be found in the first edition of Robert E. and Robert F. Cushman's *Cases in Constitutional Law* (New York, N.Y.: Appleton-Century-Crofts, 1958). What generations of law and political science students know as "Cushman and Cushman" has been updated several times, most recently with the ninth edition published in 2000 by Robert F. Cushman and Brian Stuart Koukoutchos with Susan P. Koniak, *Cases in Constitutional Law* (Upper Saddle River, N.J.: Prentice Hall, 2000), but the discussions of cases, while more current, are not nearly as extensive as in the first edition.

For First Amendment decisions, which govern the extension of federal telecommunications power into the content of what is communicated, the place to begin is with Zechariah Chafee, Jr., *Free Speech in the United States* (New York, N.Y.: Atheneum, 1969), while Floyd Abrams's *Speaking Freely: Trials of the First Amendment* (New York, N.Y.: Viking, 2005) offers a more recent treatment. Discussions of more recent cases can be found in the new "Annotated Constitution" offered over the Internet by the Legal Information Institute (LII) of the Cornell University Law School, at <http://chrome.law.cornell.edu/ancon>. Meanwhile the federal judiciary's interpretations and applications of the "commerce clause" (Article I, Section 8) have set the legal framework in which the adoption of technologies spreads into the national and global market place. When the LII Annotated Constitution (online) is completed, one can consult it for contemporary discussions of critical cases in the interpretation of constitutional grants of (or restrictions on) federal power over commerce. For discussions of the use of expert testimony (science and engineering) by the federal judiciary, see Donald Kennedy and Robert A. Merrell, "Issues in Focus:

Science and the Law," Stephen Breyer, "Science in the Courtroom," and Margaret A. Berger, "Expert Testimony: The Supreme Court's Rules," in *Issues in Science and Technology* (Washington, D.C.: National Academy of Sciences, Summer 2000).

Public administration—what federal and state bureaucracies do—is one of the most maligned and least appreciated functions of government. A succession of would-be presidents on the campaign trail has promised to "reduce" or "rid" the country of too much government and bureaucratic red tape, only to discover shortly after entering office that how they carry out their campaign promises (i.e., the administrative tools they use) can make or break the success of their policies. Unfortunately, the public administration literature rarely makes for light reading. Nonetheless the topic must be mastered to a modest extent to understand why, for example, the mandatory setting of emissions restrictions may, in the long run, prove less effective for environmental purposes than allowing firms to trade emissions "allowances."

Useful additional reading to learn more about how the U.S. policy toolkit for distributing and managing federal funds shapes the outcomes of federal support for scientific research and development includes Linda R. Cohen and Roger G. Noll, *The Technology Pork Barrel* (Washington, D.C.: The Brookings Institution Press, 1991); Hedrick Smith, "Pentagon Games: The Politics of Pork and Turf," in *The Power Game: How Washington Works* (New York, N.Y.: Random House, 1988); and Daniel S. Greenberg, *Science, Money and Politics: Political Triumph and Ethical Erosion* (Chicago, Ill.: University of Chicago Press, 2001). Two studies of federal efforts by the Department of Defense to "push" and shape computer technology provide an exceptionally close view of the intersection in the federal government of organizational politics with the challenges of technological innovation: Arthur L. Norberg and Judy E. O'Neill, *Transforming Computer Technology: Information Processing for the Pentagon, 1962–1986* (Baltimore, Md.: The Johns Hopkins University Press, 1996, 2000) and Alex Roland and Philip Shiman, *Strategic Computing: DARPA and the Quest for Machine Intelligence, 1983–1993* (Cambridge, Mass.: The MIT Press, 2002).

A major watershed was reached during the 1980s' presidency of Ronald Reagan, whose eagerness to turn federal programs over to the private sector (manifest in the intellectual property policies of the 1980s) were not reversed during the centrist presidency of his Democratic successor, Bill Clinton. A good sketch of the Reagan redesign of federal policy tools can be found in Haynes Johnson's chapters "Privatizing" and "Deregulation" in his *Sleepwalking Through History: America in the Reagan Years* (New York, N.Y.: W. W. Norton & Co., 1991). See also W. H. Schacht, "Patent Ownership and Federal Research and Development: A Discussion on the Bayh-Dole Act and the Stevenson-Wydler Act" (U.S. Congressional Research Service Report RL30320, 2002); Barry Bozeman, "Technology Transfer and Public Policy: A Review of Research and Theory, *Research*

Policy (2002); B. Hall and J. Van Reenen, "How Effective are Fiscal Incentives for R&D? A Review of the Evidence," *Research Policy* (2002); and Nathan Rosenberg and R. R. Nelson, "American Universities and Technical Advance in Industry," *Research Policy* (1994).

The distribution of federal funds for scientific (academic) research is dominated by the process of peer review by largely non-federal scientists. A good summary of this process and the policy issues it raises can be found in Richard C. Atkinson and William A. Blanpied, "Peer Review and the Public Interest," *Issues in Science and Technology*, Vol. 1, No. 4 (Washington, D.C.: National Academy of Sciences, 1985). See also H. Zuckerman and Robert Merton, "Institutionalized Patterns of Evaluation in Science," in Robert K. Merton and Norman W. Storer, eds., *The Sociology of Science: Theoretical and Empirical Investigations* (Chicago, Ill.: University of Chicago Press, 1973); Daryl Chubin and E. J. Hackett, *Peerless Science: Peer Review and U.S. Science Policy* (Albany, N.Y.: State University of New York Press, 1990); Roy Rustom, "Funding Science: The Real Defects of Peer Review and an Alternative to It" and Sheila Jasanoff, "Peer Review in the Regulatory Process," both in *Science, Technology and Human Values*, Vol. 10, No. 3 (1985); Robert K. Merton, "The Matthew Effect in Science," *Science*, Vol. 159, No. 3810 (January 5, 1968); and Charles W. McCutchen, "Peer Review: Treacherous Servant, Disastrous Master," *Technology Review* (October 1991).

For international comparisons with the administration of science and technology policy in the United States see David Mowery, "The Practice of Technology Policy," in Paul Stoneman, ed., *Handbook of the Economics of Innovation and Technological Change* (Oxford, UK: Blackwell Publishers, 1995). Set in the broader context of economic policies which today have technology at their center, such comparisons are richly explored in Daniel Yergin and Joseph Stanislaw, *The Commanding Heights: The Battle Between Government and the Marketplace That is Remaking the Modern World* (New York, N.Y.: Simon & Schuster, 1998). The international Organisation for Economic Cooperation and Development (OECD) maintains an excellent website that has numerous links to aggregate as well as country-specific information about such topics as science and innovation, science and technology policy, energy, environment, biotechnology, and information and communication technologies. See <www.oecd.org>.

THE LITERATURE ON the science and technology policy issues highlighted in chapters 6 through 9 does not observe any principle of parity. Some issues have inspired more publications (not necessarily a measure of quality) than others. That said, the following studies can be especially valuable in filling out the historical or political backdrop against which communications, health, biotechnology, space, energy, and environmental policy decisions must be assessed. For space policy, see R. Cargill Hall's essay, "Origins of U.S. Space Policy: Eisenhower, Open Skies, and Freedom of Space," in John M. Logsdon, ed., *Exploring the Unknown: Selected Documents in the History of the U.S. Civil Space Program, Vol. I, Organizing for Exploration* (Washington, D.C.: National Aeronautics and Space Administration, 1995); Nathan C. Goldman, *American Space Law: International and Domestic* (San Diego, Calif.: Univelt, Inc., 1996); Robert A. Divine's essay, "Lyndon B. Johnson and the Politics of Space," in Robert A. Divine, ed., *The Johnson Years, Vol. II: Vietnam, the Environment, and Science* (Lawrence, Kans.: University Press of Kansas, 1987); Walter A. McDougall, . . . *The Heavens and The Earth: A Political History of the Space Age* (Lawrence, Kans.: University of Kansas Press, 1985); W. Henry Lambright, *Powering Apollo: James E. Webb of NASA* (Washington, D.C.: National Aeronautics and Space Administration, 1995); and Sylvia D. Fries, "2001 to 1994: Political Environment and the Design of NASA's Space Station System," *Technology and Culture,* Vol. 29, No. 3 (July 1988).

For health policy questions—which occasion as much heat as light, thanks to their many-faceted complexity—see David Culter and Alan Garber, *A Disease Based Comparison of Health Systems* (Paris, France: OECD, 2003); Judith W. Leavitt and Ronald L. Numbers, *Sickness and Health in America: Readings in the History of Medicine and Public Health* (Madison, Wisc.: University of Wisconsin Press, 1997); Roy Porter, *The Greatest Benefit to Mankind: A Medical History of Humanity* (New York, N.Y.: W. W. Norton & Co., 1997), David J. Rothman, *Beginnings Count: The Technological Imperative in American Health Care* (New York, N.Y.: Oxford University Press, 1997) and *Strangers at the Bedside: A History of How Law and Bioethics Transformed Medical Decision-making* (New York, N.Y.: Basic Books, 1991); Paul Starr, *The Social Transformation of American Medicine* (New York, N.Y.: Basic Books, 1983); and M. L. Tina Stevens, *Bioethics in America: Origins and Cultural Politics* (Baltimore, Md.: The Johns Hopkins University Press, 2000).

Among the best introductions to the digital age are Paul E. Ceruzzi's *A History of Modern Computing* (Cambridge, Mass.: The MIT Press, 2000) and Janet Abbate's *Inventing the Internet* (Cambridge, Mass.: The MIT Press, 2000). For broader perspectives, see Alfred D. Chandler and James W. Cortada, eds., *A Nation Transformed by Information* (New York, N.Y.: Oxford University Press, 2000); Michael E. Hobart and Zachary S. Schiffman, eds. *Information Ages: Literacy, Numeracy, and the Computer Revolution* (Baltimore, Md.: The Johns Hopkins University Press, 1998); and James E. Katz and Ronald E. Rice, *Social Consequences of Internet Use: Access, Involvement, and Interaction* (Cambridge, Mass.: The MIT Press, 2002). And for thoroughly engaging writing, see Tracy Kidder's now classic and intimate look at the early hours of the digital age, *The Soul of a New Machine* (New York, N.Y.: Little, Brown and Company, 1981); and Ken Auletta's *World War 3.0: Microsoft vs. the U.S. Government, and the Battle to Rule the Digital Age* (New York, N.Y.: Broadway Books, 2001).

More than any other area of science and technology policy, issues in energy and environmental policy are driven by quantitative questions of "how much" and "for how long?" Thus anyone venturing into this area should be familiar with the following widely consulted, though not always consistent, data sources. Among government sources, see the Energy Information Administration (<http://www.eia.doe.gov>), International Energy Agency (<www.iea.org>), Nuclear Regulatory Commission Information Digest (<http://www.nrc.gov/reading-rm/>), and the Federal Energy Regulation Commission (<http://www.ferc.gov>) websites. Among organizational and commercial sources, see the American Petroleum Institute's website's (<http://api-ec.api.org/>) link to "Industry Statistics," and the American Automobile Association's "Daily Fuel Gauge Report" (<www.fuelgaugereport.com>). Data on motor vehicles use and other forms of transportation can be found at the Department of Transportation's Bureau of Transportation Statistics' website (<www.bts.dot. gov>). The most credible sources of information and policy discussions at the crossroads of energy and environmental issues are the websites of the U.S. Environmental Protection Agency (<http://www.epa.giv>), which have links to pertinent laws and policy documents, as well as data on a broad range of environmental issues; the Natural Resources Defense Council (<www.nrdc.org>) and Resources for the Future (<http://www.rff.org>).

Good introductions to both energy and environmental policy issues are David H. Guston, ed., *Science, Technology and the Environment* (Washington, D.C.: The Policy Studies Organization, 1997) and Otis L. Graham, Jr., *Environmental Politics and Policy, 1960s–1990s* (University Park, Pa.: Pennsylvania State University Press, 2000). A nearly complete education in the role of carbon-based fuels in the U.S. economy and politics can be had from Daniel Yergin's, *The Prize: The Epic Quest for Oil, Money, and Power* (New York, N.Y.: The Free Press, 1991). Also valuable for its comparably rich discussion of the post–World War II origins of U.S. energy policy is Richard K. Vietor's *Energy Policy in America Since 1945: A Study of Business-Government Relations* (New York, N.Y.: Cambridge University Press, 1984, 1987). For a forward look, see Howard Geller, *Energy Revolution: Policies for a Sustainable Future* (Washington, D.C.: Island Press, 2002).

NOTES

1. Or from $6.2 billion to $35.9 billion (current dollars), roughly half of which was allocated for defense and space related R&D. Source: Bureau of the Census, U.S. Department of Commerce.
2. Robert B. Reich, ed., *The Power of Public Ideas* (Cambridge, Mass.: Harvard University Press, 1988), 10. Italics in the original.

CHAPTER 1   INTRODUCTION

1. Office of Naval Research, Naval Research Advisory Committee, "Open Systems Architecture for Command, Control and Communications" (September 1991); Executive Summary available at <http://nrac.onr.navy.mil/webspace/exec_sum/92opensysc3.html>; U.S. Under Secretary of Defense for Acquisitions, "Open Systems Acquisitions of Weapons Systems," Memorandum dated July 10, 1996, available at <http://www.acq.osd.mil/osjtf/html/implement_dirweapons.html>. For the longer view, see Merritt Roe Smith, ed. *Military Enterprise and Technological Change* (Cambridge, Mass.: The MIT Press, 1985).
2. See, for example, Serhy Yekelchyk, "Celebrating the Soviet Present: The Zhdanov-shchina Campaign in Ukrainian Literature and the Arts," in Donald J. Raleigh, ed., *Provincial Landscapes: Local Dimensions of Soviet Power, 1917–1953* (Pittsburgh, Pa.: University of Pittsburgh Press, 2001).
3. Transcript, ABC World News Tonight broadcast for September 28, 2000.
4. Aaron Brown, "News Night with Aaron Brown," *Cable News Network* (broadcast December 15, 2003).
5. The force of gravity between two objects is the product of their respective masses divided by the square of the distance between them.
6. *Journal of the Continental Congress* (ed. 1800), I, 57; quoted by Chief Justice Hughes in *Near v. Minnesota*, 283, U.S. at 717 (1931).
7. Thomas M. Cooley, *A Treatise on the Constitutional Limitations Which Rest Upon the Legislative Power of the States of the American Union* (1868), quoted in Zechariah Chafee, Jr., *Free Speech in the United States* (New York: Atheneum, 1969), 11.
8. *Abrams v. United States*, 250 U.S. 616 (1919). The case arose out of the arrest under the Espionage Act of 1917 of the defendants for distributing leaflets criticizing the U.S. government's dispatch of troops to Vladivostok in 1918. Holmes had upheld the constitutionality of the Espionage Act of 1917 in *Schenck v. the United States* (1919), but did not believe that the defendants' actions in Abrams posed the "clear and present

danger" which he held, in *Schenck*, to be the only legitimate justification for government interference with free speech.

9. Richard Feynman (1918–1988) was a theoretical physicist with a gift for expressing the quandaries and excitement of science to audiences of all backgrounds. He shared the 1965 Nobel Prize for Physics with Julian Schwinger and Sin-Itiro Tomonaga for their fundamental work in quantum electrodynamics.

10. Richard P. Feynman, *The Pleasure of Finding Things Out* (Cambridge, Mass.: Perseus Books, 1999), 172–173, 184–185, 188, 248.

11. Drawing from interviews with Nobel laureates and personal writings of other scientists, Merton argued that scientists distribute professional recognition among their peers on the basis of psychosocial processes, with the result that, as the Bible attributes to Saint Matthew, "For unto every one that hath shall be given, and he shall have abundance; but from him that hath not shall be taken away even that which he hath." Robert K. Merton, "The Matthew Effect in Science," *Science*, Vol. 159, No. 3810 (January 5, 1968), 56–63.

CHAPTER 2 TECHNOLOGY AND THE IDEOLOGY OF FREE MARKETS

1. Jonathan Edwards, "Sinners in the Hands of an Angry God," delivered at Enfield, Connecticut, July 8, 1741. Online publication by Center for Reformed Theology and Apologetics, <http://www.reformed.org>. Downloaded April 26, 2004.

2. Witherspoon's students included not only Madison, but also six members of the Continental Congress, nine cabinet officers, twenty-one senators, thirty-nine congressmen, three Supreme Court justices, twelve governors, and thirteen college presidents. He was a delegate to the Continental Congress and an advocate in 1776 of independence. See Arthur Herman, *How the Scots Invented the Modern World* (New York, N.Y.: Crown Books, 2004).

3. D. D. Raphael and A. L. Macfie, "Introduction" to Adam Smith, *The Theory of Moral Sentiments* (Glouchestershire, England, UK: Clarendon Press, 1976), 32–33.

4. "The Theory of Moral Sentiments," in Herbert W. Schneider, ed., *Adam Smith's Moral and Political Philosophy* (New York, N.Y.: Harper & Row, 1970), 239, 251, 247.

5. Alexander Hamilton, James Madison, and John Jay, *The Federalist*, Benjamin Fletcher Wright, ed., "10. The Size and Variety of the Union as a Check on Faction" (Cambridge, Mass.: The Belknap Press, 1966), 129–136.

6. Otis L. Graham, Jr., *Losing Time: The Industrial Policy Debate* (Cambridge, Mass.: Harvard University Press, 1992), 114, 189.

7. Adam Smith, *The Wealth of Nations* (New York: Alfred A. Knopf, 1991), 374.

8. Ibid., 619–620.

9. Colbert's actual titles should be the envy of every ambitious government official. Within a few years of being named France's comptroller of finances in 1661, he acquired the additional posts of secretary of state for the King's Household, secretary of state for the Navy, superintendent of commerce, director of royal buildings, vice protector of the French Academy, and minister of state (member of the King's inner council of trusted advisors).

10. The American Revolution was provoked in part by the mercantilist policies of the British crown, attempting to recover the costs of protecting the colonies during the Seven Years War (which ended with the Treaty of Paris in 1763). The crown imposed colonial import duties on raw sugar, potash, iron, hides, glass, lead, paint, paper, tea, as well as direct taxes on newspapers, pamphlets, legal papers (Sugar Act of 1764, Stamp Act of 1765, and Townshend Acts of 1767).

11. Smith, *Wealth of Nations*, 301.

12. Ibid., 225.

13. Ibid., 12–13.

14. Ibid., 620.

15. Ibid., 107.

16. Ron Chernow, *Alexander Hamilton* (New York, N.Y.: Penguin Press, 2004), 26–33, 347.

17. Morris, born in Liverpool, England in 1734, became a successful Philadelphia merchant. As financier of the Revolution, he was responsible for raising the funds and making the purchases necessary to the success of the struggling Revolutionary armies. Hamilton, born in 1755 in the West Indies, emigrated to New York in 1773, where he became a self-made man, an effective leader of an artillery company, lawyer, and eloquent author, with James Madison, of *The Federalist Papers*.

18. Section 8, which reserves to the Congress the "Power To . . . regulate Commerce with foreign Nations, and among the several States." The decision in *Gibbons v. Ogden* struck down as unconstitutional a monopoly granted by the New York state legislature to Robert Fulton and Robert Livingston (and their successors, one of whom was Aaron Ogden) to operate a steamship between New York and New Jersey.

19. *Federalist*, Hamilton (Publius), "The Value of Union to Commerce and the Advantages of a Navy," 136–142.

20. Under Articles of Confederation (1781–1789), Congress lacked the power of taxation or control over interstate and foreign commerce and, as a result, in 1790 the government of the United States was bankrupt. Its total indebtedness to foreign (mostly French) and domestic creditors exceeded $50 million, but it had only been able to collect $2 million from the states. Nonetheless, the Congress agreed to assume state war debts totaling $21.5 million. The measure was bitterly opposed by the southern states, which finally agreed to it in exchange for the relocation of the national capital to Washington, D.C.

21. Roughly 25 percent of the federal government's debt was owed to New England merchants.

22. Smith, *Wealth of Nations*, 285.

23. The Federalists' fiscal program was bitterly opposed by the Jeffersonian Republicans, who correctly recognized in it a strategy that would inextricably connect the mercantile interests of the northeastern United States with the fortunes of the new federal government. The Congress's constitutional authority to establish a national bank was affirmed in Chief Justice Marshall's decision in *McCulloch v. Maryland* (1819). But the Congress did not always act on that authority. During Jefferson's administration the Congress (1802) refused to authorize funds to pay federal debts to the bank, requiring Jefferson to sell the government's remaining 2,000 shares to Baring Brothers of London. The financial chaos resulting from the lack of a uniform currency during and after the war of 1812 led to the recreation of a national bank in 1816.

24. Smith, *Wealth of Nations*, 9–10.

25. Ibid., 620.

26. Alexander Hamilton, "Report on the Subject of Manufactures," in Arthur Harrison Cole, ed., *Industrial and Commercial Correspondence of Alexander Hamilton, Anticipating His Report on Manufactures* (Chicago, Ill.: A. W. Shaw Company, 1928), 318–320.

27. The late colonial period is well known for its interest in technological innovation. First, of course, there was Benjamin Franklin, with his enthusiasm for labor-saving mechanical inventions. Both Boston and Philadelphia saw societies established to promote "useful knowledge," viz., the American Philosophical Society in Philadelphia,

first proposed in 1743 and the American Academy of Arts and Sciences, established in Boston in 1780. The Continental Congress had proposed forming similar societies throughout the colonies. "Report on . . . Manufactures," 320.

28. So called by Henry Clay (1777–1852), senator from Kentucky, secretary of state under President John Quincy Adams, and unsuccessful Whig candidate in the presidential election of 1832, when he was defeated by Andrew Jackson, partly because of his own support for the 2nd Bank of the United States.

29. David A. Hounshell, *From the American System to Mass Production, 1800–1932* (Baltimore: The Johns Hopkins University Press, 1984), 39.

30. Ibid., 15–65.

31. Chernow, *Alexander Hamilton*, 362–388; George Tice, *Paterson* (New Brunswick, N.J.: Rutgers University Press, 1972); and Rev. Samuel Fisher, D.D., *Census of Paterson, New Jersey, 1827–1832* (Salem, Mass.: Higginson Book Co., 1997).

32. Joseph A. Schumpeter, *The Theory of Economic Development: An Inquiry Into Profits, Capital, Credit, Interest, and the Business Cycle*, Redvers Opie, trans. (Cambridge, Mass.: Harvard University Press, 1949), 64. For a critical discussion of Schumpeter's treatment of the endogeneity of science and technology to the modern economy, see Nathan Rosenberg, *Schumpeter and the Endogeneity of Technology: Some American Perspectives* (London: Routledge, 2000).

33. Schumpeter, *The Theory of Economic Development*, 65.

34. Ibid., 66.

35. Joseph A. Schumpeter, *Capitalism, Socialism, and Democracy* (New York and London: Harper & Brothers, Publishers, 1942), 67–68.

36. Kuznets demonstrated slackening of "technological progress" by examining the ratio of patents during the most recent year to the average of patents in the preceding four years, in typewriters (1872–1923), sewing machines (1849–1923), plows and plow sulkys (1850–1923), "electrical field" (1866–1924), and "various electric appliances" (1872–1921). All show declining ratios. Simon S. Kuznets, *Secular Movements in Production and Prices* (Boston: Houghton Mifflin Co., 1930); reprinted 1967 by Augustus M. Kelley, New York, 54.

37. Ibid., 47.

38. "The ratio of net returns to capital invested is larger in the early periods of growth. . . . The chief reasons . . . seem to lie in the rapid rate of technical change, rapid improvement of the product, and lowering of costs. The stability that comes eventually is a reflection of the retarded rate of technical changes and of their economic effects." Kuznets, *Secular Movements*, 49–51.

39. Jacob Schmookler, *Invention and Economic Growth* (Cambridge, Mass.: Harvard University Press, 1966), 4–5.

40. The "important inventions" were in agriculture (234), petroleum refining (284), paper making (185), and railroading (230).

41. Schmookler, *Invention and Economic Growth*, 65–67. Italics in the original.

42. Ibid., 70–71, 201.

CHAPTER 3   THE IDEOLOGIES OF SCIENCE

1. President's Committee of Advisors on Science and Technology, "Wellspring of Prosperity: Science and Technology in the U.S. Economy—How Investments in Discovery Are Making Our Lives Better" (Washington, D.C., Spring, 2000).

2. Ibid., iv. (Italics added.)

3. Ibid., iv. (Italics added.)
4. Ibid., iv.
5. Ibid., 1. (Italics added.)
6. Title 47, Code of Federal Regulations, "Federal Acquisitions Regulations," Subpart 2.1—Definitions.
7. The International Organisation for Economic Co-operation and Development (OECD) uses comparable categories.
8. National Science Board, *Science and Engineering Indicators 2004*, 2 vols. (Washington, D.C.: National Science Foundation, 2004), Sidebar to Chapter 4: "U.S. and International Research and Development: Funds and Technology Linkages," at <http://www.nsf.gov/sbe/srs/seind04/c4/c4s1.htm>.
9. *Indicators 2004*, chapter 5, at <http://www.nsf.gov/sbe/srs/seind04/c4/c4s1.htm>. In its comments on the precision of these definitions, the NSF authors acknowledge Donald E. Stokes's *Pasteur's Quadrant: Basic Science and Technological Innovation* (Washington, D.C.: The Brookings Institution, 1997) as offering an alternative taxonomy.
10. *Indicators 2004*, Vol. 1, 5–7, fn. 3.
11. Ibid., fn. 4.
12. Office of Management and Budget, Executive Office of the President, Circular A-110, "Uniform Administrative Requirements with Institutions of Higher Education, Hospitals and Other Non-Profit Organizations," November 19, 1993; further amended September 30, 1999; Sect. 2 (dd).
13. Office of Management and Budget, Executive Office of the President, Circular A-21, "Cost Principles for Educational Institutions," August 8, 2000; Clarification Memorandum, January 5, 2001.
14. See, for example, David H. Guston and Kenneth Keniston, eds., *The Fragile Contract: University Science and the Federal Government* (Cambridge, Mass.: The MIT Press, 1994). This characterization of the report is attributed to Bush by his biographer, Pascal Zachary, in *Endless Frontier: Vannevar Bush: Engineer of the American Century* (New York: The Free Press, 1997), 222.
15. G. Pascal Zachary, *Endless Frontier: Vannevar Bush, Engineer of the American Century* (New York: The Free Press, 1997), 3–14.
16. Vannevar Bush, *Pieces of the Action* (New York: William Morrow and Company, Inc., 1970), 246.
17. For example, the Coast and Geodetic Survey, the Naval Observatory, the agricultural research stations of the Department of Agriculture, research conducted by the Bureau of Mines and the Forestry Service, and federal support of research in state institutions as a result of the Land Grant College acts of the post–Civil War era. See A. Hunter Dupree, *Science in the Federal Government: A History of Policies and Activities* (Baltimore, Md.: The Johns Hopkins University Press, 1986).
18. Francis Bacon, *The Advancement of Learning and New Atlantis* (London: Oxford University Press, 1960). In his autobiography Bush makes scant mention of his famous report. *Pieces of the Action*, 64.
19. See Daniel Kevles's discussion of the aspirations of Henry and others like him during the mid-nineteenth century in *The Physicists: The History of a Scientific Community in Modern America* (Cambridge, Mass.: Harvard University Press, 1997), chapters 1 and 2.
20. Guggenheim Foundation, Rockefeller Foundation, Bell Telephone Labs, Polaroid Corporation, Standard Oil of Indiana, and E. I. DuPont.
21. Vannevar Bush, *Science—The Endless Frontier: A Report to the President* (Washington: U.S. Government Printing Office, 1945), 5.

22. Ibid., 17, 27.

23. Ibid., 28.

24. Ibid., 27–28.

25. Ibid., 31–33.

26. Ibid., 28.

27. Ibid., 29–33. Bush and his committee envisioned that the foundation would initially have five divisions: medical research, natural sciences, national defense, scientific personnel and education, publications and scientific collaboration.

28. Ibid., 32.

29. These characterizations appear repeatedly throughout the text.

30. See Perry Miller, ed., *The American Transcendentalists: Their Prose and Poetry* (Garden City, N.Y.: Doubleday Anchor Books, 1957).

31. See Kevles, chapter 17, "The New Deal and Research," 252–266.

32. The others were Secretary of War Henry L. Stimson, then U.S. Army Chief of Staff General George C. Marshall, and Vice President Henry A. Wallace. *Pieces of the Action*, 292, *passim*.

33. Quoted in J. Merton England, *A Patron for Pure Science: The National Science Foundation's Formative Years, 1945–57* (Washington, D.C.: National Science Foundation, 1982), 29.

34. Plutarch's characterization of materialism in the *Morals*, Book VII, Question II; William W. Goodwin, ed. *Plutarch's Morals*, Vol. 3 (Boston, Mass.: Little, Brown, and Company, 1870), 402–404. Quoted by Ralph Waldo Emerson in his "Introduction" to the 1870 Goodwin edition of *Plutarch's Morals*, Vol. I, xvi, xxii.

35. The receipt of research grants and contracts is critical to a successful career in science, and is thus an integral part of the profession's reward system. For a classic discussion of the allocation of rewards in science, see Merton, "The Matthew Effect in Science".

36. Thomas F. Gieryn interprets the issue of including the social sciences in the NSF's portfolio as an instance of "boundary setting," a phase in the sociological process undergone by the sciences to establish their credibility in modern American culture. The issue reemerges periodically, as it did in 1995, when the chair of the House Science Committee (Robert Walker, R-Penn.) proposed that a 3 percent increase in the NSF's proposed budget not be extended to the foundation's program in the social behavioral and economic sciences. See Thomas F. Gieryn, "The U.S. Congress Demarcates Natural and Social Science (Twice)," in *Cultural Boundaries of Science: Credibility on the Line* (Chicago, Ill.: University of Chicago Press, 1999), 65–114.

37. The Bureau of the Budget was created by the Budget and Accounting Act of 1921; its name was changed to the Office of Management and Budget (OMB) in 1970. Appointments as director and deputy director of OMB require Senate confirmation; the incumbents are normally among the president's principal assistants. The U.S. Patent Office's name was expanded to the U.S. Patent and Trademark Office in 1975.

38. Quoted in England, *A Patron for Pure Science*, 71–72.

39. Milton Lomask, *A Minor Miracle: An Informal History of the National Science Foundation* (Washington, D.C.: National Science Foundation, 1976), 65.

40. See Alexandra Oleson and John Voss, *The Organization of Knowledge in Modern America, 1860–1920* (Baltimore, Md.: The Johns Hopkins University Press, 1979).

41. Lomask, *A Minor Miracle*, 119–214.

42. NOAA, an agency of the Department of Commerce, combined the much older National Ocean Service and Coast and Geodetic Survey (both est. 1807) with the National Weather Service (est. 1870s).

43. See Dupree, *Science in the Federal Government*.

44. Amazon.com lists over 2,000 titles for which Lewis Mumford is either an author or a subject. One of his best known works is *Technics and Civilization* (New York, N.Y.: Harcourt Brace, 1963). Roszak coined the term "counter culture" in *The Making of a Counter Culture* (Berkeley, Calif.: University of California Press, 1968), in which he identified the touchstone of student radicalism of the 1960s as "technocracy."

45. Roszak, *The Making of a Counter Culture*, 7–8.

46. See chapter 7, U.S. Space Policy.

47. Robert K. Merton, *Science, Technology and Society in Seventeenth-Century England* (New York, N.Y.: Fertig/Harper & Row, 1970).

48. Robert K. Merton, *The Sociology of Science: Theoretical and Empirical Investigations,* Norman W. Storer, ed. (Chicago, Ill.: University of Chicago Press, 1979).

49. See, for example, Ronald S. Burt, "Structural Holes: The Social Structure of Competition" (Cambridge, Mass.: Belknap Press, 1995), and Michael Erard, "Where to Get a Good Idea: Steal It Outside Your Group," *The New York Times* (May 22, 2004).

50. K. R. Popper, *Logik der Forschung* (Vienna, 1935); *The Logic of Scientific Discovery* (London, UK: Taylor & Francis, 1959).

51. Thomas S. Kuhn, *The Structure of Scientific Revolutions* (Chicago, Ill.: University of Chicago, 1962), 146.

52. Ibid., 148–151.

53. *The Logic of Scientific Discovery,* first published in German in 1934, argued that empirical falsification of scientific hypotheses, rather than their verification through inductive empiricism—which could never be absolute—constituted the essential means of progress in science.

54. Karl Popper, *The Open Society and Its Enemies,* 2 vols. (New York: Harper & Row, 1962).

CHAPTER 4   THE SCIENCE AND TECHNOLOGY POLICY TOOLKIT

1. "Commercialization" is often used interchangeably with "privatization." In the case of commercialization, a product or service is produced and marketed by private companies in the commercial market place. Privatization occurs when a public service contracts with a private sector firm to deliver its services, though the ownership of the service's facilities may remain in public hands.

2. See chapter 2.

3. A Department of Science was proposed during the Nixon administration; its advocates had to settle for the statutory establishment in the White House of the Office of Science and Technology Policy, intended to provide government-wide coordination of federal programs and, as necessary, study of national-level science and technology policy issues. The idea resurfaced unsuccessfully during the Clinton administration.

4. For a good account of the ebb and flow of the ideal of comprehensive government social and economic policy see Otis L. Graham, Jr., *Toward a Planned Society: From Roosevelt to Nixon* (New York: Oxford University Press, 1976) and Otis L. Graham, Jr., *Losing Time: The Industrial Policy Debate* (Cambridge, Mass.: Harvard University Press, 1992).

5. See Bruce L. R. Smith, *American Science Policy Since World War II* (Washington, D.C.: The Brookings Institution, 1990).

6. Linda R. Cohen and Roger G. Noll, *The Technology Pork Barrel* (Washington, D.C.: Brookings Institution, 1991) is an excellent study of the political and economic dimensions of selected federally funded and/or conducted projects to commercialize research and development in particular areas of technology, e.g., supersonic transportation, synthetic fuels.

7. For an overview see Harold Orlans, *Contracting for Knowledge* (Hoboken, N.J.: Jossey-Bass, 1973).

8. Johnson's successful "Great Society" legislative initiatives included the Civil Rights Act of 1964, the Voting Rights Act of 1965, Medicare, Medicaid, creation of a Job Corps for the unemployed, and Head Start for pre-school children.

9. The NSF's current program areas include biology, computer and information sciences, education, engineering, environmental research and education, geosciences, international, math and physical sciences, polar research, and social, behavioral and economic sciences.

10. See Hounshell, *From the American System to Mass Production.*

11. The White House, "Fact Sheet: Federal Laboratory Reform" (Washington, D.C.: Office of Science and Technology Policy, September 25, 1995), 1.

12. See, for example, Thomas P. Hughes, *Networks of Power: Electrification in Western Society, 1880–1930* (Baltimore, Md.: The Johns Hopkins University Press, 1993).

13. Ross M. Robertson, *History of the American Economy* (New York: Harcourt, Brace & World, Inc., 1964), 283–284.

14. Daniel Yergin and Joseph Stanislaw, *The Commanding Heights: The Battle Between Government and the Marketplace That is Remaking the Modern World* (New York, N.Y.: Simon & Schuster, 1998), 25–26, passim.

15. For a good account of the industrial policy controversy in U.S. politics, see Graham, Jr., *Losing Time.*

16. Smith, *American Science Policy Since World War II*, 88–89.

17. *Science Indicators, 2002,* 4–35.

18. Ibid., 4–36.

19. David Mowery, "The Practice of Technology Policy," in Paul Stoneman, ed. *Handbook of the Economics of Innovation and Technological Change* (Oxford, UK: Blackwell Publishers, 1995), 513–557.

20. *Science and Engineering Indicators, 2002,* Vol. I, 4-15–4-17. Alan Goodacre and Ian Tonks, "Finance and Technological Change"; Stan Metcalfe, "The Economic Foundations of Technology Policy: Equilibrium and Evolutionary Perspectives"; and David Mowry, "The Practice of Technology Policy," in Stoneman, *Handbook of the Economics of Innovation and Technological Change,* 313–315, 440–442, 529–531.

21. These policies and regulations are summarized in Office of Management and Budget (OMB) Circular A-76 and the Federal Acquisitions Regulations (FAR).

22. Norman R. Augustine, *Augustine's Laws and Major System Development Programs* (New York, N.Y.: American Institute of Aeronautics and Astronautics, 1983), 8.

23. National Science and Technology Council, *Renewing the Federal Government-University Research Partnership for the Twenty-first Century,* NSTC Presidential Review Directive-4 (Washington, D.C.: Executive Office of the President, Office of Science and Technology Policy, April 1999).

24. The five recommendations that did not pertain to the procedures by which federal research grants are awarded and administered called for the National Science and Technology Council and universities to develop a statement of principles to guide their future interactions, efforts to strengthen the "linkages between research and education," uniform or government-wide policies and practices to address research misconduct, improved environmental protection practices in research laboratories, and the creation of a task force to "provide continuing dialogue and review."

25. Merton, "The Matthew Effect in Science."

26. Charles W. McCutchen, "Peer Review: Treacherous Servant, Disastrous Master," *Technology Review* (October 1991), 28–40.

27. National Science and Technology Council, *Renewing the . . . Partnership*, 8.

28. This statute, the grounding precedent in both British and American patent law, voided all existing monopolies but authorized the issuance of patents, valid for fourteen years, to the true inventors of new manufactures for the exclusive "working" of their inventions.

29. Lawrence M. Friedman, *A History of American Law* (New York: Simon and Schuster, 1973), 224–226; Bruce W. Bugbee, *Genesis of American Patent and Copyright Law* (Washington, D.C.: Public Affairs, 1967).

30. Section 8.

31. 1 Stats. 109–110 (Act of April 10, 1790).

32. 5 Stats. 117, 119, 120 (Act of July 4, 1836).

33. Bureau of the Census, U.S. Department of Commerce, *The Statistical History of the United States: From Colonial Times to the Present* (Fairfield, Conn.: Fairfield Publishers, 1965), series W 66–76.

34. Patents issued in the U.S. are of four kinds: utility patents for inventions, design patents, plant patents, and reissue patents. For technology policy purposes, utility patents are the most significant; they also far outnumber all other kinds. U.S. Patents are generally valid for twenty years. Source of data: U.S. Patent and Trademark Office, <www.uspto.gov>.

35. Statistical Abstract of the United States, Foreign Commerce and Aid (Washington, D.C.: US Census Bureau, 2002), Table No. 1262, U.S. International Transactions by Type of Transaction: 1990–2001.

36. For a good overview, see Wesley M. Cohen and Stephen A. Merrill, eds., *Patents in the Knowledge-Based Economy* (Washington, D.C.: The National Academies Press, 2003).

37. John L. King, "Patent Examination Procedures and Patent Quality," in Cohen and Merrill, *Patents in the Knowledge-Based Economy*, 54–73.

38. Adam Jaffe, "The US Patent System in Transition: Policy Innovation and The Innovation Process," *Research Policy Symposium on Technology Policy* (July 1999), 4.

39. *Diamond, Commissioner of Patents and Trademarks v. Diehr and Lutton*, 450 U.S. 175 (1981). Both the majority and minority opinions in this case are well worth reading, for they illustrate the arguable nature of both the legal and technological aspects of the question.

40. U.S. Court of Appeals for the Federal Circuit (July 23, 1998).

41. The minority opinion in *Diamond v. Diehr* observed that "cases considering the patentability of program-related inventions do not establish rules that enable a conscientious patent lawyer to determine with a fair degree of accuracy which, if any, program related inventions will be patentable. Second the inclusion of the ambiguous concept of 'algorithm' within the 'law of nature' category of unpatentable subject matter has given rise to the concern that almost any process might be so described and therefore held unpatentable." *Diamond v. Diehr and Lutton*.

42. "The right of the people to be secure in their persons, houses, papers, and effects, against unreasonable searches and seizures, shall not be violated, and no Warrants shall issue, but upon probable cause, supported by Oath or affirmation, and particularly describing the place to be searched, and the persons or things to be seized."

43. *Diamond, Commissioner of Patents v. Ananda Chakrabarty*, 447 U.S. 303 (1980).

44. Ibid.

45. S. Rep. No. 1979, 82nd Cong., 2d Sess., 5 (1952); H.R. Rep. No. 1923, 82d Cong., 2d Sess., 6 (1952).

46. Federal agencies do not appear as litigants in federal courts on their own behalf; while they may be named litigants, their cases are argued by the Department of Justice on the behalf of the U.S. government.

47. *Diamond v. Chakrabarty.*

48. Jonathan King and Doreen Stabinsky, "Patents on Cells, Genes, and Organisms Undermine the Exchange of Scientific Ideas," *The Chronicle of Higher Education* (February 5, 1999), B6–B7.

49. *Pallin v. Singer and Hitchcock Associates of Randall*, Civil Action No. 5: 93-CV-202 (United States District Court for the District of Vermont); Susan Leach DeBlasio, "Patents on Medical Procedures and the Physician Profiteer," Tillinghast Licht Perkins Smith & Cohen, LLP (<www.tlslaw.com/publications/medpatents.html>; downloaded November 22, 2004); and H.R. 1127, 104th Cong., 1st Session, § 2 (1995), "Medical Procedures Innovation and Affordability Act."

50. See, for example, Merritt Roe Smith, *Harper's Ferry and the New Technology: The Challenge of Change* (Ithaca, N.Y.: Cornell University Press, 1977).

51. *Statistical History*, Series WW 66–76.

52. Established in 1947 and incorporated into the Department of Energy created in 1977.

53. Established in 1915.

54. Long cited in *The Patent, Trademark and Copyright Journal of Research and Education*, Vol. 3, No. 3 (Fall, 1959), 237–238.

55. Russell B. Long, "Federal Contract Patent Policy and the Public Interest," *Federal Bar Journal*, Vol. 21, No. 1 (Winter, 1961), 20.

56. "John F. Kennedy, Memorandum and Statement on Government Patent Policy," *Federal Register*, Vol. 28, No. 200 (October 12, 1963), 10943–10946.

57. Richard M. Nixon, Memorandum for Heads of Executive Departments and Agencies, Statement of Government Patent Policy (August 23, 1971).

58. The view that modern government policy in general has alternated between government and private management of market economies is elaborated with impressive detail in Yergin and Stanislaw, *The Commanding Heights.*

59. Amendments to the Patent and Trademark Laws, Public Law 96-517, named for their sponsors, Senators Birch Bayh and Robert Dole. Exceptions were limited to instances when retaining title might impede U.S. foreign intelligence or counter-intelligence activities, and Awhen the funding agreement includes the operation of a Government-owned, contractor-operated facility of the Department of Energy primarily dedicated to that Department's naval nuclear propulsion or weapons related programs. The federal agency under whose funding agreement the invention had been made was given "march-in" rights to issue an exclusive license to a third party, "upon terms that are reasonable under the circumstances," if the original contractor or assignee failed to take, "or is not expected to take within a reasonable time, effective steps to achieve practical application of the subject invention." U.S.C., Title 35, Part II, chapter 18.

60. Authorized by the Federal Technology Transfer Act of 1986.

61. Office of Technology Policy, U.S. Department of Commerce, *Tech Transfer 2000: Making Partnerships Work* (May 2000), 10.

62. As quoted in *The Washington Post* for July 30, 1965. See also Sylvia Katharine Kraemer, "Federal Intellectual Property Policy and the History of Technology: The Case of NASA Patents," *History and Technology*, Vol. 17, No. 3 (2001), 183–216.

63. Author's analysis of USPTO patent counts by patent class, 1977–1999.

64. Fred Warshofsky, *The Patent Wars: The Battle to Own the World's Technology* (New York: John Wiley & Sons, Inc., 1994), 230–244.

65. "Taxol: How the NIH Gave Away the Store," Public Citizen's Health Research Group *Health Letter* (August 2003).

66. The original patent owners as well as the firms to which patents were transferred were: the Boeing Company, Boeing-Rocketdyne, David Sarnoff Research Center, GenCorp Aerojet, General Dynamics, General Electric Company, ITT Corporation, Johnson Controls, Inc., Lockheed Martin Corporation, McDonnell Douglas Corporation, Martin Marietta Corporation, Northrup Grumman Corporation, Oceaneering International, Inc., Olin Corporation, Orbital Sciences Corporation, Space Systems/Loral Inc., and United Technologies Corp. Kraemer, "Federal Intellectual Property Policy."

67. Jong-Tsong Chiang, "Technology Policy Paradigms and Intellectual Property Strategies: Three National Models," *Technological Forecasting and Social Change*, 49, 35–48 (1995). Patent counts are available from the USPTO website, <www.uspto.gov>, and include utility, design, plant, and reissued patents.

68. This frequently used expression is usually meant to have broader meaning than the more traditional "technology," which might, or might not, depending on who was using it, include increased understanding of phenomena in the natural sciences or, at the other extreme, cumulative business acumen.

69. James Love, representing the Consumer Project on Technology, USPTO Roundtable on International Patent Harmonization held December 19, 2002 in Arlington, Va. The report is available from the USPTO website at <www.uspto.gov>.

70. Statement of the National Association of Manufacturers, USPTO Roundtable on International Patent Harmonization held December 19, 2002 in Arlington, Va. at <www.uspto.gov>.

CHAPTER 5   SCIENCE, TECHNOLOGY, AND POLITICAL AUTHORITY

1. The Supreme Court's abandonment of the "separate but equal" doctrine in *Plessy v. Ferguson* (1896) relied extensively on psychological and sociological studies.

2. 293 F. 1013 (D.C. Circuit, 1923).

3. *Frye v. United States*, 293 F. 1013 (D.C. Circuit 1923).

4. *Federal Rules of Evidence*, Rules 401 and 702. See <http://www.law.cornell.edu/rules/fre/rules.htm>.

5. The Federal Judicial Center was created by the Congress in 1967 "to promote improvements in judicial administration in the courts of the United States." See <www.fjc.gov>.

6. (92-102) 509 U.S. (1993).

7. 522 US 136 (1997).

8. 526 US 137 (1999).

9. Discussions of science use various expressions interchangeably, i.e., "fundamental research," "academic research," "basic research," and "university based research," which are contrasted with "commercial research," "industrial research," "research and technology," and "research and development." For the purposes of this book, "academic research" refers primarily to research in the natural sciences conducted in colleges and universities.

10. Traceable to Frederick Taylor's late-nineteenth-century time and motion studies of

factory workers, a managerial approach relying on quantitative or statistical process management for quality control was popularized in the 1940s by W. Edward Deming.

11. "The Government Performance and Results Act of 1993," PL 103-62.

12. An organization of roughly 145 research universities.

13. See National Academy of Sciences, Committee on Science, Engineering, and Public Policy, *Evaluating Federal Research Programs: Research and the Government Performance and Results Act* (Washington, D.C.: National Academy Press, 1999) and National Academy of Sciences, Committee on Science, Engineering, and Public Policy, *Implementing the Government Performance and Results Act for Research: A Status Report* (Washington, D.C.: National Academy Press, 2001).

14. A philosophy of scientific management developed by Frederick W. Taylor (1856–1915), which taught that manufacturing efficiency could be improved by disaggregating and measuring the work processes of men and machines, thus enabling rational steps toward greater productivity.

15. *Science and Engineering Indicators 2002,* Vol. 1, 4–13.

16. The FOIA also requires federal agencies to publish in the *Federal Register* descriptions of their procedures and general rules, and to respond within ten working days to a request for information resulting from their activities (that is, information "created by" them), charging requesters only a reasonable cost for assembling and providing the information. First enacted in 1966 as one of the consequences of the Vietnam War era, the federal FOIA was amended in 1997 to include electronic information.

17. 33 U.S. (8. Pet) 591.

18. 17 U.S. Code, Sect. 105.

19. Executive Office of the President, Office of Management and Budget, "Circular No. A-130, Management of Federal Information Resources," available at <http://www.whitehouse.gov/omb/circulars/a130trans4.html>.

20. Fiscal Year 1999 Omnibus Appropriations Act, Public Law 105–277.

21. For example, Alabama Power, Birmingham Steel Corporation, Gulf States Steel, Sonat Inc. (natural gas), and Southern Natural Gas Company.

22. Much of that data had been generated by a Harvard Medical School study (funded by the EPA and the National Institute of Environmental Health Sciences) of mortality in six cities, which concluded that air quality was one of four principal predictors of survival. The other three factors were age and sex; smoking, obesity, and socioeconomic status; and occupation. The other contested studies, Shelby's explanation of the reasons behind his amendment, and summaries as well as original statements submitted to the OMB both in support of, and opposition to, the amendment can be found in Linda R. Cohen and Robert W. Hahn, "Should Researchers Be Required to Share Data Used in Supporting Regulatory Decisions?" Regulatory Analysis 99-1, American Enterprise Institute-Brookings Joint Center for Regulatory Studies (May 1999). See also Senator Richard Shelby, Letter to Mr. Jacob Lew (Director, Office of Management and Budget), September 10, 1999 (Downloaded from <http://Shelby. senate.gov/press/psrsrs283.htm> on March 1, 2002); Public Citizen, Comments on OMB's Proposed Revision of OMB Circular A-110, "Uniform Administrative Requirements For Grants and Agreements With Institutions of Higher Education, Hospitals, and Other Non-Profit Organizations" (downloaded from <http://www.citizen. org.litigation.ombcomment.htm>, February 22, 2001); Bruce Alberts, President of the National Academy of Sciences, Letter to F. James Charney, Policy Analyst, Office of Management and Budget (September 10, 1999); and Office of Management and Budget, "Final Revision: Circular A-110, Uniform Administrative Requirements for Grants

and Agreements with Institutions of Higher Education, Hospitals, and Other Non-Profit Organizations" (Washington, D.C.: Executive Office of the President, November 6, 1999).

23. Administrative Procedure Act (5 U.S.C. 553, et. seq.)

24. National Research Council, *Bits of Power: Issues in Global Access to Scientific Data* (Washington, D.C.: National Research Council, 1997).

25. Public Law 106-554; H.R. 5658. Section 515. Also referred to as the Data Quality Act.

26. Letter to OMB director Joshua Bolton, December 15, 2003, signed by Henry Waxman (D-Calif.), ranking minority member, Committee on Government Reform; John F. Tierney (D-Mass.), ranking minority member, Subcommittee on Energy Policy, Natural Resources and Regulatory Affairs, committee on Government Reform; Sherrod Brown (D-Ohio), ranking minority member, Subcommittee on Health, Committee on Energy and Commerce; Eddie Bernice Johnson (D-Tex.), ranking minority member, Subcommittee on Research, Committee on Science; Mark Udall (D-Colo.), ranking minority member, Subcommittee on Environment, Technology, and Standards, Committee on Science; Brian Baird (D-Wash.), member, Committee on Science; and Michael M. Honda (D-Calif.), member, Committee on Science.

27. The complete list of respondents is available from the OMB website, as well as the responses themselves. See <http://www.whitehouse.gov/omb/inforeg/2003iq/iq_list.html>. Downloaded August 14, 2004.

28. PL 92-463.

29. Congressional Research Service, "Introduction and Legislative History of the Federal Advisory Committee Act," in *Federal Advisory Committee Act (Public Law 92-463) Source Book: Legislative History, Texts and Other Documents* (Washington, D.C.: U.S. Government Printing Office, 1978), 3–4.

30. Committee on Government Operations, U.S. House of Representatives, "Report" accompanying H.R. 7390, Amending the Administrative Expenses Act of 1946, and For Other Purposes, reprinted in ibid., 46.

31. United States Code, Title 18—Crimes and Criminal Procedure; Section 208. Acts affecting a personal financial interest.

32. *Association of American Physicians and Surgeons, Inc. et al. v. Hillary Rodham Clinton et. al.,* U.S. District Court for the District of Columbia, No. 93-0399 (March 10, 1993);

33. *Richard B. Cheney, Vice President of the United States et al., Petitioners v. United States District Court for the District of Columbia et. al.* U.S. Supreme Court 542 U.S. (2004).

34. Championed by Congressman Emilio Q. Daddario (D-Conn.), the Office of Technology Assessment was established in 1972 by the Technology Assessment Act (PL 92-484). Over time OTA's studies acquired a reputation of technical authority and political balance. However, after the Republican ascendancy in the mid-term elections of 1994, OTA was denied funding and was required to close down.

35. Charles E. Lindblom, "The Science of 'Muddling Through,'" *Public Administration Review,* Vol. 19 (1959), 79–88.

CHAPTER 6   OPEN SYSTEMS IN THE DIGITAL WORLD

1. "Cyberspace" is that vast region through which electrons travel, storing and transporting digital data, images, and sound. One can think of it as an ocean composed of electrons rather than water, over which we travel or into which we descend in our computer-ships, powered by software engines and electronic fuel.

2. For a good overview, see Paul E. Ceruzzi, *A History of Modern Computing* (Cambridge, Mass.: The MIT Press, 1998).

3. In this context, "bus" architecture refers to a system's internal design and structure.

4. Much has been written about the Internet; an excellent account of its development is Janet Abbate's *Inventing the Internet* (Cambridge, Mass.: The MIT Press, 1999). For a participant's account see Tim Berners-Lee's (with Mark Fischetti) *Weaving the Web: The Original Design and Ultimate Destiny of the World Wide Web* (New York: Harper Collins, 1999).

5. The best single source for quantitative data about US trends in science and technology is *Science and Engineering Indicators*, published annually by the National Science Board.

6. This need was not limited to meteorology or biology, of course. All disciplines that relied on quantitative data—such as economics or sociology—wanted to be able to collect and analyze increasing volumes of information.

7. Massmutual Corporate Investors, which would later advertise itself as providing the "world's farthest-reaching global Internet backbone, [connecting] 6 continents, over 140 countries, over 2,800 cities."

8. Needless to say, this policy is not followed absolutely. When the Congress decides that a technological system or economic resource should survive for whatever reason, notwithstanding "market failure," it will authorize continued spending and work on that system or resource. The survival of passenger rail systems both regionally and nationally is a good example, as is state- and federally funded unemployment insurance.

9. Such as MOSAIC, which was released gratis in 1993 by the Illinois National Center for Computing Applications. MOSAIC was compatible with any computer's operating system, and enabled transmission over the Internet of images and sound as well as alpha-numeric data. Only a year later, MOSAIC's Marc Andreessen ventured out on his own and released the commercial web browser, NETSCAPE. ISPs are companies that allowed individual computers to connect to the telephone or cable wires (and more frequently wireless transmission companies) that served as the Internet's highway system or "backbone."

10. The First Amendment protects freedom of speech and of the press, while the Fourth Amendment guarantees the "right of the people to be secure in their persons, houses, papers, and effects, against unreasonable searches and seizures" without warrants issued on the basis of "probable cause, supported by Oath or affirmation, and particularly describing the place to be searched, and the persons or things to be seized."

11. *Near v. Minnesota*, 283 U.S. 697. This case involved a Minnesota statute allowing state officials to seek court injunctions against "producing, publishing or circulating, having in possession, selling or giving away, (a) an obscene, lewd and lascivious newspaper, magazine, or other periodical, or (b) a malicious, scandalous and defamatory newspaper, magazine or other periodical." Robert F. Cushman and Brian Stuart Koukoutchos, *Cases in Constitutional Law*, ninth ed. (Saddle River, N.J.: Prentice Hall, 2000), 416–420. See also *Gitlow v. New York*, 268 U.S. 652 (1925) and *Whitney v. California*, 274 U.S. 357 (1957).

12. *Schenck v. United States*, 249 U.S. 47 (1919).

13. *Chaplinski v. New Hampshire*, 315 U.S. 568 (1942); see Robert E. Cushman and Robert F. Cushman, *Cases in Constitutional Law* (New York: Appleton-Century-Crofts, Inc., 1958), 686.

14. Ronald Dworkin, "The Coming Battles Over Free Speech," *New York Review of Books,* Vol. 39, No. 11 (June 11, 1992); and Anthony Lewis, *Make No Law: The Sullivan Case and the First Amendment* (New York: Random House, 1992).

15. See, for example, *Roth v. the United States* (1957) and *Kingsley Books v. Brown* (1957), discussed in Cushman and Koukoutchos, *Cases in Constitutional Law*, 421–428.

16. As summarized by Justice Warren Burger in *Miller v. California*, 413 U.S. 15 (1973). See, for example, *Roth v. the United States* (1957) and *Kingsley Books v. Brown* (1957), discussed in Cushman and Koukoutchos, *Cases in Constitutional Law*, 424.

17. The plaintiff in this case was Marvin Miller, not the writer Henry V. Miller, who gained some notoriety during the post-war period as an author of erotic fiction.

18. Chief Justice Warren Burger, majority opinion in *Miller v. California*, 413 U.S. 15 (1973), Cushman and Koukoutchos, *Cases in Constitutional Law*, 427.

19. See Hugh R. Slotten, *Radio and Television Regulation: Broadcast Technology in the United States, 1920–1960* (Baltimore, Md.: The Johns Hopkins University Press, 2000).

20. *FCC v. Pacifica Foundation* (1978) and *Sable Communications of California, Inc. v. FCC* (1989), discussed in Cushman and Koukoutchos, *Cases in Constitutional Law*, 442.

21. *Reno v. American Civil Liberties Union*, 117 S.Ct.2329 (1997).

22. Opinion excerpted in *The New York Times* (June 1, 2002). See also John Schwartz, "Court Blocks Law that Limits Access to Web in Library," *The New York Times* (June 1, 2002).

23. *Bigelow v. Virginia*, 421 U.S. 809, 829 (1975). Rehnquist was joined by Justices Sandra Day O'Connor, Antonin Scalia, Clarence Thomas, Anthony Kennedy, and Stephen G. Breyer. Justices John Paul Stevens, III and David H. Souter filed dissenting opinions; Justice Ruth Bader Ginsberg joined in Justice Souter's opinion. *United States et al. v. American Library Association, Inc., et al.*, 539 U.S. (2003).

24. Spurred largely by the horrific passenger aircraft assaults on the twin towers of New York's World Trade Center and the Pentagon in Washington, D.C. of September 11, 2001, and the unsuccessful attempt that same day to crash a third airliner into either the White House or the Capitol.

25. For a thorough analysis of the USA Patriot Act's (2001) electronic surveillance provisions, see the website maintained by the Electronic Privacy Information Center, at <www.epic.org>.

26. See Slotten, *Radio and Television Regulation*.

27. For example, in 1998 total revenues for the electronic mass media industry (cable TV and other pay TV, TV and radio broadcasting, and TV/radio communications equipment) totaled $143.8 billion, an average growth of 40.8 percent since 1995. Source: National Telecommunications and Information Administration (NTIA).

28. The electromagnatic "highways" over which digital Internet packets travel. Technically, the "primary forward-direction path traced sequentially through two or more major relay or switching stations." (Source: NTIA.)

29. The growth of "e-commerce" over the Internet has been significant for some businesses; however, with the exception of a few highly successful companies such as the bookseller Amazon.com, it is not yet clear how great an economic impact the Internet will have in the long run.

30. National Telecommunications and Information Agency, "A Nation Online: How Americans Are Expanding Their Use of the Internet" (Washington, D.C.: Department of Commerce, February 2002).

31. So named for Heinrich Hertz (1857–1894), discoverer of radio waves, who determined their velocity, now measured in -hertz units, ranging from as low as 3 kilohertz (KHz) to as high as 300 gigahertz (Ghz). Most ship-to-shore communications occur at very low frequencies, commercial radio broadcasts at medium frequencies (300 KHz to around 30 megahertz, or MHz), while satellite communications occur at extremely

high frequencies (around 3 Ghz to 30 Ghz). (One kilo = 1 thousand units; One mega = 1 million units; one giga = 1 thousand million units.)

32. The wavelength at the very low frequency (VLF) radio band is 100 kilometers, while the wavelength at extra high frequency (EHF) band of 300 gigahertz is only 1 millimeter.

33. Such high frequency signals are, however, very vulnerable to attenuation from terrestrial or atmospheric interference, and thus are used mostly for microwave and satellite communications.

34. This is not a peculiarly American trend. For an excellent treatment of the tensions in the older industrialized democracies between active government oversight of national economies and reliance on laissez faire minimalism, see Daniel Yergin and Joseph Stanislaw, *The Commanding Heights: The Battle Between Government and the Marketplace That is Remaking the Modern World* (New York: Simon and Schuster, 1998).

35. John Markoff, "Businesses, Big and Small, Bet on Wireless Internet Access," *The New York Times* (November 18, 2002). "Wideband" or "broadband" refers to transmission at a bandwidth exceeding the 4 KHz of the normal telephone channel.

36. Nicholas Lemann, in a recent portrait of the FCC and its chairman Michael Powell, writes that the "real purpose" of the FCC is to supervise the industry, and its more specific purpose is to protect "the interests of members of Congress, many of whom regard the media companies in their districts as the single most terrifying category of interest group—you can cross the local bank president and live to tell the tale, but not the local broadcaster." See Lemann, "The Chairman," *The New Yorker* (October 7, 2002), 49–50.

37. Older televisions designed for analog reception can be retrofitted to receive digital TV with a converter box. See the FCC's website, for its news release for November 12, 1999 <http://www.fcc.gov/mb/video/files/descrip.html>.

38. Broadband operation is necessary for the development of integrated systems digital networks (ISDN), networks that use both digital transmission and digital switching (digital signals are switched without conversion to or from analog signals). Digital signal switching offers integrated access to voice, high-speed data service, video-demand services, and interactive delivery services (FCC "Glossary of Telecommunications Terms," <http://www.fcc.gov> (June 22, 2003).

39. Principally Verizon, SBC, BellSouth, and Qwest Communications International.

40. See, for example, Jennifer Lee, "F.C.C. Members Testify About New High-Speed Rules," *The New York Times* (February 27, 2003).

41. Stephen Labaton, "FCC Clears Internet Access by Power Lines," *The New York Times* (October 15, 2004).

42. Matt Richtel, "Court Rules F.C.C. Erred in Decision on Net Access," *The New York Times* (October 7, 2003).

43. Supporters of legislation reversing the FCC's new rules include Sens. Ted Stevens (R-Alaska), Ernest Hollings (D-S.C.), Byron Dorgan (D-N.Dak.), Oympia Snowe (R-Maine), and Kay Bailey Hutchison (R-Tex.). The new rules were supported by the Republican Bush administration.

44. The increased local channel service does not need to be satellite service (indeed, some local conditions might prevent satellite service), thus leaving local broadcasters to rely on antenna transmissions, as before. Frank Ahrens, "FCC Approves Murdoch Purchase of DirecTV," *The Washington Post* (December 19, 2003).

45. Useful general treatments can be found in Ceruzzi, *A History of Modern Computing*; James W. Cortada, "Progenitors of the Information Age: The Development of Chips and Computers," in Alfred D. Chandler, Jr., and James W. Cortada, *A Nation Trans-*

formed by Information (New York: Oxford University Press, 2000); and Harry Wulforst, *Breakthrough to the Computer Age* (New York: Charles Scribner's Sons, 1982).

46. Gates had already earned attention in the nascent PC market by writing BASIC, the programming language chosen for the first true personal computer, the Altair, released as a kit for hobbyists in 1974. BASIC's principal advantages over its competitors (FORTRAN and COBOL) were that it could run on relatively little memory, performed well, and was easy to learn.

47. Ceruzzi writes, "Within ten years there were over fifty million computers installed that were variants of the original PC architecture and ran advanced versions of MS-DOS. Ceruzzi, *A History of Modern Computing*, 272. Microsoft's history, as well as a detailed account of its antitrust legal troubles, are well covered in Ken Auletta's *World War 3.0: Microsoft vs. the U.S. Government, and the Battle to Rule the Digital Age* (New York: Broadway Books, 2001).

48. *Houston, East & West Texas Ry. Co. v. United States*, 234 U.S. 342 (1924); *National Labor Relations Board v. Jones & Laughlin Steel Corporation*, 301 U.S. 1 (1937).

49. Microsoft feared the possibility that with Netscape users might run software applications online produced by Microsoft's competitors, thus making its own applications superfluous. Auletta, *World War 3.0*, xx–xxi.

50. Auletta, *World War 3.0*, 52–53.

51. Users who attempted to remove Explorer from Windows received warning messages that their other programs might not work properly if they did so.

52. Auletta, *World War 3.0*, 69, 72, 298–299.

53. Jonathan Krim, "Judges Hear Challenges to Microsoft Settlement," *The Washington Post* (November 5, 2003).

54. See, for example, Steve Lohr and David D. Kirkpatrick, "Microsoft and AOL Time Warner Settle Antitrust Suit," *The New York Times* (May 29, 2003) and Steve Lohr, "Microsoft Can Leave Java Out of Windows, Court Rules," *The New York Times* (June 27, 2003).

55. Paul Meller, "Microsoft to Alter Online System to Satisfy Europe," *The New York Times* (January 31, 2003).

56. Scott Morrison of *The Financial Times*, "HP Plans Switch to Linux for New PCs," *The New York Times* (March 26, 2004).

57. Steve Lohr, "Digital Media Becomes Focus as Microsoft and AOL Settle," *The New York Times* (June 2, 2003).

58. Rob Pegoraro, "Apple Exploits Open Source to Produce an Outstanding Browser," *The Washington Post* (June 29, 2003).

59. Besides Holmes, William James, Charles Peirce, and John Dewey.

60. Louis Menand, *The Metaphysical Club* (New York: Farrar, Straus and Giroux, 2001), 441–442.

61. Challenges to a cyber-driven global market place are part of the dynamics explored in Thomas L. Friedman's *The Lexus and the Olive Tree* (New York: Farrar, Straus and Giroux, 1999); see also David Leonhardt, "Globalization Hits a Political Speed Bump," *The New York Times* (June 1, 2003).

CHAPTER 7   OPEN SYSTEMS IN OUTER SPACE

1. Years later Eisenhower recalled during an interview, "then he [Sec. of War Henry L. Stimson] told me they were going to drop it [the atomic bomb] on the Japanese. Well, I listened, and I didn't volunteer anything because, after all, my war was over in

Europe and it wasn't up to me. But I was getting more and more depressed just thinking about it. Then he asked for my opinion, so I told him I was against it on two
counts. First, the Japanese were ready to surrender and it wasn't necessary to hit them
with that awful thing. Second, I hated to see our country be the first to use such a
weapon. Well . . . the old gentleman got furious." Quoted in Richard Rhodes, *The Making of the Atomic Bomb* (New York: Simon and Schuster, 1986), 688.

2. Public Papers of the Presidents, Dwight D. Eisenhower, Farewell Address, January 17,
1961. Downloaded July 9, 2003 from <http://www.yale.edu/lawweb/avalon/presiden/
speeches/eisenhower001.htm>.

3. Stephen E. Ambrose, *Eisenhower: Soldier and President* (New York: Simon & Schuster,
1990), 461. See also Robert R. Bowie and Richard H. Immerman, *Waging Peace: How Eisenhower Shaped an Enduring Cold War Strategy* (New York: Oxford University Press, 2000).

4. A U.S. foreign policy of Soviet containment was first proposed by George F. Kennan,
head of the State Department's policy planning staff in 1947 when he published
anonymously an article in *Foreign Affairs* recommending military counter-pressure,
rather than appeasement or conciliation, as the best response to Soviet expansionism.

5. Ambrose, *Eisenhower*, 393.

6. Planned to coincide with a season of maximum physical activity on the surface of the
sun, the IGY was organized principally by the Council of Scientific Unions and the
World Meteorological Organization. Its aim was to conduct global observations from
polar and equatorial stations and orbiting satellites to collect observations on the
Earth's meteorology, geomagnetism, aurora, ionospheric physics, latitude and
longitude.

7. National Security Council, NSC 5520, "Draft Statement of Policy on U.S. Scientific Satellite Program" (Washington, D.C.: The White House, May 20, 1955).

8. "Treaty on Principles Governing the Activities of States in the Exploration and Use of
Outer Space, Including the Moon and Other Celestial Bodies" (January 27, 1967),
"Agreement on the Rescue of Astronauts, The Return of Astronauts, and The Return of
Objects Launched Into Outer Space" (April 22, 1968), "Convention on International
Liability for Damage Caused by Space Objects" (March 29, 1972), and "Convention on
Registration of Objects Launched Into Outer Space" (January 14, 1975).

9. Article I, "Treaty on Principles Governing the Activities of States in the Exploration
and Use of Outer Space, Including the Moon and Other Celestial Bodies" (January 27,
1967), reprinted in Nathan C. Goldman, *American Space Law: International and Domestic*
(San Diego, Calif.: Univelt, Inc., 1996). A fifth treaty, the "Moon Treaty" or Agreement
governing the Activities of States on the Moon and Other Celestial Bodies, was agreed
to in 1979 by UNCOPUOS through the UN's traditional process of "consensus," but has
not been ratified by the United States or any other major space-faring nation. Opponents have argued that its provisions leave little or no scope for private enterprise (v.
multilateral government-sponsored activities) in space.

10. National Security Council, NSC 5520, "Draft Statement of Policy on U.S. Scientific Satellite Program" (Washington, D.C.: The White House, May 20, 1955), reprinted in John
M. Logsdon, ed., *Exploring the Unknown: Selected Documents in the History of the U.S. Civil
Space Program, Vol. I, Organizing for Exploration* (Washington, D.C.: National Aeronautics and Space Administration, 1995), 309.

11. National Security Council, NSC 5814, "U.S. Policy on Outer Space" (The White House,
June 20, 1958), reprinted in Logsdon, *Exploring the Unknown*, 345–359.

12. Ibid.

13. Ambrose, *Eisenhower*, 454.

14. John M. Logsdon, *The Decision to Go to the Moon: Project Apollo and the National Interest* (Cambridge, Mass.: The MIT Press, 1970), 16–19.

15. Logsdon, *Decision*, 26.

16. The Senate Committee on Aeronautical and Space Sciences and the House Committee on Science and Astronautics.

17. Logsdon, *Decision*, 16–19.

18. Ibid., 35, 57–60.

19. George Kateb, "Kennedy as Statesman," Commentary (June 1966), 57, quoted in Logsdon, *Decision*, 93.

20. Johnson was minority leader from 1953 to 1955 and majority leader from 1955 to 1961.

21. Fuqua was a member of the Manned Spaceflight Subcommittee of the House Committee on Science and Technology. An excellent source for the congressional politics of the early space program is Ken Hechler's *Toward the Endless Frontier: History of the Committee on Science and Technology, 1959–79* (Washington, D.C.: U.S. House of Representatives, 1980).

22. W. Henry Lambright, *Powering Apollo: James E. Webb of NASA* (Baltimore and London: The Johns Hopkins University Press, 1995), 91, 92, 94.

23. Ibid., 115–121.

24. Executive branch department and agency heads do not customarily challenge the White House to tell them how to manage budget cuts. Tradition holds that working with given budgets is one of the obligations of public administration.

25. Robert A Divine, ed. *The Johnson Years, Vol. II: Vietnam, the Environment, and Science* (Lawrence, Kans.: The University of Kansas Press, 1987), 233–239; National Aeronautics and Space Administration, *NASA Historical Data Book, Vol. I: NASA Resources 1958–1968* (Washington, D.C.: National Aeronautics and Space Administration, 1988), 71.

26. Divine, *The Johnson Years*, 241. The "WPA" was the Works Progress Administration, a New Deal program to boost flagging employment through public sector projects.

27. Ibid., 240–245.

28. Each Saturn V was as tall as a thirty-six-story building, or thirteen times the height of the Statue of Liberty and weighed more than 3,000 tons, which is more than most large Navy destroyers.

29. NASA flew twelve Apollo missions in all between 1968 and 1972.

30. "Paid Civilian Employment in the Federal Government, by Agency, All Areas, 1965 to 1979," *Statistical Abstract of the United States* (Washington, D.C.: U.S. Government Printing Office, 1979), 277.

31. Memorandum from George M. Low to the NASA administrator (James C. Fletcher), Subject: NASA as a Technology Agency (May 25, 1971), NASA Historical Documents Collection, NASA Headquarters (hereafter NHDC).

32. In 1974 the Congress directed NASA to develop and demonstrate "solar heating and cooling technologies"; in 1975, to monitor and investigate the "chemical and physical integrity of the Earth's upper atmosphere"; in 1976, to develop "more energy efficient and petroleum conserving and environment preserving ground propulsion systems"; in 1976 to develop and demonstrate "electric and hybrid [ground] vehicle" technologies; and in 1978, to develop advanced automobile propulsion systems *and* to assist "in bioengineering research, development, and demonstration programs designed to alleviate and minimize the effects of disability." *National Aeronautics and Space Act of 1958, As Amended*, Printed for the Use of the National Aeronautics and Space

Administration (January 1990). NASA was gradually relieved of its non-aerospace responsibilities in subsequent years.

33. Low and Fletcher served in those positions during December 3, 1969–June 5, 1976 and April 27, 1971–May 1, 1977, respectively. Fletcher served an additional term after the Challenger accident, from May 12, 1986 to April 8, 1989. The quote is from the statement by the president, the White House, January 6, 1972, NHDC.

34. Memorandum from Deputy Administrator George M. Low to NASA program office administrators, May 16, NHDC.

35. Which consists of the Shuttle Orbiter, External Tank (non-recoverable) and twin recoverable Solid Rocket Boosters.

36. Mathematica, Inc., Economic Analysis of the Space Shuttle System, National Aeronautics and Space Administration Contract NASW-2081 (January 1972).

37. National Security Council, Presidential Decision Directive 37, "National Space Policy" (Washington, D.C.: The White House, May 11, 1978).

38. Jeffrey S. Banks, "The Space Shuttle," in Linda R. Cohen and Roger G. Noll, eds., The Technology Pork Barrel (Washington, D.C.: The Brookings Institution, 1991), 212.

39. Contracts went to Boeing Aerospace Co., Martin Marietta Aerospace, RCA Astro Electronics, General Electric Company Space Systems Division, Rockwell International Rocketdyne Division, TRW Federal Systems Division, Lockheed Missiles & Space Co., and McDonnell Douglas Astronautics Co.

40. In FY 1981 the DOD space budget was $4.8 billion compared to NASA's $4.9 billion (in real year dollars). By FY 1985 the DOD space budget had risen to $12.7 billion compared to NASA's budget of $6.9 billion. The DOD budget for space peaked (again, in real year dollars) to $17.9 billion in FY 1989; by the second year of the Clinton administration (FY 1995) it again dropped below NASA's budget for space ($10.6 billion vs. NASA's $12.5 billion).

41. Office of Management and Budget, Circular A-76 (Washington, D.C.: Executive Office of the President, March 1996).

42. National Security Decision Directive 94, "Commercialization of Expendable Launch Vehicles" (Washington, D.C.: Executive Office of the President, May 16, 1983); Commercial Space Launch Act of 1984 (Public Law 98-575), 98 Stat. 3055, October 30, 1984; "National Aeronautics and Space Administration Authorization Act, 1985" (Public Law 98-361, July 16, 1984).

43. Report of the Presidential Commission on the Space Shuttle Challenger Accident (Washington, D.C.: U.S. Government Printing Office, June 6, 1986).

44. "Statement by the President on the Fourth Orbiter and the Space Program," Administration of Ronald Reagan, Public Papers of the President (August 15, 1986).

45. "Space Exploration Initiative Strategy," Office of the Vice President, The White House (March 13, 1992).

46. Because of its mounting costs, the size of the ISS has been reduced to a configuration supporting a crew of only three astronauts. Critics have argued that a crew of three will be able to maintain the station, and not much else.

47. See, for example, "Report of the Space Shuttle Management Independent Review Team," ch. Christopher Kraft (February 1995), NASA Historical Reference Collection.

48. Ben Ianotta, "Cuts Into the Space Shuttle's Budget Stir Safety Concerns," Space News (March 6–12, 1995); Kathy Sawyer, "NASA Plans Privatization for Shuttle," The Washington Post (June 7, 1995; Mark Carreau, "Shuttle's Managers Say Go Slow," Houston Chronicle (March 15, 1995).

49. Gordon Adams, The Iron Triangle: The Politics of Defense Contracting (New York City: Council on Economic Priorities, 1981), 15.

50. NASA installations are located in Alabama, California, Florida, Maryland, Mississippi, Ohio, Texas, and Virginia; NASA Headquarters is located in Washington, D.C. National Aeronautics and Space Administration, "Annual Procurement Report, Fiscal Year 2002" and "Annual Procurement Report, Fiscal Year 2003" (Washington, D.C.: National Aeronautics and Space Administration, 2003 and 2004). Available online at <www.nasa.gov>.

51. The U.S. stable is dominated by the ICBM-based Atlas, Delta, and Titan series of vehicles, all of which are expendable, and the partially reusable Space Shuttle. Russia's principal launch vehicles are also ICBM based, and are dominated by the SL-series and the Proton and Energia vehicles. The heaviest payload to Earth orbit capacity is the Energia's, at roughly 100 tons, and next heaviest is the Shuttle, at roughly 30–40 tons to Earth orbit (ca. 620 miles). To reach geosynchronous orbit (ca. 22,400 miles) one can turn to the U.S. Delta, Atlas, and Titan launch vehicles, the largest of which (Titan and Atlas) can lift between 2 and 4.5 tons to geosynchronous orbit, or "GEO," depending on whether they are configured with the Centaur upper stage rocket; the Russian Proton (capable of lifting over 2 tons to GEO); ESA's Ariane 2.3 or Ariane 4 (up to 2.2 tons to GEO); and Japan's H-II vehicle (0.6 tons to GEO). Most space-based communications, data, and remote sensing systems require geosynchronous orbit.

52. For an overview of this subject, see Scott N. Pace, "The Future of Space Commerce," in W. Henry Lambright, ed., *Space Policy in the 21st Century* (Baltimore, Md.: The Johns Hopkins University Press, 2003), 55–88.

53. See <http://www.sirtf.caltech.edu>. These four large telescopes were named for renowned astronomers Edwin Hubble, Lyman Spitzer, Arthur Holly Compton, and Subrahmanyan Chandrasekhar.

54. George W. Bush, "National Space Policy Review," National Security Presidential Directive/NSPD-15 (Washington: The White House, June 28, 2002).

55. Guy Gugliotta, "Mars Mission Shows Significant Signs of Life, But Critics Skeptical," *The Washington Post* (November 29, 2004).

56. Tim Weiner, "Lockheed and the Future of Warfare," *The New York Times* (November 28, 2004), and William D. Hartung and Frida Berrigan, "Lockheed Martin and the GOP: Profiteering and Pork Barrel Politics with a Purpose" (New York: World Policy Institute, <www.worldpolicy.org>; downloaded December 16, 2004).

57. "NASA Procurement Report for 2003" (Washington, D.C.: National Aeronautics and Space Administration, 2004), available at <www.hq.nasa.gov>.

CHAPTER 8   THE CRISIS IN AMERICAN HEALTH CARE

1. For a bird's eye view of the array of social controversies and policy choices engendered by the ubiquitous American automobile, see John Tierney, "The Autonomist Manifesto (Or, How I Learned to Stop Worrying and Love the Road)," *The New York Times* (September 26, 2004).

2. In *A Social History of American Technology* (New York: Oxford University Press, 1997), 149, Ruth Schwartz Cowan describes technological systems as occurring in industrial societies when "individuals become more dependent on one another" as "they are linking together in large, complex networks that are, at one and the same time, both physical and social." For a history of U.S. medicine see Paul Starr, *The Social Transformation of American Medicine* (New York: HarperCollins, 1982).

3. Richard Rhodes, *The Making of the Atomic Bomb* (New York: Simon and Schuster, 1986), 571–572.

4. There were actually two sets of trials, the International Military Tribunal referenced here and the United States Military Tribunal. The trials were controversial, some charging that the victor's justice could have little moral or legal standing. For two contemporary views, see Herbert Wechsler, "The Issues of the Nuremburg Trial," *Political Science Quarterly*, Vol. 62 (1947) and Herman Phleger, "Nuremberg—A Fair Trial?" *Atlantic Monthly*, Vol. 177 (April 1946).

5. Extensive excerpts from the published transcripts and documents collected at Nuremberg are included in William Shirer's discussion of German medical experiments under the Third Reich. See William L. Shirer, *The Rise and Fall of the Third Reich: A History of NAZI Germany* (New York: Simon and Schuster, 1960), 971–991. The official trial record is available through the National Archives, *Trials of War Criminals before the Nuremberg Military Tribunals under Control Council Law No. 10.* Nuremberg, October 1946–April 1949 (Washington D.C.: U.S. Government Printing Office, 1949–1953); also National Archives Record Group 238, M887. The United States Holocaust Memorial Museum maintains an online exhibit on the "doctors' trial" at <http://www.ushmm.org/research/doctors/index.html>.

6. Roy Porter, *The Greatest Benefit to Mankind* (London: W. W. Norton & Co., 1997), 648–659. See also Starr, *The Social Transformation of American Medicine.*

7. "This [voluntary and informed consent] means that the person involved should have legal capacity to give consent; should be so situated as to be able to exercise free power of choice, without the intervention of any element of force, fraud, deceit, duress, over-reaching, or other ulterior form of constraint or coercion; and should have sufficient knowledge and comprehension of the elements of the subject matter involved as to enable him to make an understanding and enlightened decision. This latter element requires that before the acceptance of an affirmative decision by the experimental subject there should be made known to him the nature, duration, and purpose of the experiment; the method and means by which it is to be conducted; all inconveniences and hazards reasonably to be expected; and the effects upon his health or person which may possibly come from his participation in the experiment." The Nuremberg Code, Permissible Medical Experiments in *Trials of War Criminals.*

8. 41st World Medical Assembly, World Medical Association, "Declaration of Helsinki" (Hong Kong, September 1989). Available from the National Institutes of Health at <http://ohsr.od.nih.gov/Helsinki>.

9. Ibid. Italics added.

10. Lewis R. Thomas, *The Youngest Science: Notes of a Medicine Watcher* (New York: Penguin Books, 1995).

11. Porter, *The Greatest Benefit to Mankind*, 648.

12. Jean Heller, "Syphilis Victims in the U.S. Study Went Untreated for 40 Years," *The New York Times* (July 26, 1972). A similar story appeared the previous day in *The Washington Star*. See also James H. Jones, *Bad Blood: The Tuskegee Syphilis Experiment*, new and expanded ed. (New York: The Free Press, 1993). We know from Sinclair Lewis's novel *Arrowsmith*, the story of (among other things) a doctor torn between medical research and practice, that withholding treatment in the interest of a controlled clinical trial was a questionable practice as early as 1925, when the novel was published.

13. Office of Human Subjects Research, National Institutes of Health, "Guidelines for the Conduct of Research Involving Human Subjects at the National Institutes of Health," Appendix I. Downloaded December 16, 2003 from <http:www.nihtraining.com/ohrsite/guidelines/graybook.html>.

14. Albert R. Jonsen, Ph.D., Associate Professor of Bioethics at the University of California

at San Francisco. For general treatments of this interdisciplinary field, see M. L. Tina Stevens, *Bioethics in America: Origins and Cultural Politics* (Baltimore, Md.: The Johns Hopkins University Press, 2000), Albert Jonsen, *The Birth of Bioethics* (New York, N.Y.: Oxford University Press, 1998), and David Rothman, *Strangers at the Bedside: A History of How Law and Bioethics Transformed Medical Decision Making* (New York, N.Y.: Basic Books, 1991).

15. The Belmont Report is available from the Department of Health and Human Services website at <http://ohrp.osophs.dhhs.gov/humansubjects/guidance/Belmont.htm>.

16. Kenneth John Ryan, M.D., chairman, chief of staff, Boston Hospital for Women; Joseph V. Brady, Ph.D., Professor of Behavioral Biology, Johns Hopkins University; Robert E. Cooke, M.D., president, Medical College of Pennsylvania; Dorothy I. Height, president, National Council of Negro Women, Inc.; Albert R. Jonsen, Ph.D., Associate Professor of Bioethics, University of California at San Francisco; Patricia King, J.D., Associate Professor of Law, Georgetown University Law Center; Karen Lebacqz, Ph.D., Associate Professor of Christian Ethics, Pacific School of Religion; David W. Louisell, J.D., Professor of Law, University of California at Berkeley; Donald W. Seldin, M.D., professor and chairman, Department of Internal Medicine, University of Texas at Dallas; Eliot Stellar, Ph.D., provost of the university and Professor of Physiological Psychology, University of Pennsylvania; and Robert H. Turtle, LL.B., attorney, VomBaur, Coburn, Simmons & Turtle, Washington, D.C. (Mr. Louisell and Mr. Turtle died before the report was issued).

17. 41st World Medical Assembly, World Medical Association, "Declaration of Helsinki" (Hong Kong, September 1989), Introduction, III-1 and III-4. Available from the National Institutes of Health at <http://ohsr.od.nih.gov/Helsinki>.

18. Belmont Report.

19. Alert readers will note my preference for the term "moral" rather than "ethical" principles. Because the term "ethics" as commonly used is more often associated with arguable tenets and thus subject to casuistry, whereas "moral" is more often associated with concrete and accepted rules of conduct, I believe the latter term is more suitable to discussions of practices that may become (or should become) the subject of law and public policy.

20. The charter for the National Bioethics Advisory Commission expired October 3, 2001. A transcript of its July 14–15, 1998 meeting from Georgetown University's National Reference Center for Bioethics Literature is available at <http://www.georgetown.edu/research/nrcbl/nbac/>. http://www.georgetown.edu/research/ncbl/nbac/transcripts/jul98/belmont.html.

21. Declaration of Geneva, International Code of Medical Ethics, World Medical Association Declaration of Helsinki.

22. Belmont Report, Section C.2.

23. Title 45, Part 46, "Protection of Human Subjects" (sometimes referred to as the "Common Rule"). Administratively, compliance with 46 CFR 45 is the responsibility of the National Institutes of Health's Office of Human Subjects Research. Federal policy governing not only federally sponsored research, but research for which a federal agency has oversight responsibility (e.g., Food and Drug Administration) requires institutions performing such research to provide the sponsoring federal agency written assurances that they have in place one or more institutional review boards to review and approve the proposed research for compliance with 46 CFR 45. See <http://www.nihtraining.com/ohrsite/>.

24. Organisation for Economic Co-operation and Development, "OECD Health Data 2003

Show Health Expenditures at an All-time High," OECD website, <http://www.oecd.org>, downloaded October 10, 2004. A transcript of the October 13, 2004 presidential debate is available from the Cable News Network (CNN), *The Washington Post*, and *The New York Times* websites, among other sources.

25. Organisation for Economic Co-operation and Development, "Health at a Glance 2003—OECD Countries Struggle with Rising Demand for Health Spending," <http://www.oecd.org/>, downloaded October 31, 2005.

26. U.S. Gross Domestic Product represents the monetary measure of all goods and services produced with all factors of production physically located within the United States.

27. OECD "Health at a Glance 2003."

28. PriceWaterhouseCoopers, "The Factors Fueling Rising Healthcare Costs" (April 2002).

29. David J. Rothman, *Beginnings Count: The Technological Imperative in American Health Care* (New York, N.Y.: Oxford University Press, 1997).

30. Roy Porter, *The Greatest Benefit to Mankind* (London: W. W. Norton & Co., 1997), 628.

31. See Rothman, *Beginnings Count*, chapter 1.

32. Ibid., especially chapters 1 and 3.

33. Corporate after-tax profits for 2000 were $522.9 billion. Source: U.S. Bureau of Economic Analysis, "National Income and Product Accounts of the United States." See <www.bea.doc.gov>.

34. Bureau of Labor Statistics, U.S. Department of Labor, "The Recent Decline in Medical and Retirement Plan Coverage," *Monthly Labor Review* (September 28, 2004) and "Increases in prescription-drug copayments, 1993–2000," *Monthly Labor Review* (October 4, 2004).

35. U.S. Census Bureau, Current Population Reports, P60-226, "Income, Poverty, and Health Insurance Coverage in the United States: 2003" (Washington, D.C.: U.S. Government Printing Office, 2004).

36. Bureau of Labor Statistics, U.S. Department of Labor, "Issues in Labor Statistics: A Different Look at Part-time Employment," Summary 96-9 (April 1996).

37. Institute of Medicine, *Insuring America's Health: Principles and Recommendations* (Washington, D.C.: National Academies Press, January 2004).

38. Ibid.

39. Sheryl Gay Stolberg, "Youth's Death Shaking Up Field of Gene Experiments on Humans," *The New York Times* (January 27, 2000).

40. Ibid.

41. See, for example, NIH's guidelines for research with human subjects in the Code of Federal Regulations, Title 45, Part 46, §46.102 and §46.111.

42. Institute of Medicine, *Health Literacy: A Prescription to End Confusion* (Washington, D.C.: National Academies Press, 2004), 11, 14–16.

43. Theodore Roszak, *The Making of a Counter Culture* (Berkeley, Calif.: University of California Press, 1995), 7–8.

44. Rothman, *Beginnings Count*, chapter 6.

45. The eight presidential and congressional election years between 1980 and 2000 were distinguished by, among other things, a drop from 60 percent to slightly over 40 percent in the percentage of the voting age population that voted. The decline was most pronounced in people under the age of forty and negligible among people over sixty-five. As the age of the voting population creeps upward, the median age of the population creeps upward as well, so that by the year 2030 one in five people in the U.S. would be sixty-five or older. Source: U.S. Census Bureau, *Current Population Reports*.

46. The pharmaceutical/health products industry's rank in total campaign giving by eighty industries tracked by the Center for Responsive Politics rose from twenty-seventh ($3.2 million) in 1990 to tenth ($29.3 million) in 2002. During the 2004 election cycle the industry made $2.6 million in contributions to House and Senate candidates. A good source of information on campaign contributions is the Center for Responsive Politics' website, <www.opensecrets.org>.

47. The law could also be characterized as the health lawyers' full employment act of 2003, for much of it consists of amendments to the voluminous existing provisions of Title 42 of the U.S. Code, "The Public Health and Welfare." See E. J. Dionne, Jr., "Medicare Monstrosity," *The Washington Post* (November 18, 2003); Christopher Lee, "Medicare Bill Partly a Special Interest Care Package," *The Washington Post* (November 23, 2003); John Tierney, "A $400 Billion Purchase, All On Credit," *The New York Times* (November 30, 2003); and Robert Pear, "Medicare Law's Costs and Benefits Are Elusive," *The New York Times* (December 9, 2003).

48. Robert Pear, "Congress Weighs Drug Comparisons," *The New York Times* (August 24, 2003).

49. 42 USC 262 ("The Public Health and Welfare") defines a "biological product" as "a virus, therapeutic serum, toxin, antitoxin, vaccine, blood, blood component or derivative, allergenic product, or analogous product, or arsphenamine or derivative of arsphenamine (or any other trivalent organic arsenic compound), applicable to the prevention, treatment, or cure of a disease or condition of human beings."

    Prescription drug prices are discussed in H.R. 1, Medicare Prescription Drug, Improvement, and Modernization Act of 2003, Title I, Sec. 101. The following (author's text) references to prescription drugs shall apply as well to biological products. The Act contains numerous exceptions and sub-provisions that cannot readily be summarized without reproducing the legislation itself. The author's text only attempts top-level summaries. The full text of the Act is available at <http://frwebgate.access.gpo.gov/>.

50. H.R. 1, Medicare Prescription Drug, Improvement, and Modernization Act of 2003, Title I, Sec. 303.

51. Ibid., Sec. 1013.

52. See, for example, accounts of the marketing and clinical histories of cholesterol reducing drugs and drugs to alleviate arthritis pain, both ailments widespread among an aging population, in Marcia Angell, M.D., *The Truth About the Drug Companies: How they Deceive Us and What to Do About It* (New York: Random House, 2004), and John Abramson, M.D., *Overdosed America: The Broken Promise of American Medicine* (New York: HarperCollins, 2004). See also Mary Duenwald, "One Lesson from Vioxx: Approach New Drugs With Caution," *The New York Times* (October 5, 2004), Fran Hawthorne, *The Merck Druggernaut: The Inside Story of a Pharmaceutical Giant* (John Wiley & Sons, 2003) and Katherine Greider, *The Big Fix: How the Pharmaceutical Industry Rips Off American Consumers* (New York: Perseus, 2003).

53. Prior to the Act health maintenance organizations could contract with Medicare to serve Medicare beneficiaries. Under the Bush reform legislation Medicare beneficiaries would be offered a fixed amount of money ("premium support") which they could use to purchase traditional Medicare or a competing private health insurance plan. Supporters of the provision argued that it would promote competition, and thus reduce costs. Opponents argued that it was intended to "privatize" medicare, and thus destroy it. See Robert Pear, "Issue of Competition Causes Widest Split Over Medicare," *The New York Times* (November 10, 2003).

54. Editorial, "Global Competition in Health Care," *British Medical Journal* (September 28, 1996).

55. Families USA Foundation, "Sticker Shock: Rising Prescription Drug Prices for Seniors," Families USA Publication No. 04-103 (2004). Available at <www.familiesusa.org>.

56. Campaign for a National Health Program NOW! See <www.cnhpnow.org>. Information downloaded October 14, 2004.

57. The British and Spanish national health systems are nationally funded and rely on salaried physicians and other providers, while hospitals are publicly owned and operated. France and Germany provide universal health insurance through non-profit "sickness" funds or "mutuales," which pay physicians and hospitals uniform rates that are negotiated annually. Health care is also a universal right in Japan, which uses various insurance plans paid for by compulsory payroll deductions, taxes, and patient co-payments. Uniform fees are negotiated by a council of insurers, providers, and citizens.

58. A transcript of the October 13, 2004 presidential debate is available from the Cable News Network (CNN), *The Washington Post* and *The New York Times* websites, among other sources.

59. America's Health Insurance Plans' website, at <www.ahip.org>.

60. Center for Responsive Politics, <http://www.opensecrets.org>. Downloaded October 20, 2004.

61. Donald J. Palmisano, David W. Emmons, and Gregory D. Wozniak, "Expanding Insurance Coverage Through Tax Credits, Consumer Choice, and Market Enhancements," *Journal of the American Medical Association*, Vol. 291, No. 18 (May 12, 2004), 2237–2242. For the AMA's role in the design and passage of Medicare in 1965, see Rothman, *Beginnings Count*, 69–82.

62. John W. Wright, ed., *The New York Times Almanac for* 2004 (New York, N.Y.: Penguin Putnam Inc., 2003), 401–406.

63. For a valuable overview, see Allen Buchanan, Dan W. Brock, Norman Daniels, and Daniel Wikler, *From Chance to Choice: Genetics and Justice* (New York, N.Y.: Cambridge University Press, 2000).

64. Public Law 95-622.

65. Naturally occurring clones have been with us for millennia, witness identical twins and triplets. Dolly, a Finn Dorset Ewe, was cloned by a team led by Ian Wilmut at the Roslin Institute, near Edinburgh, Scotland. The National Reference Center for Bioethics Literature at the Joseph and Rose Kennedy Institute of Ethics, Georgetown University, maintains an excellent online bibliographical resource, "Basic Resources in Bioethics." See <http://www.georgetown.edu/research/nrcbl/>.

66. "Defining Death" (1981); "Whistleblowing in Biomedical Research" (1981), "Compensation for Research Injuries," 2 vols. (1982); "Protecting Human Subjects" (1982); "Splicing Life" (1982); "Deciding to Forego Life Sustaining Treatment" (1983); "Implementing Human Research Regulations" (1983); "Screening and Counseling for Genetic Conditions" (1983); "Making Health Care Decisions" (1982–1983); "Securing Access to Health Care" (1983); and "Summing Up" (1983). Reports are available from the National Reference Center for Bioethics Literature at the Joseph and Rose Kennedy Institute of Ethics, Georgetown University, "Basic Resources in Bioethics." See <http://www.georgetown.edu/research/nrcbl/>. See also Leon Kass, ed. *Being Human: Core Readings in the Humanities* (New York, NY: W. W. Norton & Co., 2004),

67. The seven bioethicists were Patricia Backlar, Alexander Morgan Capron, R. Alta

Charo, James F. Childress, Bette O. Kramer, Bernard Lo, and Thomas H. Murray. Arturo Brito, Eric J. Cassell, David R. Cox, Carol W. Greider, and Diane Scott-Jones represented the medical or public health faculty, and Rhetaugh Graves Dumas, Steven H. Holtzman, Lawrence H. Milke, and Laurie M. Flynn were the provost emerita, chief business officer, jurist, and spokesperson for the mentally ill, respectively.

68. For a good survey of the issues and points of view engaged in the embryonic stem cell research question, see Susan Holland, Karen Lebacqz, and Laurie Zoloth, eds. *The Human Embryonic Stem Cell Debate* (Cambridge, Mass.: The MIT Press, 2001).

69. Testimony of James F. Childress, Kyle Professor of Religious Studies and Professor of Medical Education, University of Virginia, before the Subcommittee on Labor, Health and Human Services and Education of the Committee on Appropriations, United States Senate (November 4, 1999). Available from the National Reference Center for Bioethics Literature at the Joseph and Rose Kennedy Institute of Ethics, Georgetown University, "Basic Resources in Bioethics." See <http://www.georgetown.edu/research/nrcbl/>. See also "Research Involving Human Biological Materials: Ethical Issues and Policy Guidance," Report and Recommendations of the National Advisory Commission (Rockville, Md.: August, 1999).

70. "Basic Resources in Bioethics." See <http://www.georgetown.edu/research/nrcbl/>.

71. "Research Involving Human Biological Materials: Ethical Issues and Policy Guidance," Report and Recommendations of the National Advisory Commission (Rockville, Md.: August, 1999), 51. In 1990 the California State Supreme Court ruled in favor of the University of California and co-defendants in a suit filed by John Moore, who asserted a claim to a portion of the profits derived from uses of tissue physicians had removed from his spleen to develop a cell line. *Moore v. Regents of California et al.*, 793 P.2d 479 (Cal. 1990).

72. National Bioethics Advisory Commission, "Ethical and Policy Issues in Research Involving Human Participants" (Bethesda, Md.: August 2001), 5–6. For a critical discussion of the NBAC's report, see John C. Fletcher, "NBAC's Arguments on Embryo Research: Strengths and Weaknesses," in Holland et al., *The Human Embryonic Stem Cell Debate*, 61–72.

73. Institutions represented by the Council's eighteen members include the University of Chicago (two), the University of California (two), Yale University, Washington University, the Johns Hopkins University, Dartmouth College, Princeton University, Georgetown University, Harvard University (two), and Stanford University. The only non-academic on the panel is the syndicated columnist for *The Washington Post*, Charles Krauthammer. Information on the Council and its activities is available at <http://www.bioethics.gov/>. See The President's Council on Bioethics, "Beyond Therapy: Biotechnology and the Pursuit of Happiness (Washington, D.C.: The White House, October 2003), chapter 6. Both reports are available at <http://bioethics.gov/reports/beyondtherapy>.

74. Connie Bruck, "Hollywood Science," *The New Yorker* (October 18, 2004), 62–82.

75. To examine proposed federal human cloning legislation, see <http://Thomas.loc.gov/>.

76. Human cloning has been banned in thirty-three countries, including most OECD countries. Countries where the cloning of embryos for research is not prohibited include the United Kingdom, Sweden, China, and Israel, among other countries. See the Center for Genetics and Society website, <http://www.genetics-and-society.org/policies/other/cloning.html>.

77. Department of Health and Human Services, Food and Drug Administration,

"Application of Current Statutory Authorities to Human Somatic Cell Therapy Products and Gene Therapy Products, *Federal Register*, Vol. 58, No. 197 (October 14, 1993).

78. Food and Drug Administration, "Use of Cloning Technology to Clone a Human Being," downloaded from <http:www.fda.gov/cber/genetherapy/clone.htm>.

79. Denise Grady, "With Few Suppliers of Flue Shots, Shortage Was Long in Making," *The New York Times* (October 17, 2004); Knight Ridder Newspapers, "England Acted Early, Expects No Shortage of Flu Vaccine," *The Portland Press Herald* (October 17, 2004); and Diedra Henderson, "Flu Vaccine Shortage Could Prove Costly," *Portland Press Herald* (October 22, 2004).

80. Accounts of the history of the federal regulation of drugs can be found in Peter Temin, *Taking Your Medicine: Drug Regulation in the United States* (Cambridge, Mass.: Harvard University Press, 1980) and Philip J. Hilts, *Protecting America's Health: The FDA, Business, and One Hundred Years of Regulation* (New York, N.Y.: Alfred A. Knopf, 2003).

81. See Abramson, *Overdosed America*.

82. Arnold S. Relman, "Your Doctor's Drug Problem," *The New York Times* (November 18, 2003).

83. Arnold S. Relman and Marcia Angell, "America's Other Drug Problem: How the Drug Industry Distorts Medicine and Politics," *The New Republic* (December 16, 2002), 30. Drs. Relman and Angell, both of whom are affiliated with Harvard Medical School, have each served as editor-in-chief of the *New England Journal of Medicine*.

84. Pub. L. No. 96-480, 94 Stat. 2311 and Pub. L. No. 96-517 § 6(a), 94 Stat. 3019.

85. Pub. L. No. 99-502, 100 Stat. 1785.

86. Relman and Angell, "America's Other Drug Problem," 31.

87. Ibid., 30–31.

88. BMG's sales of Taxol have declined since 2000, when the generic version paclitaxel became available on the market. See U.S. General Accounting Office, "Technology Transfer: NIH-Private Sector Partnership in the Development of Taxol," GAO-03-829 (Washington, D.C.: U.S. General Accounting Office, June 2003).

89. See 15 U.S.C. § 3710a(c)(7); 18 U.S.C. § 1905 (2000); and 5 U.S.C. § 552(b)(4)(2000), which exempts trade secrets, and commercial and financial information that is privileged or confidential, from public disclosure. Nonetheless the GAO was able to trace the research history of Taxol and the ways in which BMG was able to use existing law to prolong its monopoly sales of Taxol. See U.S. General Accounting Office, "Technology Transfer: NIH-Private Sector Partnership in the Development of Taxol," GAO-03-829 (Washington, D.C.: U.S. General Accounting Office, June 2003).

90. National Heart, Lung, and Blood Institute, "NHLBI Stops Trial of Estrogen Plus Progestin Due to Increased Breast Cancer Risk, Lack of Overall Benefit" (Bethesda, Md.: National Institutes of Health, July 9, 2002); available at <http://www.nih.gov/news/pr/Jul2002/nhlbi-09.htm>.

91. U.S. General Accounting Office (USGAO), "Adverse Drug Events: The Magnitude of the Health Risk Is Uncertain Because of Limited Incidence Data," GAO/HEHS-00-21 (Washington, D.C.: U.S. General Accounting Office, January 2000), *passim*.

92. Ibid., 3.

93. Interview with Paul Seligman, director of FDA's Office of Drug Safety, "Frontline: Dangerous Prescription: Politics, Profits & Pharma," Public Broadcasting System. Broadcast November 13, 2003. Transcripts available online at <http://www.pbs.org/wgbh/pages/frontline/>. Downloaded November 14, 2003.

94. Hilts, *Protecting America's Health*, 276–279.

95. U.S. Food and Drug Administration, "PDUFA III Five-Year Plan" (Washington, D.C.: U.S.

Department of Health and Human Services, Food and Drug Administration, July 2003); available at <http://www.fda.gov/oc/pdufa3/2003plan/defalt.htm>.

96. U.S. Food and Drug Administration, "FY 2002 Performance Report to Congress for the Prescription Drug User Fee Act of 1992," available from the Office of Planning, the Food and Drug Administration, and online at <http://www.fda.gov/oc/pdufa/report2002/>. Downloaded January 7, 2004.

97. See David Willman, "How a New Policy Led to Seven Deadly Drugs," *The Los Angeles Times* (December 20, 2000); available online at <http://www.drugawareness.org/Archives>. U.S. General Accounting Office, "Food and Drug Administration: Effect of User Fees on Drug Approval Times, Withdrawals, and Other Agency Activities," GAO-02-958 (Washington, D.C.: U.S. General Accounting Office, September 2002).

98. Eric J. Topol, "Good Riddance to a Bad Drug," *The New York Times* (October 2, 2004), Andrew Pollack, "New Scrutiny of Drugs in Vioxx's Family," *The New York Times* (October 5, 2004.

99. David M. Potter, *People of Plenty: Economic Abundance and the American Character* (Chicago, Ill.: University of Chicago Press, 1954), 90.

CHAPTER 9   FOSSIL FUELS AND CLEAN AIR

1. *Energy Policies of IEA Countries: 2002 Review*, International Energy Agency. Available at <http://www.iea.org>.

2. Marshall's reputation rests on his historic decision asserting the doctrine of national supremacy over the several states, and his associated state papers. Until he was appointed chief justice, he had served as a member of Congress from Virginia and secretary of state under John Adams. Kent had served in the New York assembly and professor of law at Columbia University and, after retiring from the bench in 1824, authored his *Commentaries on American Law* (1828–1830) (12th ed., New York, N.Y.: Hein & Company, 1989), which educated generations of lawyers in the Anglo-American heritage of the common law.

3. The best known of the government-funded expeditions to map nature's bounty in America was the 1803–1806 trans-continental reconnaissance of Meriwether Lewis and William Clark. But there were also the federally funded explorations by Lieut. Zebulon M. Pike of the sources of the Mississippi, Colorado, and New Mexico (1805–1807); the trans-Mississippi west surveys of the U.S. Army Corps of Topographical Engineers; and the 1850s soldier-civilian surveys of six possible routes for the trans-continental railway, which Secretary of War Jefferson Davis launched with a congressional appropriation of $150,000. By the 1850s individual states such as Mississippi, Illinois, California, Wisconsin, Arkansas, and Minnesota had begun their own state-funded surveys as well.

4. Endless diversity would become essential to Darwin's theory of the progressive evolution of the species by random variation. Biblical scripture, with which most nineteenth-century Americans were familiar, had taught that divine creation had filled a formless void not only with light, water, and dry land, but "every herb . . . every tree . . . the fish of the sea . . . [and] every moving thing that moveth upon the earth." For those multitudes who believed (and many who still believe) that the American prospect was divinely blessed, the notion that nature might fail to yield all that man required was and remains a difficult notion to accept.

5. David M. Potter, *People of Plenty: Economic Abundance and the American Character* (Chicago, Ill.: University of Chicago Press, 1954), 90.

6. From less than $0.12 per kilowatt hour to less than $0.8 per kilowatt hour. Congressional Research Service (CRS), "Energy: Useful Facts and Numbers," CRS Report for Congress (Washington, D.C.: Library of Congress, July 2003).

7. Technically speaking, this new access to what some would call "liquid gold" was initially pumped with a common hand water pump from the ground, where the oil had accumulated over a weekend in a hole made by Drake's employee, William A. Smith, a blacksmith who made tools for salt water drillers. Daniel Yergin, *The Prize: The Epic Quest for Oil, Money, and Power* (New York: The Free Press, 1991). Also, Harold F. Williamson and Arnold R. Daum, *The American Petroleum Industry* (Evanston: Northwestern University Press, 1959).

8. See *United States v. E. C. Knight* (1895), *United States v. Addyston Pipe and Steel Company* (1899), *United States v. United Shoe Machinery Company of New Jersey et al.* (1918) and *United States v. United States Steel Corporation* (1920).

9. Natural gas can also contain other hydrocarbons such as propane and butane, and the non-hydrocarbon gases nitrogen, carbon dioxide, hydrogen, hydrogen sulfide, helium, and argon.

10. Jad Mouawad, "Oil Demands Can Be Met, But at a High Price, Energy Agency Says," *The New York Times* (October 28, 2004).

11. Estimates of coal, oil, and natural gas reserves can be obtained from the U.S. Geological Survey, the International Energy Agency, and the Energy Information Agency of the U.S. Department of Energy.

12. From introduction to "Energy Perspectives," *Annual Energy Review* (2003) Energy Information Administration, at <www.eia.doe.gov>.

13. Department of Energy, Energy Information Administration, "Energy Perspectives: Consumption by Source," *Annual Energy Review,* 2003 (Washington, D.C.: U.S. Department of Energy, 2004), available online at <http://www.eia.doe.gov>. Hydropower's share of electricity generation has gradually declined from 30 percent in 1950 to less than 10 percent in 2000. CRS, "Energy: Useful Facts and Numbers."

14. U.S. Department of Energy, *Annual Energy Review*; US Nuclear Regulatory Commission, *Daily Power Reactor Report.*

15. CRS, "Energy: Useful Facts and Numbers."

16. The U.S. Department of Energy estimates that for every unit of electrical energy generated by natural gas that is then used by residential and commercial consumers, twice that amount has been lost during the process of generating, transmitting, and distributing electricity to those same consumers. Department of Energy, Energy Information Administration, "Energy Perspectives: Consumption by Sector," *Annual Energy Review,* 2003 (Washington, D.C.: U.S. Department of Energy, 2004), available online at <http://www.eia.doe.gov>.

17. The total U.S. consumption of petroleum, natural gas, and coal increased from roughly 42 quadrillion BTUs in 1960 to 84 quadrillion BTUs in 2002, during the same period the U.S. gross domestic product in constant 1996 dollars *quadrupled* from $2,376.7 billion to $9,439.9 billion. GDP = output of production attributable to all factors of production (labor and property) physically located within a country. Department of Energy, Energy Information Administration, *Annual Energy Review, 2003* (Washington, DC: U.S. Department of Energy, 2004), available online at <http://www.eia.doe.gov>; CRS, "Energy: Useful Facts and Numbers."

18. Oil is measured in barrels per day; coal in million tons per year; natural gas in trillion cubic feet (TCF) per year. A common measure is often used to aggregate various types of energy, the British Thermal Unit (BTU). For example, one million barrels per day of

oil is approximately equivalent to two quadrillion BTUs per year. Per capita consumption is usually given in one million BTUs per year.

19. U.S. Department of Energy, *Annual Energy Review.* Renewable sources = hydroelectric power, geothermal energy, solar energy, wind energy, and biofuels such as wood waste, landfill gases, animal waste.

20. CRS, "Energy: Useful Facts and Numbers;" Energy Information Administration, *International Energy Annual.*

21. Yergin, *The Prize*, 175.

22. Close inter-workings between the government and the oil industry were somewhat discredited during the administration of Warren G. Harding (1921–1923), when Harding's secretary of the interior, former Senator Albert B. Fall, issued a sweetheart license to exploit the national oil reserve at Teapot Dome in Wyoming to Sinclair Oil Company. For his pains Fall received $100,000 from Sinclair and one year in jail from the federal government (he only served nine months), the first cabinet member to be convicted of a felony committed while in office.

23. Cited in Yergin, *The Prize*, 254.

24. *Panama Refining Co. v. Ryan*, 293 U.S. 388; *Schecter v. United States*, 295 U.S. 495 (1935).

25. Since its founding OPEC has been enlarged by the addition of Qatar (1961), Indonesia and Libya (1962), Abu Dhabi (1967), Algeria (1969), Nigeria (1971), the United Arab Emirates (which absorbed Abu Dhabi), and Ecuador and Gabon, which later withdrew.

26. The Strategic Petroleum Reserve (SPR) helped to prevent further price increases during the Iraq invasion of Kuwait in 1990 and the Gulf War. In 1985, the SPR held enough oil for normal U.S. supply for 115 days, absent any petroleum imports. Thirty million barrels were released by Clinton in 2000, with an intent to repay, with interest; not all has been repaid.

27. Leo Marx, *The Machine in the Garden: Technology and the Pastoral Ideal in America* (New York, N.Y.: Oxford University Press, 1964), 352–353. Marx explored this theme in the work of Nathaniel Hawthorne, David Thoreau, Ralph Waldo Emerson, Herman Melville, Samuel Clemens, and Henry James.

28. *Under the Sea Wind* (1941) and *The Sea Around Us* (1951).

29. Jeff Girth, "Big Oil Steps Aside in Battle Over Arctic," *The New York Times* (February 21, 2005).

30. U.S. Department of Energy, Annual Energy Review Interactive Data Query System, <http://tonto.eia.doe.gov/aer/>.

31. In 2000 there were 209,128,094 persons over eighteen residing in the United States, and 213,299,313 registered motor vehicles (which include trucks and buses as well as cars). In 1960 there were 115,121,000 persons over eighteen residing in the United States and 73,857,768 registered motor vehicles. Note that data includes commercial vehicles. Sources: U.S. Bureau of the Census, Department of Commerce, *Current Population Reports* and *The Statistical Abstract of the United States* (annual); Ward's Communications, *Ward's Motor Vehicle Facts & Figures* (2003).

32. CRS, "Energy: Useful Facts and Numbers."

33. U.S. Environmental Protection Agency, Office of Air Quality, "The Plain English Guide to The Clean Air Act" (2005), at <www.epa.gov/oar/oaqps/>.

34. U.S. Department of Energy, Annual Energy Review Interactive Data Query System, <http://tonto.eia.doe.gov/aer/>, and U.S. Department of Energy, Energy Information Administration, "Annual; Survey of Alternative Fuel Vehicle Suppliers and Users" (2004). The number of alternative fueled vehicles available in 1992 was around

230,000. Alternative fuel vehicles currently are those fueled by compressed natural gas (CNG), electricity (EVC), Ethanol, 85 percent (E85), hydrogen (HYD), liquefied natural gas (LNG), and liquefied petroleum gas (LPG).

35. Matthew L. Wald, "Test Set on Transmission That Could Save Fuel," *The New York Times* (February 10, 2005).

36. CRS, "Energy: Useful Facts and Numbers."

37. William K. Reilly, "The New Clean Air Act: An Environmental Milestone," *EPA Journal* (January–February 1991).

38. Ibid.

39. Gregg Easterbrook, "Clear Skies, No Lies," *The New York Times* (February 16, 2005). The Senate Environment and Public Works Committee was also divided over the bill's initial lack of actual caps for carbon dioxide, and exemption of small coal-fired power plants from caps for mercury emissions.

40. The Kyoto Protocol commits its signatories—Europe, Japan, and others—to reduce by 2012 emissions (largely from fossil fuels) of carbon dioxide, methane, nitrous oxide, hydroflourocarbons, perfluorocarbons, and sulphur hexafluoride (i.e., gases that trap heat, or "global warming" gases) to 5 percent below 1990 levels.

41. William J. Broad, "Deadly and Yet Necessary, Quakes Renew the Planet," *The New York Times* (January 11, 2005).

42. Danny Hakim, "Big Pickup Trucks Eclipsing S.U.V.'s," *The New York Times* (February 8, 2005).

43. Jeffrey H. Birnbaum and Alan S. Murray, *Showdown at Gucci Gulch: Lawmakers, Lobbyists, and the Unlikely Triumph of Tax Reform* (New York, N.Y.: Random House, 1987).

44. Linda R. Cohen and Roger G. Noll, "Synthetic Fuels From Coal," in *The Technology Pork Barrel*, Linda R. Cohen and Roger G. Noll, eds. (Washington, D.C.: The Brookings Institution, 1991).

45. Richard H. K. Vietor, *Energy Policy in America since 1945: A Study in Business-Government Relations* (New York, N.Y.: Cambridge University Press, 1984).

46. Editorial, *The New York Times* (October 24, 2005).

CHAPTER 10   EPILOGUE

1. Standard Oil of Indiana was one of the seven companies formed in 1911 when the U.S. Supreme Court ordered the break-up of Standard Oil.

2. The Center for Responsive Politics' website, <www.opensecrets.org>, provides detailed information on the political contributions of all major industries. Pip Coburn of UBS Warburg quoted in Andrew Ross Sorkin and Barnaby Feder, "A Sector Where 'Merger' Can Mean the Start of Something Ugly," *The New York Times* (February 10, 2005).

3. Alan B. Kreuger, "The Farm-Subsidy Model of Financing Academia," *The New York Times* (May 26, 2005).

4. Mary L. Good, "Increased Commercialization of the Academy Following the Bayh-Dole Act of 1980," in Donald G. Stein, ed., *Buying In or Selling Out? The Commercialization of the American Research University* (New Brunswick, N.J.: Rutgers University Press, 2004), 54.

5. For discussions of the policy issues entailed by the patenting of scientific research, as well as possible ways of resolving them, see Dorothy Nelkin, *Science as Intellectual Property: Who Controls Research?* (New York: Macmillan Publishing Company, 1984); Derek Bok, *Universities in the Marketplace: The Commercialization of Higher Education*

(Princeton, N.J.: Princeton University Press, 2003), Sheldon Krimsky, *Science in the Private Interest: Has the Lure of Profits Corrupted Biomedical Research?* (Lanham, Md.: Rowman & Littlefield, 2003); and Stein, *Buying In or Selling Out?*

6. David Blumenthal, Nancyanne Causino, and Eric Campbell, "Relationships between Academic Institutions and Industry in the Life Sciences: An Industry Survey," *New England Journal of Medicine*, Vol. 334, No. 6 (February 8, 1996); David Blumenthal, Nancyanne Causino, and Eric Campbell, "Participation of Life Science Faculty in Research Relationships with Industry," *New England Journal of Medicine*, Vol. 335, No. 23 (December 5, 1996); National Institutes of Health, *Report of the National Institutes of Health (NIH) Working Group on Research Tools* (June 4, 1998); Eric G. Campbell, et al., "Data Withholding Academic Genetics," *Journal of the American Medical Association*, Vol. 287 (January 2002), 473–480; Sheldon Krimsky and L. S. Rothenberg, "Conflict of Interest Policies in Science and Medical Journals: Editorial Practices and Author Disclosure," *Science and Engineering Ethics*, Vol. 7, No. 2 (2001), 205–218.

7. *Prometheus Radio Project v. Federal Communications Commission*, U.S. Court of Appeals for the Third Circuit, June 24, 2004; Stephen Labaton, "U.S. Backs Off Relaxing Rules for Big Media," *The New York Times* (January 28, 2005).

8. John M. Broder, "California's New Stem-Cell Initiative is Already Raising Concerns," *The New York Times* (November 27, 2004).

9. Robert Pear and James Dao, "States' Tactics Aim to Reduce Drug Spending," *The New York Times* (November 21, 2004); and Patricia Barry, "The Real Value of Drugs: Evidence-based Research Is Fast Emerging As an Important Tool to Assess which Medicines Are Most Effective," *AARP Bulletin* (March 2005). See <http://www. crbestbuydrugs. org>, <http://www.aarp.org/researchRx>, and <http://www.oregonrx.org>.

10. Paul Krugman, "The Chinese Connection," *The New York Times* (May 20, 2005); Chris Buckley, "China Says It Will No Longer Peg Its Currency to the U.S. Dollar," *The New York Times* (July 21, 2005); and Keith Bradscher, "China Adopts Opague Currency Policy," *The New York Times* (July 22, 2005).

11. *The Lexus and the Olive Tree* (New York, N.Y.: Farrar, Straus and Giroux, 1999) and *The World is Flat* (New York, N.Y.: Farrar, Straus and Giroux, 2005).

12. CAFTA involves Costa Rica, El Salvador, Guatemala, Honduras, Nicaragua, and the Dominican Republic, as well as the United States.

13. Robert J. Samuelson, "The World Is Still Round," *The Washington Post* (July 22, 2005).

INDEX

ABOUT THE AUTHOR

Sylvia Kraemer served as chief historian and director of policy development at the National Aeronautics and Space Administration, representing the agency during the Clinton administration's development of its space policy issued in 1996, as well as its formulation of federal policy for academic research. She has taught courses in U.S. science and technology policy at the George Mason University School of Public Policy and Colby College. Prior to her involvement in science and technology policy in Washington, she received her doctorate in the history of ideas from The Johns Hopkins University and taught American and European history at Vassar College, Southern Methodist University, and the University of Maine at Orono. Kraemer is also the author of *The Urban Idea in Colonial America* (1977) and *NASA Engineers in the Age of Apollo* (1993).